solid-state device theory

with illustrative problems

Phillip Cutler
Education Research Associates
and
Orange Coast College
Costa Mesa, California

McGraw-Hill Book Company
New York
San Francisco
St. Louis
Düsseldorf
Johannesburg
Kuala Lumpur
London
Mexico
Montreal
New Delhi
Panama
Rio de Janeiro
Singapore
Sydney
Toronto

Other books by Phillip Cutler

Electronic Circuit Analysis: Passive Networks
Semiconductor Circuit Analysis
Outline for DC Circuit Analysis with Illustrative Problems

This book was set in Times Roman by Holmes Typography, Inc., and printed and bound by Edwards Brothers, Inc. The illustrations were done by John Foster. The editors were Alan W. Lowe and Michael A. Ungersma. Charles A. Goehring supervised production.

solid-state device theory with illustrative problems

Printed in the United States of America.

Library of Congress catalog card number: 78–179079

234567890 EBEB 79876543

07–015002–8

contents

appendixes

To Betty
my wife,
my heart,
my right arm

preface

This text provides the initial framework for a course in electronic devices and circuit applications, with heavy emphasis on illustrative problems—some completely worked out for the student and many for the student to solve. The subject matter is presented in a twofold manner designed to develop the student's ability to analyze electronic circuits. First, a "feel" for the physical behavior of a device or circuit is imparted, and second, a brief and practical mathematical analysis is given. Although the subject matter is more diverse than the author's "Outline for DC Circuit Analysis with Illustrative Problems," it relentlessly pursues the same philosophy of encouraging the student to examine the physical world (a circuit or system in this case) and express it in mathematical terms. Wherever possible the theory is interwoven with practical applications.

In general, this text may be used as a primary source of information or as an effective supplement to existing works. It is also the prerequisite for the next book in this series, which will explore linear electronic circuits in depth. The text level is applicable for the serious technician who seeks a solid foundation in linear electronic circuits or the engineer who needs a practical review.

This background in devices and linear circuits provides a logical stepping-stone to a subsequent course in nonlinear electronics, involving waveshaping, timing, digital circuits, and similar subjects.

Although this text has been written in terms of electron flow rather than conventional current flow, this presentation should pose no problem for the reader who is more at home with the latter form. It is only necessary to first reverse the current arrow directions and then proceed as usual. The reader should refer to Appendixes A and B for a more detailed description of the current and voltage notation used in this text, as well as the others in this series.

In general, each chapter concludes with approximately 30 problems divided into three groups of 10 problems each. The first 10 include detailed solutions that thoroughly illustrate the concept and application of the lesson material. The next 10 are problems with answers, so that the student may test himself, and the last 10 are published without answers for the instructor's convenience in assigning homework. A separate solutions manual is available.

The text is structured as follows: Chapter 1 develops the concept of an ideal diode and the breakpoint on the volt-ampere characteristic. Chapter 2 introduces the practical solid-state diode with a minimum of physics and imparts a solid feel for the electrical and thermal characteristics of a *PN* junction. Chapter 3 introduces diode models while Chapter 4 outlines some useful techniques for single- and multiple-diode circuit analysis.

Chapters 1 through 4 set the stage for understanding

and analyzing solid-state devices whose unique characteristics relate to *PN* or similar junction properties. Chapter 5 introduces the transistor by building upon the properties of the *PN* junction developed in Chapter 2. Great pains are taken to promote an insight into transistor action. Chapters 6 and 7 introduce the common-base and common-emitter models respectively. The application of these models and the interrelation between their parameters is vigorously stressed. Chapter 9 presents some useful small-signal equivalent circuits of the transistor. The relationships between the most often used parameter sets are developed from a physical viewpoint based on measurement concepts, rather than on a *purely* mathematical approach. This manner of presentation breeds familiarity with the incremental models and is clear to the student possessing lesser mathematical skill. Chapter 10 introduces the hybrid-π model, which is useful in evaluating the high-frequency performance of the transistor. Chapter 11 outlines the characteristics of the basic transistor amplifier configurations. Although the various gain and impedance relations are presented, the emphasis is upon developing a physical "feel" for the equivalent circuit seen by looking into a pair of input or output terminals. The subsequent circuit analysis is then considerably simplified. Chapters 12, 13, 14, and 15 develop the field-effect transistor in a format similar to that used for the junction transistor. Both JFET's and IGFET's are covered. The text material, in conjunction with the drill problems, has proven most effective in providing the FET neophyte with a solid grasp of the subject.

The bulk of the material in this book has been used with outstanding success in industrial training programs and at Orange Coast College. Although at first glance this level of presentation may impress some instructors as being too formidable for technicians, it has been repeatedly demonstrated that with the proper prerequisites the average technician can thoroughly digest it and be thrilled with his sense of accomplishment. In fact, the student who just masters "Outline for DC Circuit Analysis with Illustrative Problems" can handle 80 percent of this text.

Audio-visual aids applicable to classroom presentation and/or individualized study to supplement the material in these texts are available through the author. For further information contact Phillip Cutler, 13181 Balboa, Apt. 1, Garden Grove, California 92640.

The author heartily recommends the use of an electronic calculator in lieu of a slide rule or sophisticated computer terminal. The accuracy and savings in time are enormous, and one can truly focus one's attention on the problem.

I should like to thank my dear friend, Mrs. Bertha K. Rapaport, for guidance and consideration in promoting the completion of this text, Mr. Robert Decker for his bloodhound ability to sniff out errors, Mr. Theodore Hoffman for invaluable feedback, Tektronix, Inc., for their valuable assistance, Mrs. Darlene Hormann, who typed this manuscript, and my wife, Betty, who contributed mightily to this work in so many, many different ways. And I shall always be indebted to the more than 10,000 students of industrial and college training programs whose expressions of pain and delight guided me through the years.

Phillip Cutler

1 the ideal diode

Diodes are two-terminal electronic devices with the unique property of permitting an electric current to flow more readily in one direction than the other. This property renders diodes exceedingly useful in electronic applications, as we shall soon see. To facilitate our understanding of diodes, we will first discuss the concept of an ideal diode.

1-1 The ideal diode

The concept of the ideal diode may be more clearly understood if we first compare its characteristics with those of a linear resistor as shown in Fig. 1-1. Both the resistor and diode are two-terminal devices. The linear resistor *V-I* characteristic indicates that it obeys Ohm's law for any polarity of applied voltage. By that we mean that it conducts equally well in both directions.

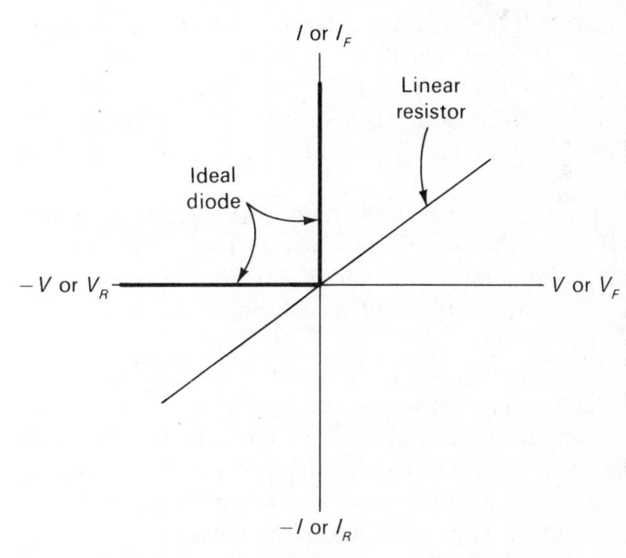

FIGURE 1-1

A resistor is therefore said to be a *bilateral* circuit element. An ideal diode, as we shall see, is a *unilateral* device. That is, it conducts perfectly in one direction and not at all in the other, depending upon the polarity of the applied voltage.

The schematic representation of an ideal diode is shown in Fig. 1-2. This is also the symbol for a practical solid-state diode and, therefore, the discussion should indicate whether an ideal or practical diode is being considered. However, in most cases the practical solid-

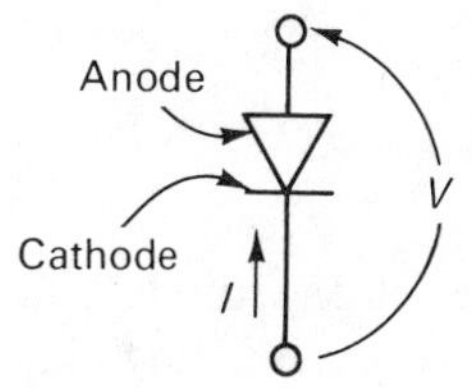

Anode
Cathode
V
I

FIGURE 1-2

state diode so closely approximates the ideal diode characteristic that the two are interchangeable.

Correlating the *V-I* curve of Fig. 1-1 with Fig. 1-2, we conclude that the ideal diode has the following characteristics. If we attempt to bias the anode positive with respect to the cathode, a condition called forward bias, the diode exhibits zero resistance since we see from the *V-I* curve that $V = 0$ for any I. The current I is then limited only by circuitry external to the diode. This is called the "ON," or *forward-bias* or *forward-conducting* state of the diode. Any voltage in the first quadrant which tends to turn the diode ON is also called the *forward voltage* (V_F) and the corresponding current is the *forward current* (I_F). Conversely, for an ideal diode, $V_F = 0$ when some forward current I_F is forced through it as from a current source. We see that a forward-biased ideal diode simulates a closed switch.

If V is a negative quantity (anode negative with respect to cathode), V is called a reverse voltage (V_R) and I is correspondingly negative and called the *reverse current* (I_R). When reverse-biased the ideal diode is turned OFF and it simulates an open switch, since no current can flow in the reverse direction. Hence, $I_R = 0$ for all values of V_R or $-V$.

The point at which the diode just changes state from OFF to ON or vice versa is called the *breakpoint*. At the breakpoint $V = 0$ and $I = 0$. This concept of the breakpoint is very useful in analyzing circuits containing several diodes.

Summarizing: (1) If the current directed into or the voltage impressed across the ideal diode tends to bias the anode positive relative to the cathode, the diode becomes forward-biased (ON) and simulates a closed switch with zero resistance and zero voltage drop. (2) If the voltage impressed across the ideal diode tends to bias the anode negative relative to the cathode, the diode becomes reversed-biased (OFF) and simulates an open switch with infinite resistance and zero current through it. (3) At the breakpoint, $V = 0$, $I = 0$.

PROBLEMS WITH SOLUTIONS

PS 1-1 Determine E and I in the circuit of Fig. PS 1-1. Assume an ideal diode.

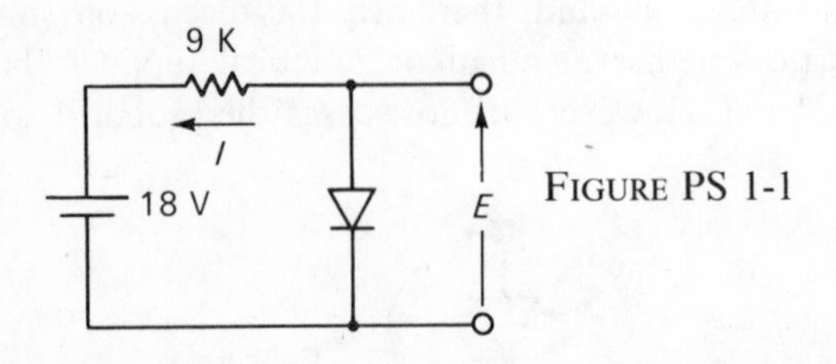

FIGURE PS 1-1

SOLUTION The state of the diode must first be established. This may be done by imagining the diode temporarily removed from the circuit and determining the polarity of the open-circuit voltage V_{OC} that the diode looks into. Clearly, V_{OC} is of such a polarity that the anode of the diode will tend to become positive relative to the cathode. Hence, the diode will be forward-biased, which implies that it acts like a closed switch. Therefore, $E = 0$ and $I = 18$ volts/9 kilohms $= 2$ ma.

PS 1-2 Determine E and I in Fig. PS 1-2. Assume an ideal diode.

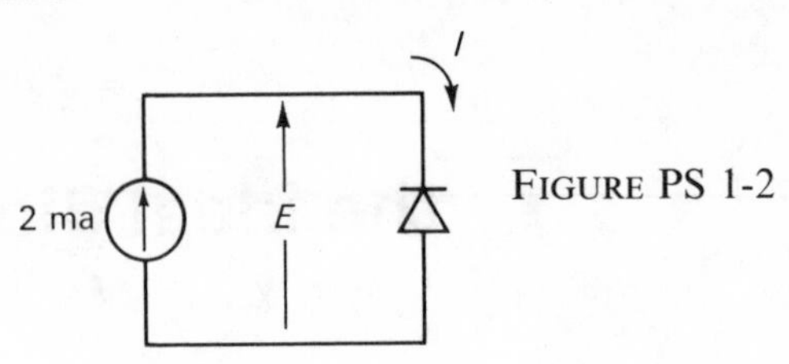

FIGURE PS 1-2

SOLUTION Clearly, the current source is trying to force the current I in the forward-bias direction, that is, from cathode to anode. Hence, the diode will be turned ON and simulate a closed switch. Therefore, $E = 0$ and $I = 2$ ma as determined by the current source.

PS 1-3 Determine E and I in Fig. PS 1-3. Assume an ideal diode.

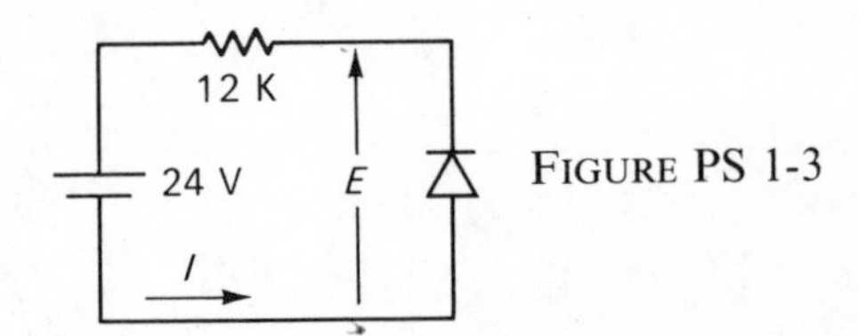

FIGURE PS 1-3

SOLUTION The open-circuit voltage the diode looks into is of such a polarity that the anode tends to be biased negative relative to the cathode. This is a condition of reverse-bias and hence the diode will be OFF and simulate an open switch. Therefore, $I = 0$ and $E =$ 24 volts.

PS 1-4 Determine V_{BC} in Fig. PS 1-4.

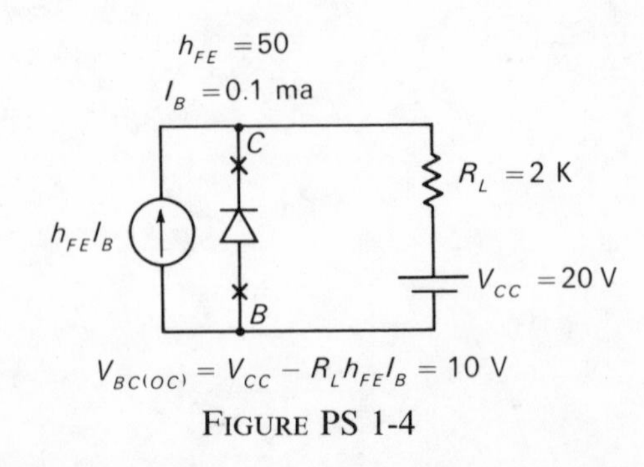

FIGURE PS 1-4

SOLUTION To determine V_{BC} we must first determine the state of the diode. This can be done by imagining the diode removed from the circuit and then computing the open-circuit voltage that the diode looks into.

$$V_{BC(OC)} = V_{CC} - R_L(h_{FE}I_B)$$
$$= 20 \text{ volts} - 2 \text{ kilohms } (5 \text{ ma}) = 10 \text{ volts}$$

Therefore the diode is OFF and $V_{BC} = 10$ volts. This particular problem is directly applicable to the analysis of transistor circuits. The quantity h_{FE} is a current gain. It has no dimensions and is called a numeric.

PROBLEMS WITH ANSWERS (assume ideal diodes)

PA 1-1 Determine I and E in Fig. PA 1-1.

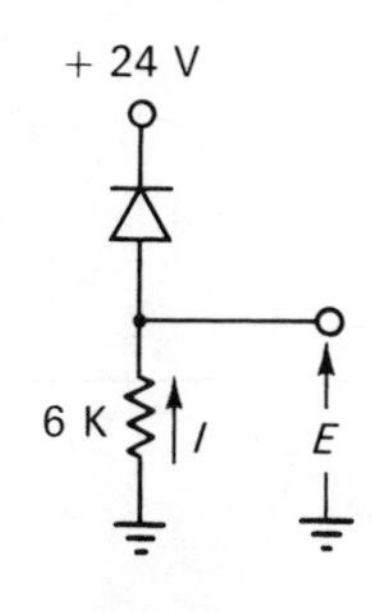

ANSWER $I = 0$,
$E = 0$

PA 1-2 Sketch the e_0 wave form.

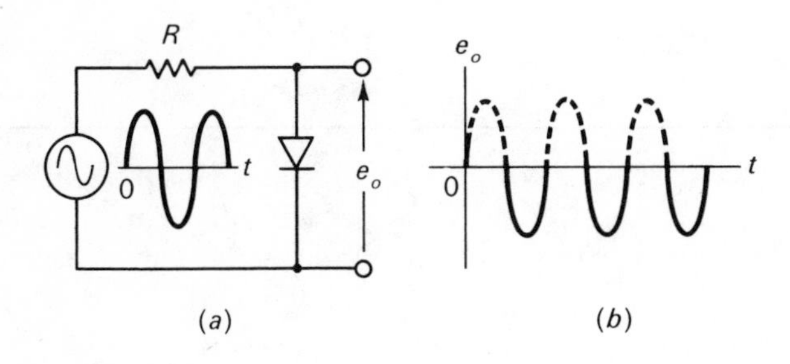

(a) (b)

ANSWER See Fig. PA 1-2*b*.

PA 1-3 Determine E_{AB} in Fig. PA 1-3.

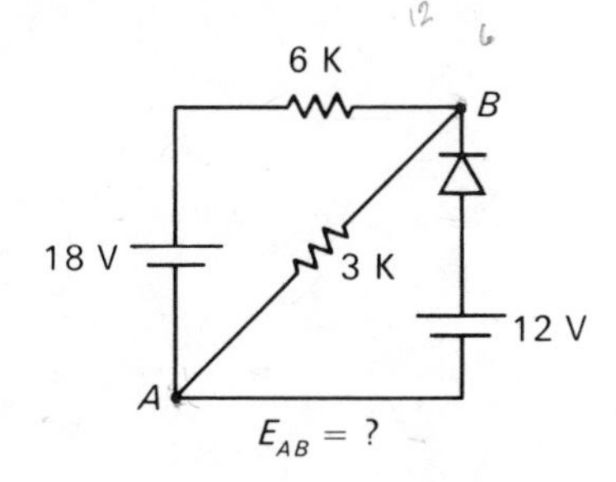

ANSWER $E_{AB} = 12$ volts

PA 1-4 Sketch the volt-ampere characteristic seen looking into the terminals of the circuit shown in Fig. PA 1-4*a*. Be sure to observe the proper reference directions

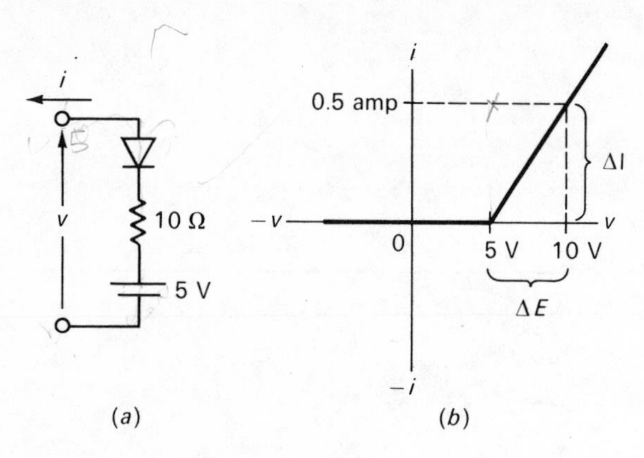

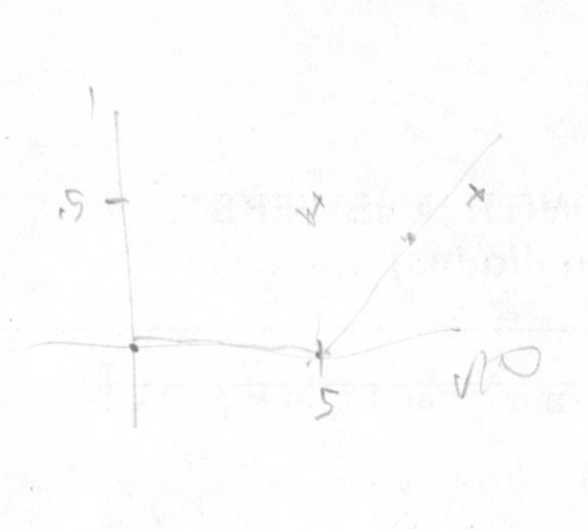

ANSWER See Fig. PA 1-4*b*.

PA 1-5 What value of current I_x is required for the diode to just be at its breakpoint?

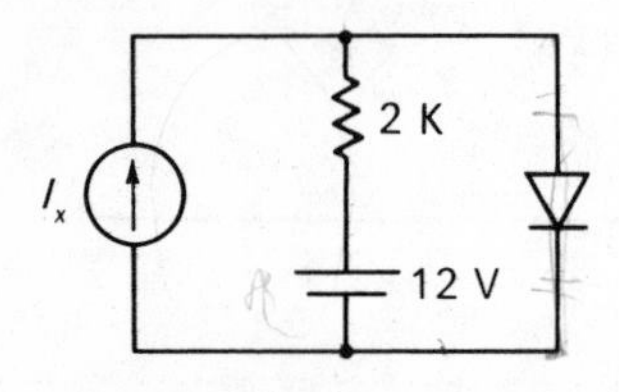

ANSWER $I_x = 6$ ma

PROBLEMS WITHOUT ANSWERS (assume ideal diodes)

P 1-1 How would you connect a diode between the terminals shown in Fig. P 1-1 so that current flows through the load R_L?

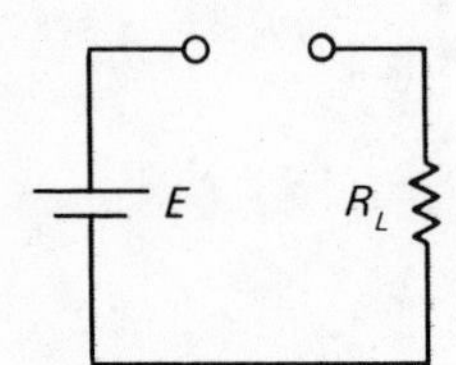

P 1-2 What is the current I equal to in Fig. P 1-2?

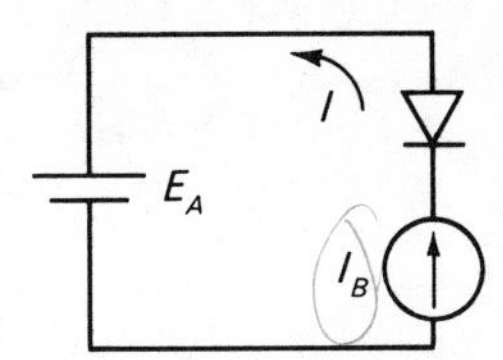

P 1-3 Determine I in Fig. P 1-3.

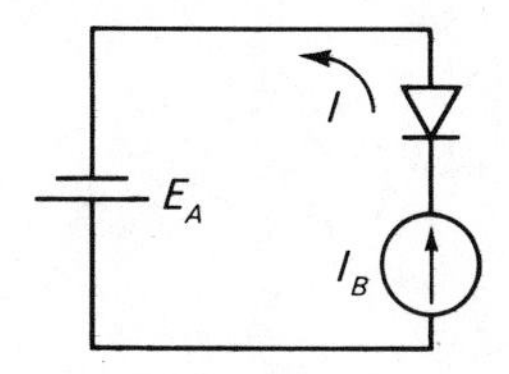

P 1-4 Determine I in Fig. P 1-4.

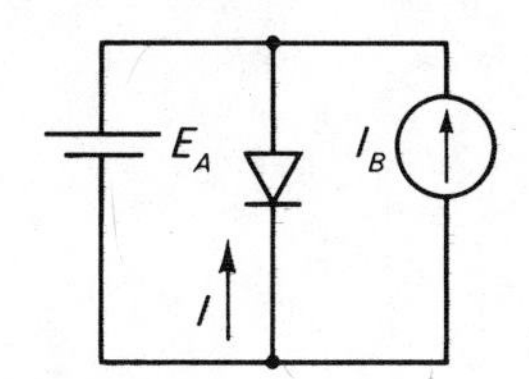

P 1-5 Determine I in Fig. P 1-5.

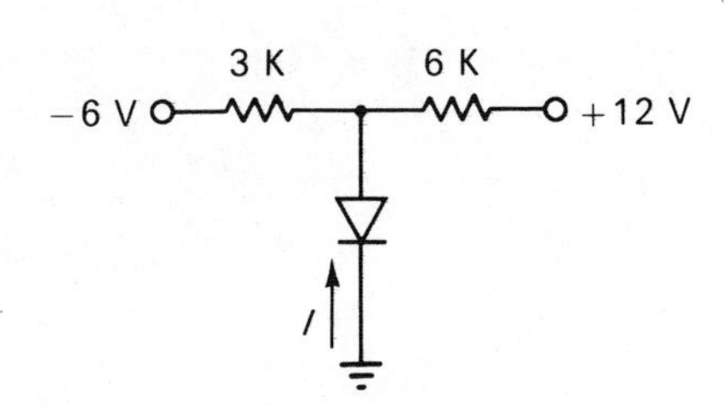

P 1-6 Determine I and V in Fig. P 1-6.

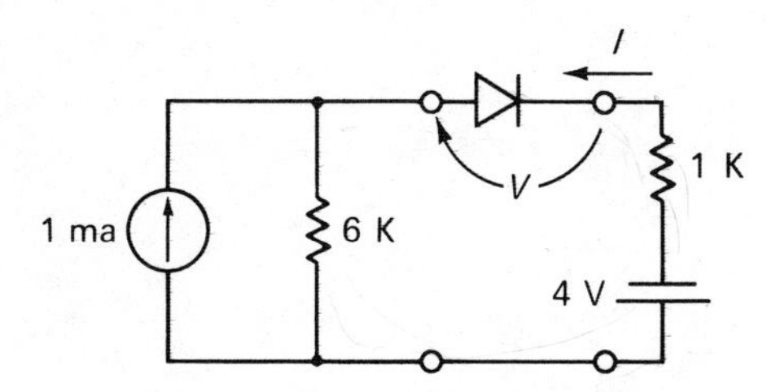

---

2 the solid-state junction diode

We previously studied the ideal diode. In practice, there is no ideal diode, but fortunately its characteristics are closely approximated by the solid-state junction diode in most cases.

2-1 The junction diode

Most junction diodes are primarily composed of the element silicon, and to a lesser extent, germanium. Both silicon and germanium are called *semiconductors* because the resistance of these elements lies somewhere between the low resistivity of metallic conductors and the high resistivity of insulators.

A significant characteristic of the atomic structure of these materials is that they possess four valence electrons. You will recall that it is the valence electrons that largely determine the chemical and electrical properties of an element. Now if these elements are prepared in a crystalline form, the valence electrons in the outer ring of any one atom align themselves with the valence electrons of adjacent atoms to form pairs of shared electrons called *covalent bonds*. These covalent bonds bind the silicon or germanium atoms into an orderly geometric pattern within the crystal.

For our purposes the individual silicon or germanium atom may be represented by the schematic form shown in Fig. 2-1. Each line emanating from the nucleus

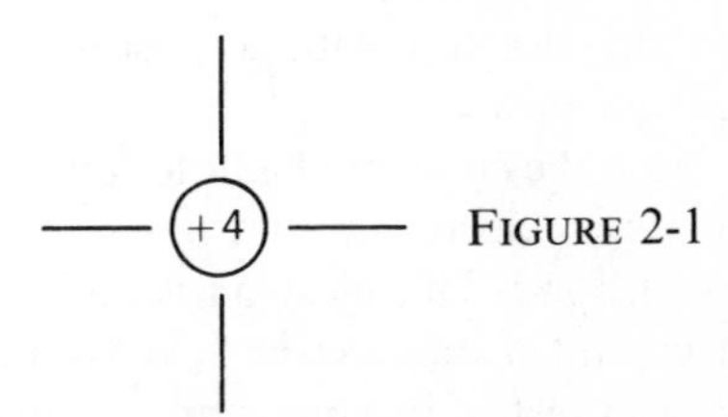

FIGURE 2-1

represents a valence electron. We will consider only the four protons necessary to balance the four valence electrons in order to keep the normal atom electrically neutral. Figure 2-2 illustrates how each silicon or germanium atom is linked by covalent bonds to each adjacent atom.

2-2 Intrinsic conduction

In a pure semiconductor specimen few electrons are available for conduction at low or moderate temperatures, because most of the valence electrons are tightly bound in covalent bonds. The semiconductor specimen would therefore exhibit a rather high resistance. However, if the temperature of the crystal is increased, we would find that the ohmic resistance of the specimen had decreased. This negative temperature coefficient of resistance is due to the temperature rise imparting additional kinetic energy to the valence electrons. The

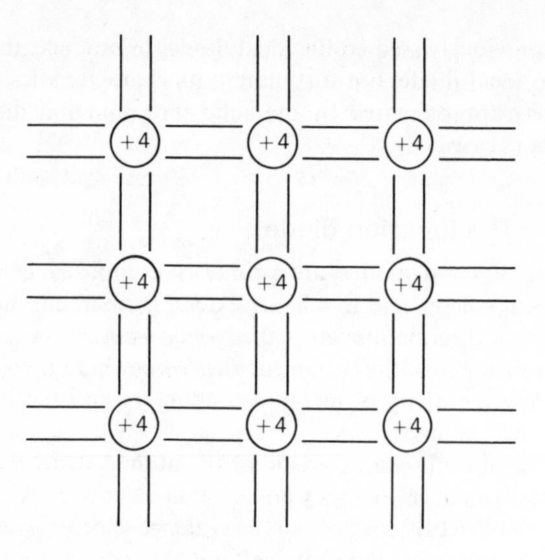

FIGURE 2-2

additional energy enables some of the valence electrons to break free of their covalent bonds. These free electrons are then available to act as current carriers under the influence of an applied electric field.

The vacancy in the covalent bond where the free electron used to be is called a *hole*. It turns out that the hole behaves as an additional and positively charged type of current carrier. Although there are some significant differences, we may consider the hole as being equivalent to an electron with a positive charge for most practical purposes.

To better grasp the concept of a hole consider Fig. 2-3 which illustrates an electron *e* that has been torn loose from a covalent bond. This electron is now free to drift through the crystal if an electric field is applied. The field could be established by connecting a voltage source to each end of the crystal. Now, the electron constitutes

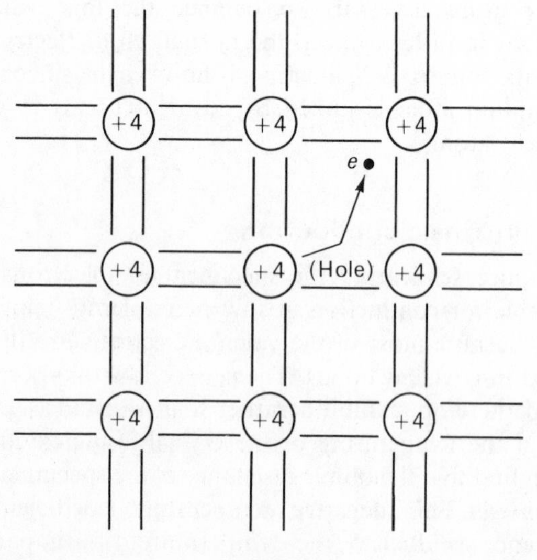

FIGURE 2-3

one current carrier when considered by itself. Notice that where the covalent bond was ruptured there remains an electron vacancy or hole. If we consider that the removed electron was balanced by an equal and opposite charge due to the surrounding nuclei, we should expect to find that this hole has an equal but opposite charge to that of the electron. This seems reasonable since the region was electrically neutral prior to the removal of the electron. This hole will then attract an electron from a neighboring atom's covalent bond, and it, too, will leave a hole in its former position. A third electron will then fill the hole left by the second electron, and so on. If an electric field is applied, the holes, which behave as mobile positive charges, drift towards the negative end of the crystal, while the electrons drift towards the positive end. The hole is also a current carrier, and on reaching the negative end of the crystal it is neutralized by an electron from the wiring attached to the negative terminal of the voltage source. Thus we see that breaking a covalent bond produces not one but two current carriers.

It should be noted that the holes do not flow in the external circuit but only within the semiconductor. This type of conduction, involving the generation of hole-electron pairs, is known as *intrinsic conduction.* Later we shall see that the presence of hole-electron pairs is usually undesirable, and that we want just holes or just electrons. It should also be noted that we have taken many liberties in order to simplify the solid-state physics involved. Nevertheless, the presentation is of sufficient depth to impart a good intuitive feel for understanding diodes and transistors.

The conductivity of silicon or germanium will increase as the number of current carriers increases. Conversely, it will decrease as the number of current carriers decreases. We could, therefore, control the conductivity by disrupting covalent bonds, but this is undesirable for two reasons. First, our control source must furnish a good deal of energy to break a covalent bond. Second, current carriers of both types (holes and electrons) are produced in equal amounts. Other means were therefore sought to control conductivity in semiconductors; it was found that the addition of certain impurities would reduce these undesirable features.

2-3 Doped semiconductors

To see how the addition of certain impurity atoms affects the conductivity of a semiconductor, consider the addition of a small amount of arsenic to a batch of pure, molten silicon. Figure 2-4 is a schematic representation of an arsenic atom which has five valence electrons. When the melt solidifies into a crystal, the arsenic atoms will be uniformly dispersed throughout the crystal structure. Also, since there are so many

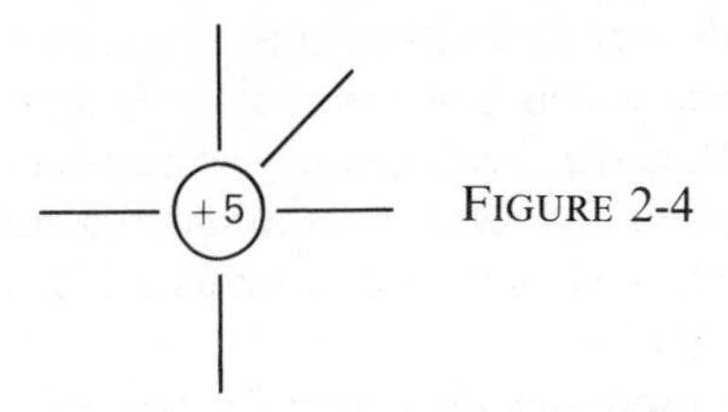

FIGURE 2-4

silicon atoms compared with arsenic atoms, virtually every arsenic atom will be surrounded by silicon atoms. This is illustrated in Fig. 2-5. Now arsenic has five valence electrons compared with four for silicon or germanium. Thus, in the crystal, four of arsenic's five

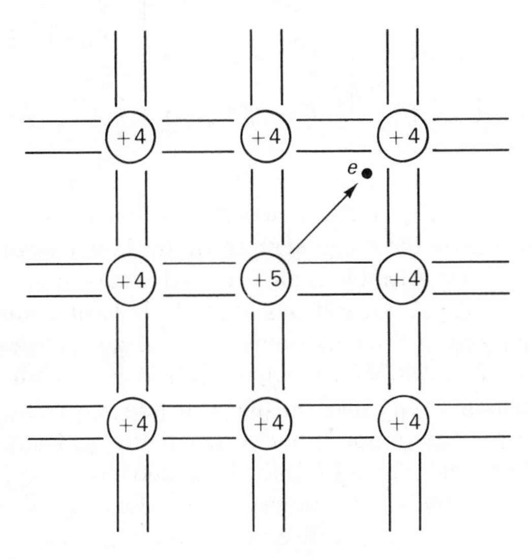

FIGURE 2-5

valence electrons will enter into covalent bonds with the surrounding silicon (or germanium) atoms. This, however, leaves a fifth, loosely-bound arsenic valence electron with no particular place to go. If this electron drifts away from the parent arsenic atom due to thermal agitation or other causes, there remains a positively charged arsenic ion which is rigidly bound into the crystal structure. This must be so since this region was electrically neutral prior to the loss of the electron.

Since this fifth valence electron of the arsenic atom is so loosely bound to its parent atom, it requires only a small amount of energy to break it loose and make it available for conduction under the influence of an electric field. Most significantly, note that this electron did not come from a covalent bond. This means that no hole is left behind. A hole, you will recall, is an electron vacancy in a covalent bond. Therefore, by adding a pentavalent (five valence electrons) impurity, we have overcome the two objections associated with varying conductivity obtained by controlling production of electron-hole pairs. Other pentavalent impurities such as phosphorus or antimony could have also been used to give us a form of silicon or germanium which is abundant in loosely bound electrons. This electron-rich semiconductor is known as *N*-type, and the pentavalent impurities are called *donor atoms.* The process of adding impurity atoms to a pure semiconductor is called *doping.*

It is also possible to produce *P*-type silicon, which is rich in holes, by adding minute amounts of a trivalent (three valence electrons) impurity to the melt. Typical trivalent impurities (*acceptor atoms*) are indium, aluminum, and gallium. Figure 2-6 illustrates a schematic

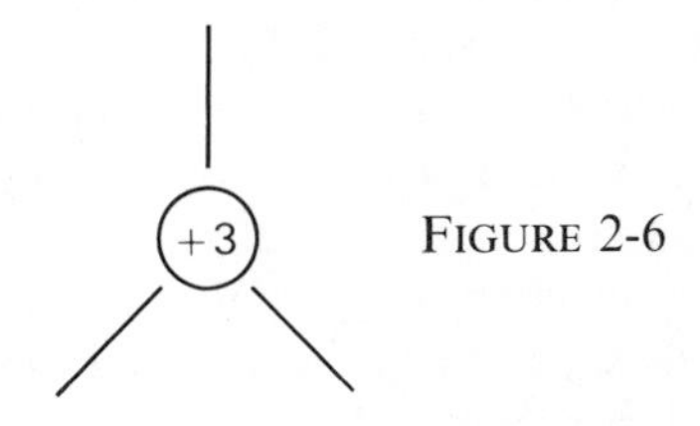

FIGURE 2-6

representation of a trivalent impurity atom. Figure 2-7 shows the effect of a trivalent impurity within the lattice structure. Note that we now have an incomplete covalent bond which is short an electron. This is in

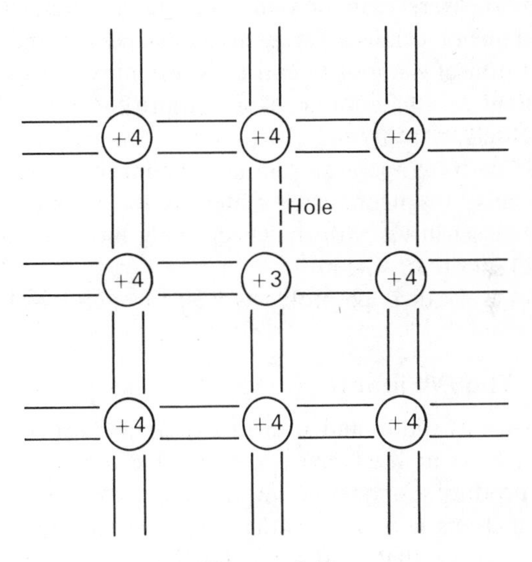

FIGURE 2-7

agreement with our previous definition of a hole. In this case, the hole is the only current carrier, because no loosely bound electrons are present. Here we find an abundance of free positive charges and hence the material is called *P*-type silicon or germanium as the case may be. Note in this figure that the impurity atom becomes a bound negative ion as soon as an electron from an adjacent covalent bond drifts into the hole. This original hole is then neutralized, but another hole appears where the neutralizing electron came from.

If small but equal amounts of *P*- and *N*-type impurities were mixed in the melt, we would have a crystal which behaved very much like a pure semiconductor. The reason for this is that the *N*-type impurity electrons fill the *P*-type impurity holes. It should be realized that no practical specimen is solely *N*-type or *P*-type but controlled addition of impurities (doping) causes one or the other to predominate.

More energy is required to break up a covalent bond in silicon than germanium. This means that, at a given temperature, pure Si has fewer current carriers available than pure Ge. Although the resistivity of pure Si is higher, which may prove a disadvantage in some cases, the overall effect of temperature upon Si transistors will be less than upon Ge. For this reason silicon transistors are usually preferred.

Electrons in an *N* region or holes in a *P* region are for obvious reasons called *majority carriers*, whereas electrons in a *P* region or holes in an *N* region are called *minority carriers*. Due to rupture of covalent bonds no semiconductor material is purely *N* or *P*. Instead it contains both majority and minority carriers. The majority carriers are continuously recombining with minority carriers and neutralizing them. The minority carriers, however, do not get used up as they are being continuously recreated by thermal energy. Neither do the majority carriers get used up, since the thermal generation of a minority carrier is accompanied by the simultaneous generation of a majority carrier. Unfortunately, the thermal generation of minority carriers varies with temperature in an exponential manner. This causes semiconductor materials to be quite temperature sensitive, with the effect being more severe in germanium than silicon.

Now work through problems P2-1 through P2-11.

2-4 The *PN* junction

The various diode and transistor manufacturing techniques have at least one common objective, and that is to produce a crystal in which there are one or more PN junctions present. It is the unique properties of the PN junction that make rectification and transistor action possible.

If a semiconductor crystal is so prepared that there exists a layer of *P*-type material adjacent to a layer of *N*-type material, the interface between the two is known as a *PN junction*. This situation is illustrated in Fig. 2-8.

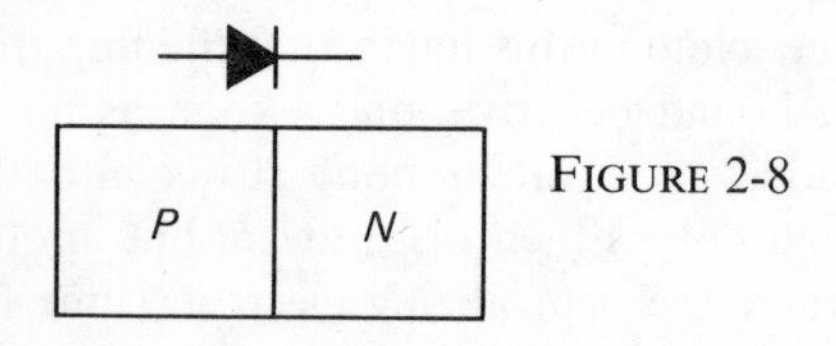

FIGURE 2-8

If some leads are attached to this structure we have a junction diode which is schematically illustrated above the crystal. Note that the anode (arrowhead) corresponds to the *P* material, whereas the cathode corresponds to the *N* material. Memorize this figure as it will prove invaluable later.

To understand the properties of the *PN* junction, consider Fig. 2-9. For purposes of illustration, assume that at time zero we somehow form a single crystal in which a *P*-type region interfaces with an *N*-type region.

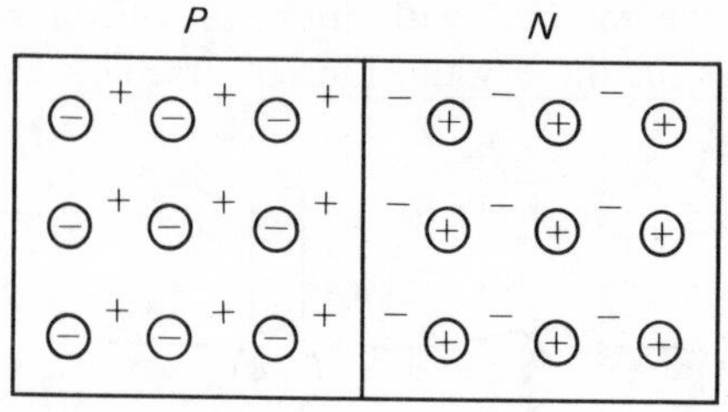

FIGURE 2-9

This interface is a *PN* junction. The circles with the minus signs in the *P* region represent the bound negative impurity ions. These ions are negatively charged because they have captured electrons to fill the holes introduced by the acceptor (*P*-type) atoms. The plus signs represent the free holes drifting throughout the *P* region. Similarly, the encircled plus signs in the *N* region represent the bound positive donor impurity ions that have lost their loosely bound electron. Both the *P* and the *N* regions are electrically neutral, however, since there are just as many free holes as negative ions in the *P* region and as many free electrons as positive ions in the *N* region.

Assume that the *PN* junction has just been formed, that the temperature is constant, and that no voltage is impressed across the crystal. The *P* side is loaded with free holes and bound negative ions, while the *N* side abounds in free electrons and immobile positive ions. Since the hole concentration on the *P* side is so much greater than that on the *N* side, the holes will diffuse into the *N* region. The diffusion mechanism is similar to the uniform distribution of ink molecules in a glass of water after an ink drop has been introduced. The ink molecules try to distribute themselves uniformly. In technical parlance we say that a hole concentration gradient exists from the *P* to *N* region. Similarly, an electron concentration gradient exists from the *N* to *P* region and results in electrons diffusing across the junction.

At first glance, it would seem that the holes and electrons would keep diffusing across the junction and recombine with each other until no current carrier remained or until just one or the other kind of charge carrier remained. However, this is not the case. For each hole that crosses the junction from the *P* to the *N* side, there remains an unneutralized immobile negative ion on the P side. Similarly, each electron that crosses from the N to the P region leaves an unneutralized positive ion. These unneutralized, immobile ions on each side of the junction are called *uncovered charges*, and the electric field between them can be conveniently represented by a battery placed across the junction, as shown in Fig. 2-10.

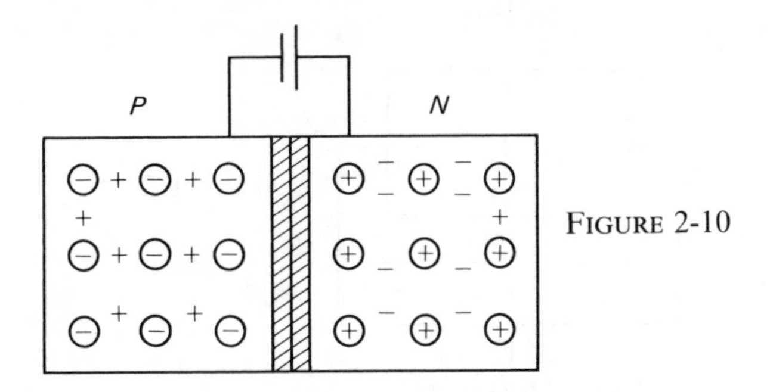

FIGURE 2-10

This internal barrier potential tends to restrain the diffusion of holes from the P side to the N side and electrons from the N side to the P side. Also, this internal barrier potential tends to sweep minority carriers, that is, electrons in the P material or holes in the N material, across the junction. Therefore, the region adjacent to and on each side of the junction is relatively free of holes and electrons. This essentially charge-free region is called a *depletion region.* The width of the depletion region is a function of the manner in which the crystal is prepared. The width of the depletion region on the P side need not be the same as on the N side. The side made of the higher resistivity material (fewer impurity atoms) will sustain a wider depletion region.

The holes that cross from the P to the N region recombine with electrons on the N side. Similarly, electrons from the N region recombine with holes on the P side. This flow of holes from the P to N side and electrons from the N to the P side constitutes a recombination current across the junction. This recombination current does not, however, persist at some constant value. Instead it falls to some very low value because the recombination process keeps uncovering charges in the vicinity of the junction. The uncovered negative ions on the P side start repelling the electrons from the N side while the wall of uncovered positive ions on the N side repels the holes from the P side. The battery in Fig. 2-10 therefore represents the barrier potential set up by the uncovered charges, which inhibits the recombination current. Thus it seems that a condition of equilibrium is established between the diffusive potential of the concentration gradient and the barrier potential of the uncovered charges.

If thermal agitation caused all the mobile charge carriers to have exactly the same kinetic energy, this simple explanation for equilibrium conditions at the barrier would suffice. However, the thermal energy imparted to the mobile charge carriers is randomly distributed. Statistically speaking, some holes and electrons have only a small amount of kinetic energy whereas others have a very large amount. Some of the high-energy carriers will from time to time be capable of overcoming the barrier potential. If this were the only action, it would seem that the barrier height would keep increasing in an effort to compensate for those high-energy carriers that manage to hurdle it. Ultimately, we might expect the last of the mobile charges to cross the barrier, leaving some large barrier potential.

This, however, is an incomplete although improved picture of conditions at the junction. The thing we are neglecting is that no material is perfectly P or N. The P material will have some free electrons in it caused by the rupture of covalent bonds by thermal agitation. The hole which is produced is no different from any other hole in the P side, where holes are obviously the majority current carriers. The electron in the P material constitutes a minority carrier, and it will have some average time (called lifetime) before it combines with one of the numerous holes available. The lifetime of a minority carrier clearly depends upon the number of surrounding majority carriers, which in turn is determined by the number of impurity atoms introduced into the crystal lattice.

If this electron in the P region survives long enough to drift into the vicinity of the junction, it will come under the influence of the electric field existing there. The direction of the field is such that the electron will be swept across the depletion region (region containing the uncovered charges) since it is attracted by the uncovered positive ions on the N side. Another way of visualizing this is to imagine the barrier battery in Fig. 2-10 forcing electrons from the P to the N side.

By similar reasoning we see that a thermally generated hole in the N material constitutes a minority carrier that would be swept across the depletion region from the N to the P side. The flow of thermally generated minority carriers across the junction is *aided* by the potential barrier.

We now have a complete picture as shown in Fig. 2-11 below.

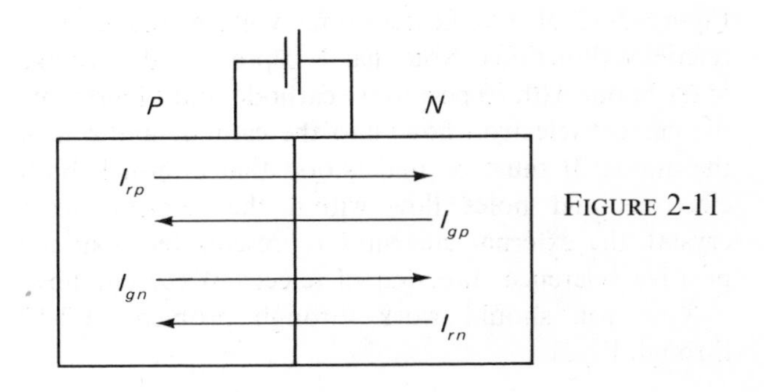

FIGURE 2-11

With no external voltage applied, the actual equilibrium conditions are as follows. There will be a net recombination current I_r across the junction which consists of holes I_{rp} climbing the barrier from the P to the N side and electrons I_{rn}, which climb the barrier in the opposite direction. Since a hole going from left to right is equivalent to an electron going from right to left, the two recombination current components are additive, and we may write

$$I_r = I_{rp} + I_{rn}$$

At the same time the rupture of covalent bonds will cause a net thermally generated current I_g because the minority carriers are swept across the barrier. This thermally generated current will also have two components, a hole component I_{gp}, which flows from the N to the P region, and an electron component I_{gn}, which flows from the P to the N region. Therefore,

$$I_g = I_{gp} + I_{gn}$$

The thermally generated current I_g depends solely upon temperature and is sometimes called I_s, the saturation current. Under equilibrium conditions the carriers which cross the junction due to I_g replace those due to I_r, which has components flowing oppositely to those of I_g. The net result is that the total junction current is zero, which it must be, since shorting a PN junction with a piece of wire does not result in a current flow through the wire. The barrier height will assume a potential of such value that it permits the recombination current to just equal the thermally generated current.

The schematic representation of a junction diode is again shown in Fig. 2-12. As shown, the anode consists of the P material and the cathode consists of the N material.

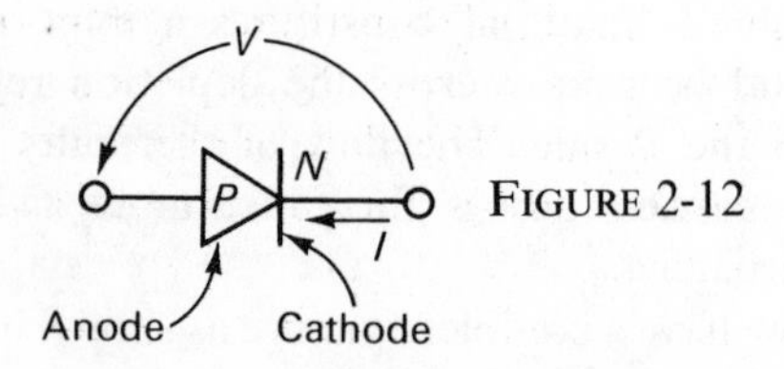

FIGURE 2-12

Figure 2-12 also indicates some voltage and current reference directions. Note that V represents the voltage at the anode with respect to the cathode, and I represents the current (electron flow) into the cathode and out of the anode. It must be understood that although both electrons and holes flow within the semiconductor crystal, the external current I represents the assumed positive reference direction of (electron) current flow.

Now you should work through problems P2-12 through P2-21.

2-5 Forward bias

If a voltage source is placed across the diode as shown in Fig. 2-13, the diode is said to be forward-biased. Note that when forward-biased, V is a positive quantity which means the P material is maintained at a positive

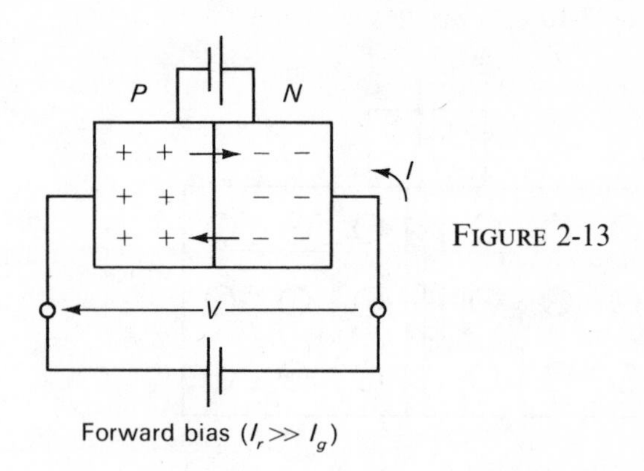

FIGURE 2-13

potential with respect to the N material. The external voltage V sets up an electric field across the junction which opposes the barrier potential and thereby reduces its effect. Consequently, the recombination (majority) current is increased. Intuitively, we can see that the voltage V will tend to push holes from the P side to the N side and electrons from the N side to the P side, which vastly increases the current across the junction. In fact, it is necessary to either keep V quite small or insert a current limiting resistor in series with the voltage source to keep the diode current at a reasonable value. However, the thermally generated current is virtually unaffected since this current depends, in theory at least, only upon the temperature rather than upon the applied voltage.

A theoretical relationship between the diode current and the applied external voltage is

1) $$I \quad I_S(\epsilon^{V/\eta T} - 1) \qquad (2\text{-}1)$$

In this equation I_S is the thermally generated leakage current, sometimes called a saturation current. The voltage V represents the voltage of the P material relative to the N material and could be a positive or negative quantity depending upon the polarity of the applied voltage. If V is a positive quantity, it is sometimes called the forward voltage V_F. If V is a negative quantity, it is sometimes called a *reverse voltage*, V_R. Correspondingly, the current I in the reference direction shown can be a positive or a negative quantity, depending upon the polarity of the applied voltage. If the current is a positive quantity, it is sometimes called a forward current I_F, the condition corresponding to the application of a forward voltage V_F. On the other hand, if I is a negative quantity, it is sometimes called a reverse current I_R which corresponds to a condition in which V is a negative quantity or a reverse voltage. The quantity V_T has the dimensions of volts and is given by

2) $$V_T = \frac{T}{11{,}600} \text{ volts} \qquad (2\text{-}2)$$

T is the temperature in degrees Kelvin and is related to the temperature in degrees centigrade by

3) $$T \text{ in °Kelvin} = \text{°C} + 273$$

4) $$\eta = \text{a number between 1 and 2}$$

The quantity η is a relative constant which is generally taken as 1 for germanium. For silicon, η may vary from approximately 2 at low current levels, say less than 0.2 ma to approximately 1 at forward currents greater than 0.2 ma.

At normal room temperature, 27°C, $V_T = 26$ mv. Hence, we see that in equation 2-1 with η assumed $= 1$ and for $V \geqq 100$ mv, $I \approx I_S(\epsilon^{V/V_T})$. By similar reasoning, $I \cong -I_S$ if V is a negative quantity (reverse bias) in excess of 100 mv.

The slope of the forward-current versus forward-voltage characteristic represents the dynamic (ac) conductance of the diode when it is forward-biased. The theoretical value of this conductance g_f is readily obtained by differentiating equation 2-1 to obtain

5) $$g_f = \frac{dI}{dV} \approx \frac{\Delta I}{\Delta V} = \frac{I_S\epsilon^{V/\eta V_T}}{\eta V_T} \qquad (2\text{-}3)$$

From equation 1 we see that for values of $V > 100$ mv, $I \approx I_S\epsilon^{V/\eta V_T}$ and consequently

6) $$g_f \approx \frac{I}{\eta V_T}$$

The reciprocal of the forward conductance is the *forward resistance.* Therefore, the forward dynamic resistance is

7) $$r_f = \frac{1}{g_f} = \frac{\eta V_T}{I} \qquad (2\text{-}4)$$

The current I is the static (dc) current through the diode. Equation 2-4 indicates that for an assumed value of $\eta = 1$, a forward-biased diode will exhibit a dynamic resistance of 26 ohms at a forward current of 1 ma and 2.6 ohms at $I = 10$ ma. The actual forward dynamic resistance will always be somewhat higher than this theoretical value due to the effect of bulk and lead resistance which appear in series with r_f.

Figure 2-14 illustrates the typical forward characteristic of a low-power silicon junction diode. The forward current is plotted for three different values of temperature. Evidently, as the temperature increases, the forward voltage decreases. Specifically, if the diode current is held constant, the forward voltage will typically decrease at the rate of 2 to 3 mv for each °C of temperature rise. Mathematically, this is usually expressed as

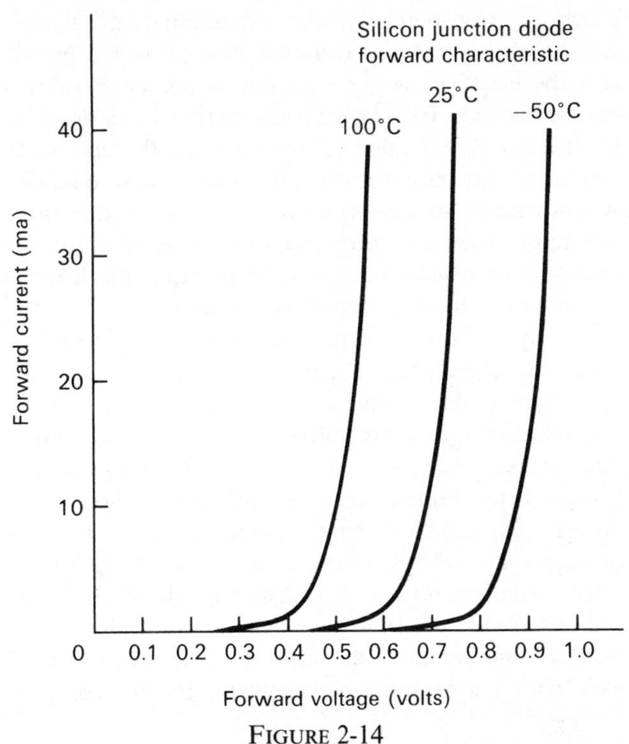

FIGURE 2-14

8) $$\frac{dV_F}{dT} \approx -2.5 \text{ mv/°C} \qquad (2\text{-}5)$$

Note that this temperature coefficient is negative, which implies the forward voltage decreases with a rise in temperature. In practice this temperature coefficient tends to become less negative as the forward current increases and that at high current levels may even become slightly positive.

2-6 Reverse bias

If, as shown in Fig. 2-15, V is made a negative quantity so that the P material is held negative with respect to the N material, the diode is said to be reverse-biased. When reverse-biased, the electric field established by V is of such a polarity that it aids the internal barrier

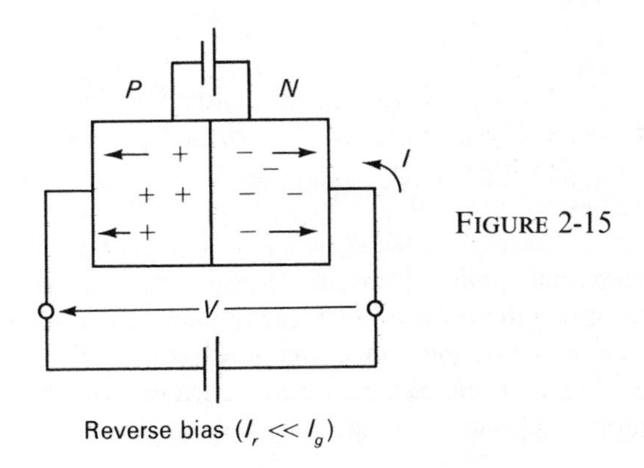

FIGURE 2-15

potential. Consequently, the recombination current, which consists of holes going from the P to the N side and the electrons going from the N to the P side, is drastically reduced. The external current I is a negative quantity and it is due to the flow of the thermally generated carriers across the junction. Ideally, the diode would exhibit no reverse current, and hence this small thermally generated reverse current is commonly referred to as a *leakage current.* In practice, the leakage current actually consists of two components; I_S, the thermally generated component which is not dependent upon the magnitude of the reverse bias (it only depends upon temperature), and a component due to surface leakage effects, which manifests itself where the junction terminates at the edges of the crystal. The component of leakage current due to these surface effects is somewhat voltage sensitive so that the leakage current in practice does increase with an increase in reverse bias. This is illustrated in the reverse characteristic of the silicon diode illustrated in Fig. 2-16. Note that I_S shown in the dashed line is some small and constant value which would vary with temperature but not with the reverse bias.

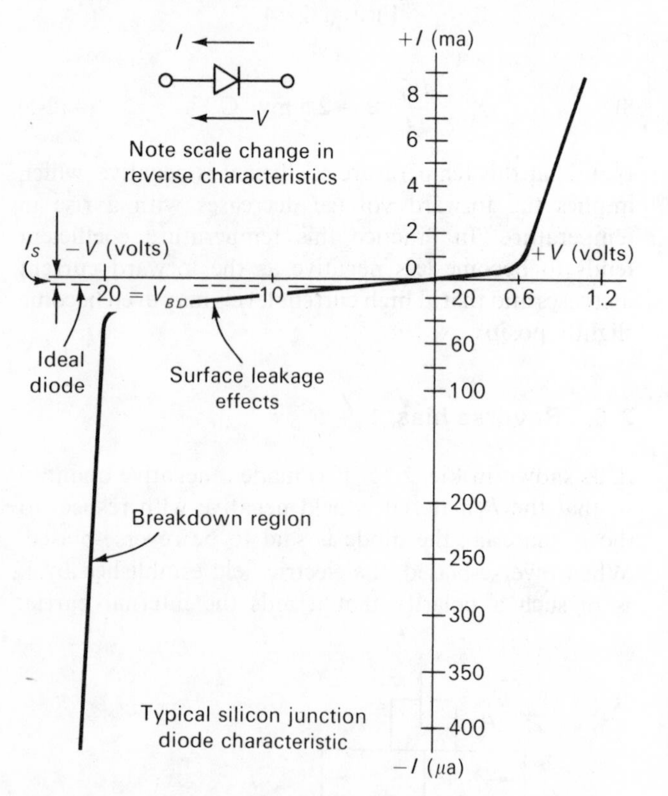

FIGURE 2-16

The actual reverse current, due to surface leakage effects, exhibits an ohmic component of resistance across the junction, since an increase in reverse bias does cause an increase in reverse current. As a rule of thumb, the thermally generated component may be assumed to roughly double for every 10°C rise in temperature.

When a PN junction is reverse-biased, most of the externally applied reverse voltage appears across the depletion region because it is in this carrier-free region that the highest resistance of the circuit manifests itself. In other words, the voltage drop across the depletion region will, in general, almost be equal to the total applied voltage. The depletion region is very narrow, being typically on the order of 0.0001 in. or less. Hence, a reverse voltage as little as, say, 6 volts would develop an electric field strength of 60,000 volts/in. This high field intensity may cause the junction to break down. When a junction breaks down, its impedance drops considerably due to the production of many additional current carriers by an ionization and/or secondary emission phenomena.

Diodes deliberately designed to be used in the breakdown region are commonly referred to as zener or avalanche diodes. Another term sometimes used for these diodes is regulator diodes. This term arises from the fact that in the breakdown region (note Fig. 2-16) the reverse voltage in this example is maintained at a substantially constant value of 20 volts. If the diode power dissipation is limited to a safe value by restricting the current through it in the breakdown region, then operation in the breakdown region is not destructive. The breakdown voltage, which may range from 3 volts to several hundred volts, is determined by the manufacturing process.

The reverse breakdown voltage V_{BD} is somewhat temperature sensitive. For diodes which break down around 5 volts or less, V_{BD} exhibits a negative temperature coefficient. That is, V_{BD} decreases with increasing temperature. On the other hand, diodes which exhibit a V_{BD} of about 6 volts or more tend to exhibit a positive temperature coefficient. That is, V_{BD} increases with increasing temperature. In the range between 5 and 6 volts, it is possible to find diodes which exhibit an almost zero temperature coefficient, which makes them highly valuable as voltage reference sources.

2-7 Barrier capacitance

We previously mentioned that the depletion region, which is relatively free of mobile charge carriers, looks like a high resistance. In fact, it looks more like an insulator which may be considered the dielectric of a capacitor. The P and N regions on each side of the depletion region are loaded with mobile charge carriers and look like conductors which may be considered the plates of a capacitor. This apparent capacity of the junction is called the *barrier* or *junction capacitance.* This capacitance is voltage sensitive, and it varies inversely with the junction voltage raised to some power, usually between ⅓ and ½. Hence, as the magnitude of the reverse voltage increases, the depletion region width increases and the junction capacitance decreases. The voltage sensitivity of this capacitance is due to the fact that an external reverse bias aids the internal barrier potential in establishing the width of the depletion region. This phenomenon is put to use in the design of frequency selective circuits which are voltage controlled.

2-8 Diffusion capacitance

If a *PN* junction is forward-biased, the *N* side will contain some holes that were injected from the *P* side, but which had not yet combined with electrons. Likewise, the *P* side will contain electrons injected from the *N* side. The diode is then carrying some net forward current as shown in Fig. 2-17*a*. At some instant of time, t_0, we

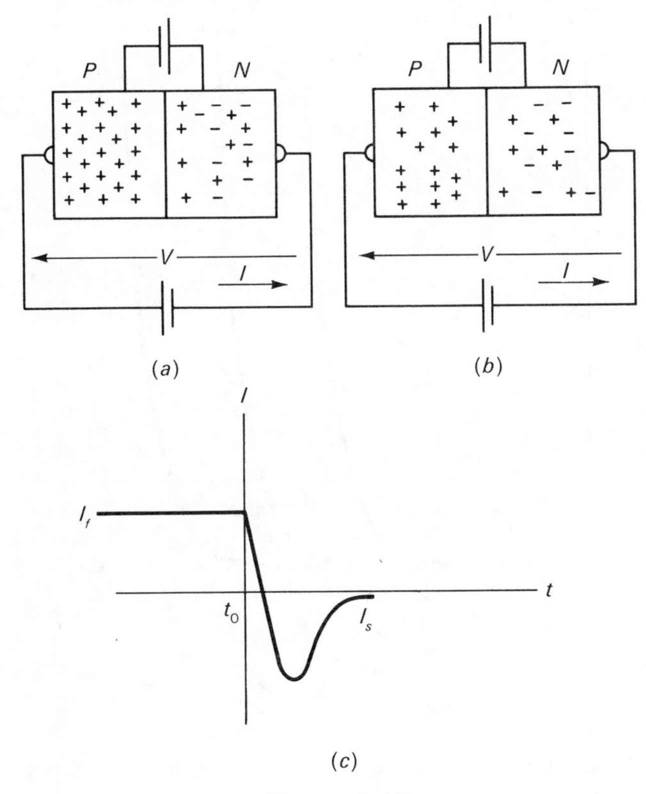

FIGURE 2-17

suddenly reverse-bias the diode as shown in Fig. 2-17*b*. At first glance, we might suspect that the diode current would immediately fall to the small value of reverse current which normally flows in the diode when reverse-biased. However, we would find that actually a large pulse of reverse current flows before the reverse current stabilizes at its saturation value. This is illustrated in Fig. 2.17*c*. Evidently, there is a recovery time involved before the diode assumes its high-impedance state when reverse-biased. The reason for this large pulse of reverse current is the presence of those holes in the *N* region and electrons in the *P* region which did not have a chance to recombine with majority carriers in the region they were injected into before the reverse bias was applied. Those holes which were injected into the *N* region will be swept back into the *P* region when the reverse bias is applied. Similarly, electrons trapped in the *P* region will be swept back into the *N* region. This constitutes a reverse current. Since it takes time for this charge to be moved, the diode cannot assume its OFF (nonconducting) state immediately. This effect is of significance only when the amount of forward bias is changed or when the diode is switched between forward and reverse bias.

A convenient way to regard this charge-storage effect is to visualize a capacitor C_D, called *diffusion capacitance*, as being in parallel with the junction barrier capacitance. When reverse-biased, the recombination current is negligibly small, and consequently there are virtually no injected carriers. Therefore, the diffusion capacitance appears very small. When forward-biased, the injected carriers result in a large diffusion capacitance. The diffusion capacitance varies directly with the forward current. The product of diode forward resistance and diffusion capacitance is a time constant $\tau_D = r_f C_D$ which limits the usefulness of the diode at high frequencies.

PROBLEMS WITH SOLUTIONS

PS 2-1 Estimate the output voltage in Fig. PS 2-1.

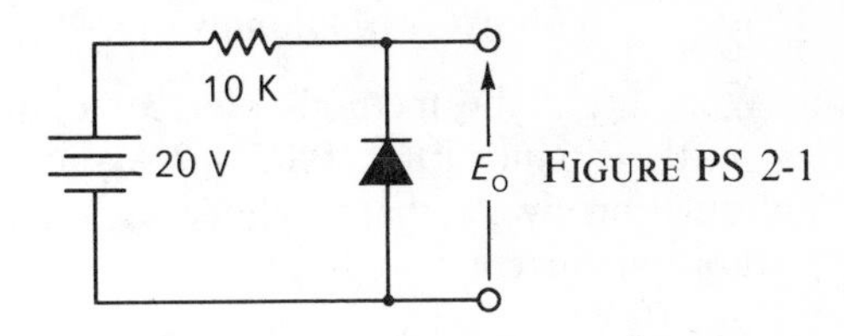

FIGURE PS 2-1

SOLUTION Since the diode is reverse-biased it may be assumed to exhibit a very high reverse (leakage) resistance, say, several hundred kilohms or more. Hence, treating the circuit as a voltage divider, we see that virtually the full 20 volts appears at the output. Thus E_0 = 20 volts.

PS 2-2 Estimate the output voltage in Fig. PS 2-2.

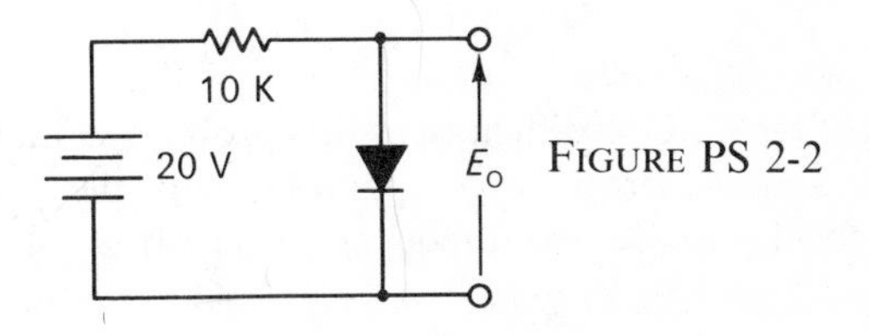

FIGURE PS 2-2

SOLUTION Plainly, the diode is forward-biased. For a germanium diode the forward voltage will be about 0.2 to 0.3 volts and for silicon about 0.6 to 0.7 volts. The accuracy of these estimates, of course, depends on the forward current, which is $I = (20 \text{ volts} - V_F)/10$ kilohms. Since V_F will (even though it isn't exactly known) be so small compared to 20 volts, we can safely assume $I = 2$ ma. For typical diodes, a forward current of 2 ma is consistent with the previously assumed voltage values. In most cases we could even assume the diode is ideal so that $V_F = 0 \text{ volts} = E_O$.

PS 2-3 Estimate the ac output voltage e_o in Fig. PS 2-3*a*.

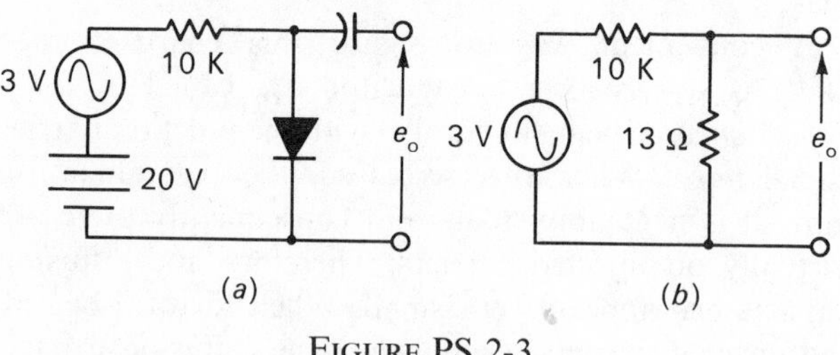

FIGURE PS 2-3

SOLUTION The ac output voltage will be the voltage which appears across the diode ac resistance, r_f, which, in turn, is a function of the dc diode current. The dc diode current is approximately 20 volts/10 kilohms = 2 ma. Since $r_f = V_T/I = 26 \text{ mv}/2 \text{ ma} = 13$ ohms (room temperature assumed) we arrive at the ac equivalent circuit of Fig. PS 2-3*b*. Therefore,

$$e_o = (3 \text{ volts}) \frac{13 \text{ ohms}}{13 \text{ ohms} + 10 \text{ kilohms}} = 3.9 \text{ mv}$$

PS 2-4 What are some methods you could use to estimate V in the circuit of Fig. PS 2-4*a*? Assume that the only information given is that the diode is germanium and the saturation current $I_s = 10\ \mu a$.

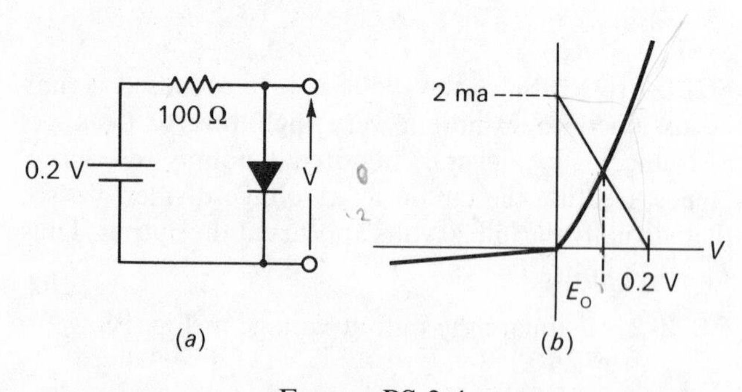

FIGURE PS 2-4

SOLUTION Our difficulty in this problem is that the diode voltage drop is comparable with the source voltage. Therefore the diode characteristics will significantly affect the circuit current. One method of solution is to assume values of V in the diode equation (equation 2-1) and calculate the corresponding value of I. We can assume $\eta = 1$, $V_T = 26$ mv, and $I_s = 10\ \mu a$. The V-I curve may then be plotted and a load line constructed on it as shown in Fig. PS 2-4*b*. The intersection of the load line and the diode curve is the *quiescent operating point*. For a review of load line techniques see Chapter 16 of "Outline for DC Circuit Analysis" in this series of texts.

PS 2-5 Determine the average temperature coefficient over the range from −50°C to 100°C for a forward-biased diode having the characteristics shown in Fig. PS 2-5. Assume the forward current is held constant at 20 ma.

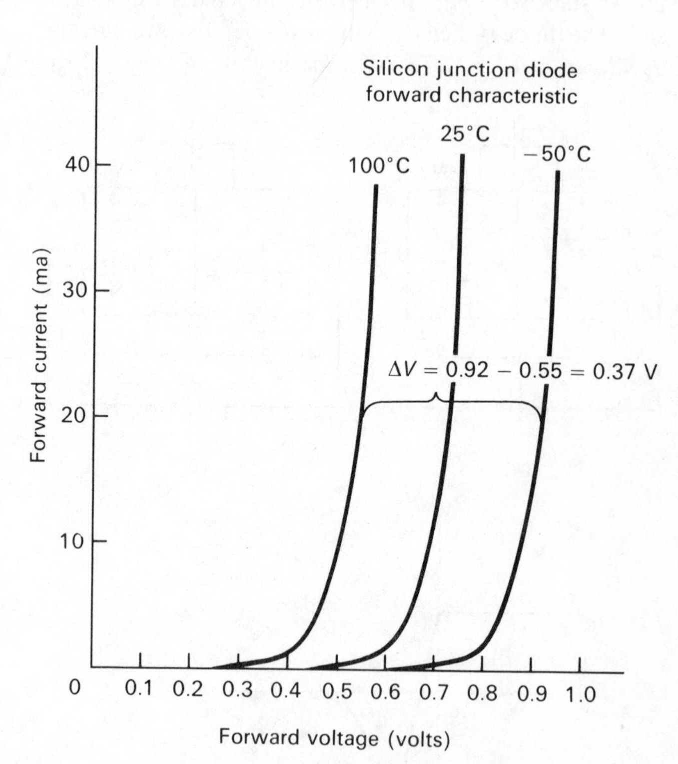

FIGURE PS 2-5

SOLUTION The temperature coefficient (T.C.) as determined from Fig. PS 2-5 is given by

$$\text{T.C.} = \left.\frac{\Delta V_F}{\Delta T}\right|_{\Delta I_F = 0} = \frac{0.92 \text{ volt} - 0.55 \text{ volt}}{-50°\text{C} - 100°\text{C}}$$

$$= -2.47 \text{ mv/°C}$$

PS 2-6 What is the static (dc) diode resistance at 25°C and at an operating point (Q point) of 20 ma for the diode represented by Fig. PS 2-6?

SOLUTION At $I_F = 20$ ma and at 25°C, $V_F = 0.73$ volt. Therefore, $R_F = 0.73 \text{ volt}/20 \text{ ma} = 36.5$ ohms.

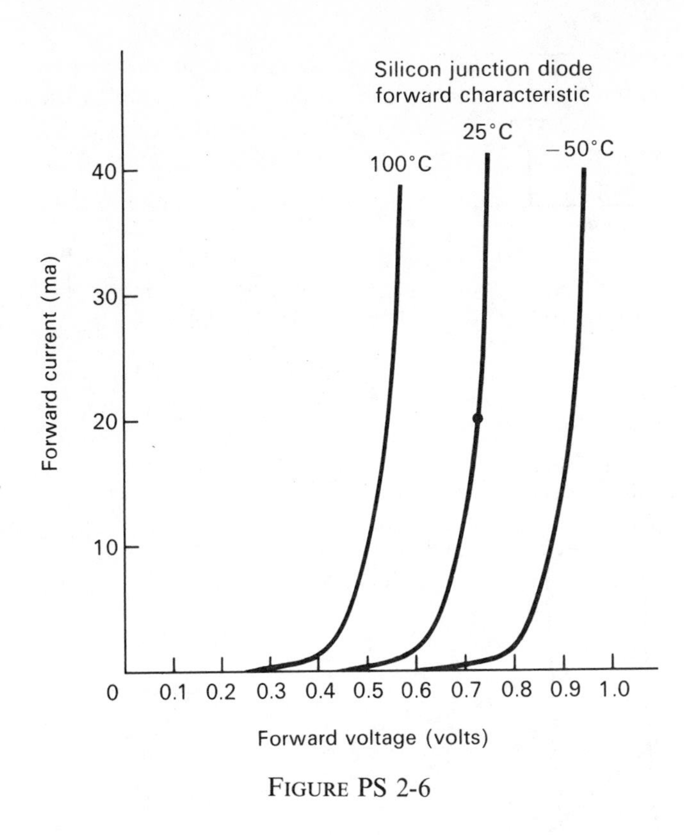

FIGURE PS 2-6

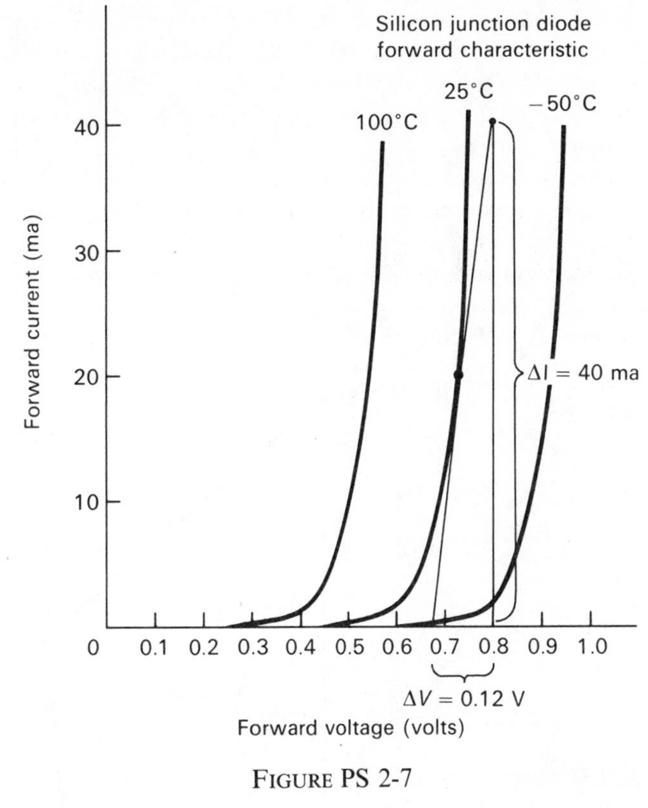

FIGURE PS 2-7

PS 2-7 What is the dynamic resistance at the 20 ma, 25°C operating point for the diode in Fig. PS 2-6?

SOLUTION Constructing a tangent line to the Q point as shown in Fig. PS 2-7, and then determining the reciprocal of its slope, we have

$$r_f = \frac{\Delta V_F}{\Delta I_F} = \frac{0.12 \text{ volt}}{40 \text{ ma}} = 3 \text{ ohms}$$

The theoretical dynamic junction resistance is (at 27°C)

$$r_f = 26 \text{ mv}/40 \text{ ma} = 0.65 \text{ ohms}$$

The bulk and lead resistance, which equals 3 ohms − 0.65 ohms = 2.35 ohms, primarily accounts for the discrepancy between the measured and theoretical values.

PS 2-8 Determine the Q point for a diode having the forward characteristics of Fig. PS 2-6 when it is used in the circuit of Fig. PS 2-8*a*.

SOLUTION Construct a load line on the diode V-I curve as shown in Fig. PS 2-8*b*. The load line is based on the equation $V_F = 1.5$ volts − (30 ohms)I_F. This is a linear equation requiring only two points to plot it. Therefore, if we first assume $I_F = 0$ we have $V_F =$ 1.5 volts as one point on the load line equation. Next assume $V_F = 0$ and calculate the corresponding I_F. Thus with $V_F = 0$, $I_F = 1.5$ volts/30 ohms = 50 ma. The intersection of the load line with the 25°C diode curve indicates that at the Q point $I_F = 25$ ma, $V_F = 0.74$ volt.

PS 2-9 The forward-voltage temperature coefficient for the diode of PS 2-8 was determined in PS 2-5 as T.C. = −2.47 mv/°C. How could this information be used to predict the Q point of the diode in the circuit of Fig. PS 2-8*a* at a temperature of −50°C? Assume $T = 25°$C initially.

SOLUTION The total temperature change is $\Delta T =$ (−50°C) − 25°C = −75°C. From the load line we see that the initial $I_F = 25$ ma at 25°C and $V_F = 740$ mv. At the final temperature, V_F will be given by

$$\begin{aligned} V_{F(\text{final})} &= V_{F(\text{initial})} + \text{T.C.}\,(\Delta T) \\ &= 740 \text{ mv} + (-2.47 \text{ mv/°C})(-75°\text{C}) \\ &= 740 \text{ mv} + 185 \text{ mv} = 925 \text{ mv} \end{aligned}$$

The final current is then

$$I_{F(\text{final})} = \frac{1500 \text{ mv} - 925 \text{ mv}}{30 \text{ ohms}} = 19.2 \text{ ma}$$

Checking the final current value with the intersection of the load line and the −50°C curve in Fig. PS 2-8*b*, we find excellent correlation between the calculated and graphic solutions.

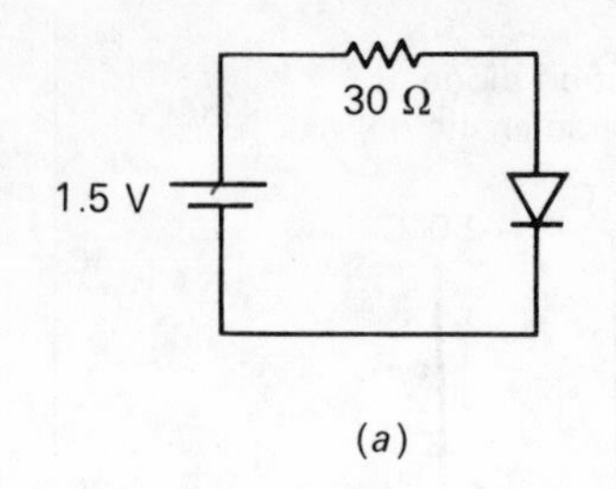

(*a*)

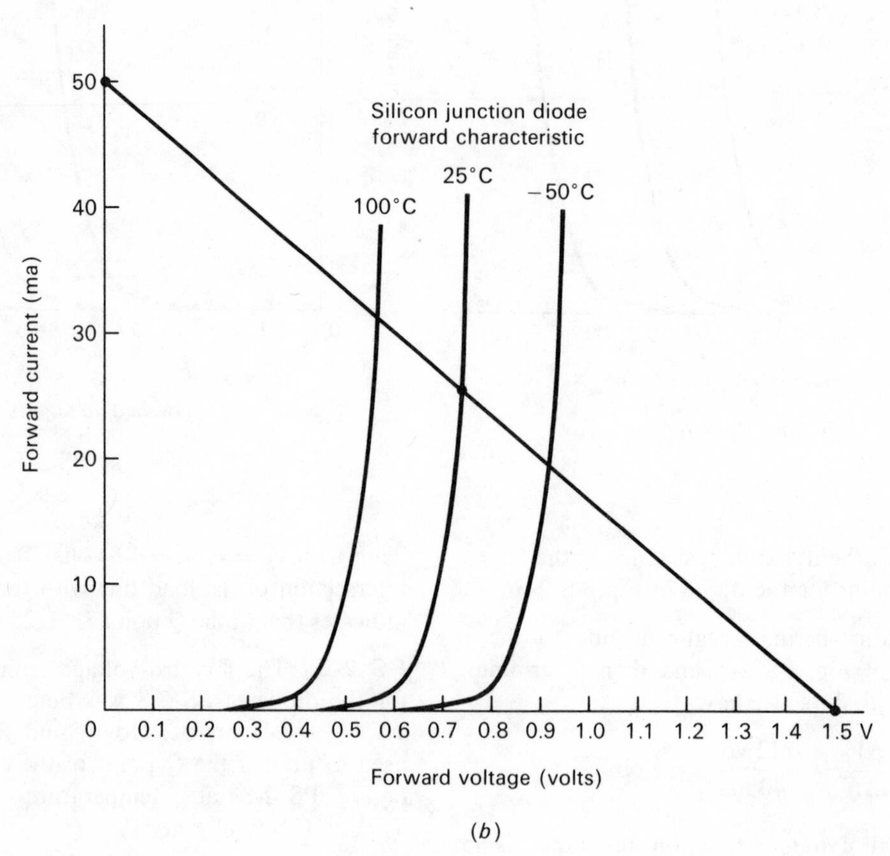

(*b*)

FIGURE PS 2-8

PROBLEMS WITH ANSWERS

PA 2-1 What are majority carriers in an N region?
ANSWER Electrons

PA 2-2 How is the recombination current in a diode controlled?
ANSWER By varying the forward bias.

PA 2-3 What happens to the width of the depletion region as the reverse voltage is varied?
ANSWER An increase in V_R widens the depletion region and a decrease in V_R will narrow it.

PA 2-4 How is the dynamic resistance related to the forward current in an ideal diode? Assume $T = 27°C$.
ANSWER $r_f = 26 \text{ mv}/I_F$

PA 2-5 What effect does decreasing the forward current have upon diffusion capacitance?
ANSWER The diffusion capacitance decreases.

PA 2-6 At a forward current of 12 ma and at 25°C, the static (dc) voltage drop across a diode is 0.31 volt. If the forward current is held constant but the temperature rises to 50°C, what would the new forward voltage approximately be?
ANSWER 0.248 volt

PA 2-7 The temperature coefficient of a zener diode is 0.08 percent/°C. If the diode exhibits voltage breakdown at 18 volts at 30°C, what would the breakdown voltage be at 80°C?
ANSWER 18.72 volts

PA 2-8 Determine E in Fig. PA 2-8.

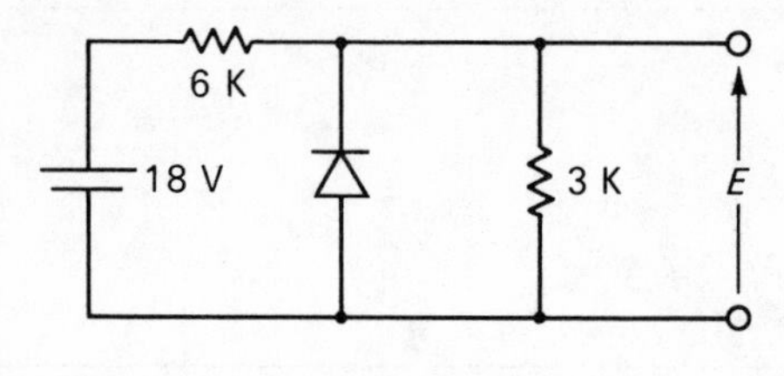

ANSWER $E = 6$ volts

PA 2-9 Determine E_1 and E_2 in Fig. PA 2-9.

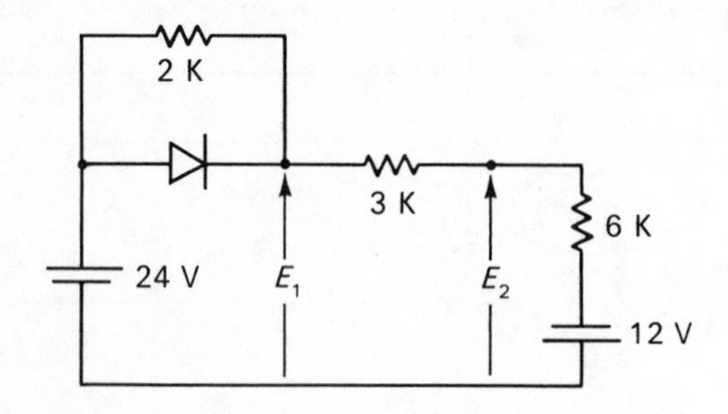

ANSWER $E_1 = 24$ volts
$E_2 = 12$ volts

PA 2-10 Determine e_o at room temperature in the circuit of Fig. PA 2-10. Assume the capacitors have negligible reactance and that the transformer has twice as many turns on the primary as on the secondary.

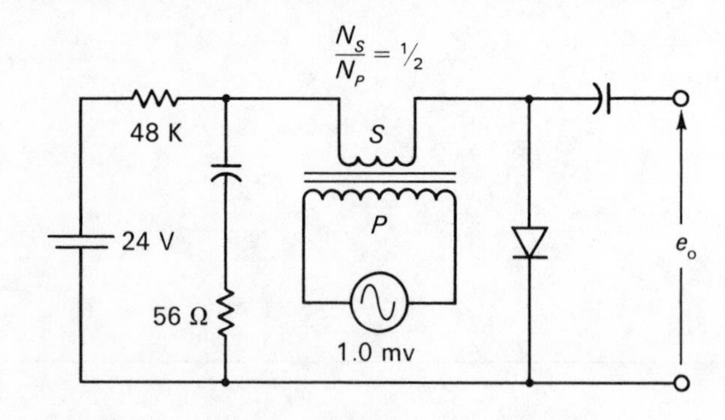

ANSWER $e_o = 0.25$ mv

PROBLEMS WITHOUT ANSWERS

P 2-1 Semiconductor diodes are commonly made from ________ or ________.

P 2-2 The *P* region is rich in ________.

P 2-3 ________ in the *N* region are majority carriers.

P 2-4 ________ in the *P* region are minority carriers.

P 2-5 In a semiconductor, the positive charge carriers are called ________.

P 2-6 In a semiconductor, the negative charge carriers are called ________.

P 2-7 The *P* region contains many ________ and few ________.

P 2-8 The *N* region contains many ________ and few ________.

P 2-9 Recombination is continuously occurring between __________ and __________.

P 2-10 Thermal agitation causes the creation of new __________ and __________.

P 2-11 Silicon is preferred to germanium at high temperatures because __________.

P 2-12 The boundary between the *P* and *N* regions in a semiconductor crystal is called a __________ __________.

P 2-13 The recombination of __________ from the *P* region to the *N* region and electrons from the __________ region to the *P* region is inhibited by the presence of the internal barrier potential.

P 2-14 The polarity of the internal barrier potential is such that the *P* region appears __________ with respect to the *N* region.

P 2-15 The polarity of the internal barrier potential is such that it tends to restrict the flow of __________ from the *N* region to the __________ region, but it enhances the flow of __________ from the *N* region to the __________ region and __________ from the *P* region to the *N* region.

P 2-16 The region about the junction which is relatively free of charge carriers is called the ________ region.

P 2-17 The width of the ________ region will be larger on the side of the junction which has the ______ resistivity.

P 2-18 The ________ field of the internal barrier potential manifests itself across the ________ region.

P 2-19 The recombination current consists of holes going from the ________ region to the ________ region and electrons going from the ________ region to the ________ region.

P 2-20 The thermally generated current consists of electrons going from the ______ region to the ______ region and holes going from the ________ region to the ________ region.

P 2-21 Sketch the schematic symbol for a junction diode. Identify the cathode and the anode. Which is the *P* material and which is the *N* material? What are the voltage and current reference directions that we have agreed upon for the diode?

P 2-22 What are minority carriers in a *P* region?

P 2-23 What effect does forward-biasing a diode have upon the thermally generated current?

P 2-24 What is meant by the depletion region?

P 2-25 How does the capacity of a reverse-biased junction relate to the applied voltage?

P 2-26 Is voltage breakdown necessarily destructive?

P 2-27 Explain diffusion capacitance.

P 2-28 How would you measure the thermally generated component of diode reversed current?

P 2-29 A diode exhibits a dynamic forward resistance of 13 ohms at 27°C. If the temperature rises to 50°C, determine the new forward resistance. Assume the forward current remains constant.

P 2-30 If the diode in the circuit of Fig. P 2-30*a* has the forward characteristic shown in Fig. P 2-30*b*, determine I and V at 25°C.

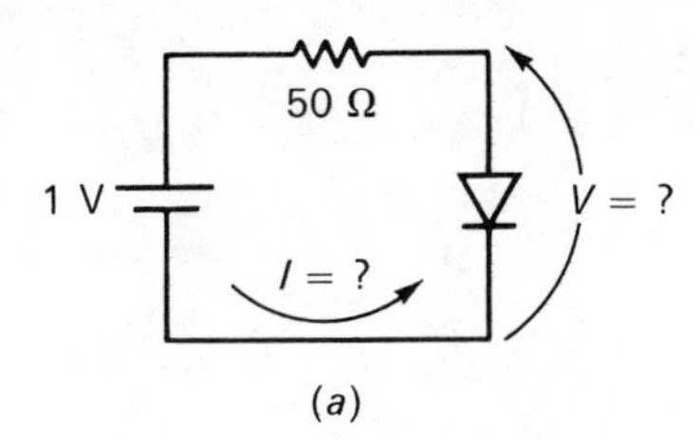

(*a*)

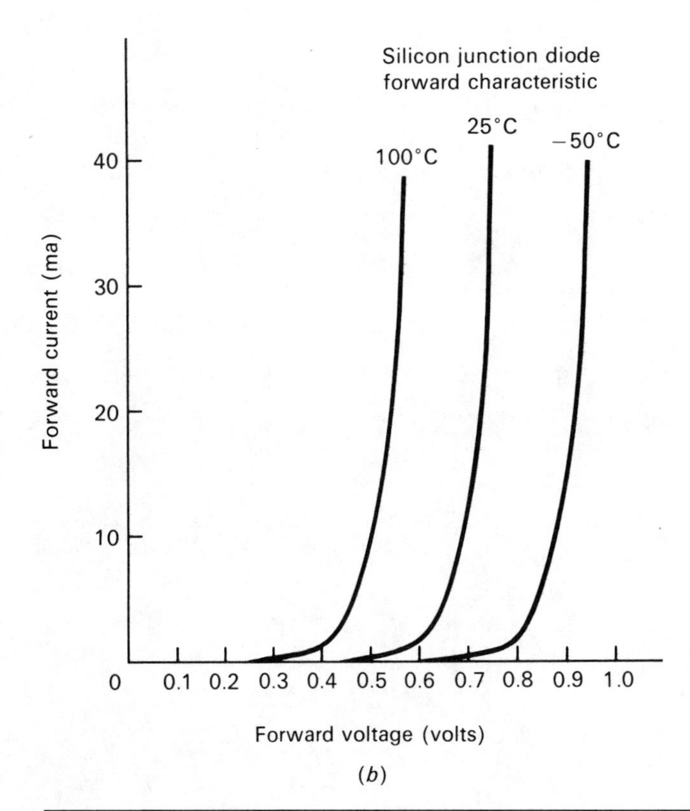

(*b*)

P 2-31 A diode has a reverse leakage current of 1.2 ma at 25°C. Estimate the leakage current at 50°C.

P 2-32 Determine E and I in Fig. P 2-32.

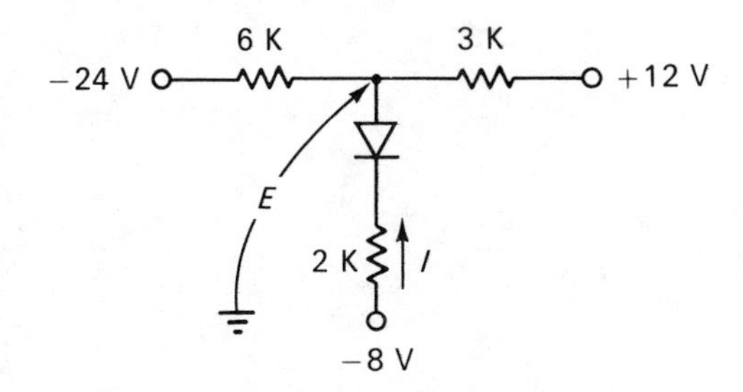

P 2-33 Determine the range of e_o if R is varied from 1 kilohm to 12 kilohms in the circuit of Fig. P 2-33.

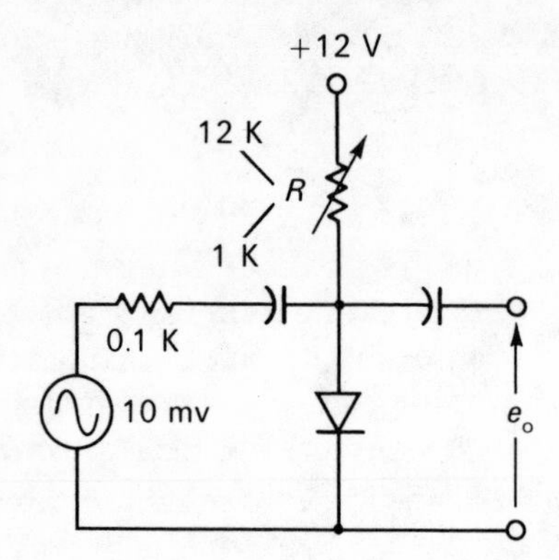

3 diode equivalent circuits

Since diodes and diode-like devices are commonly encountered in electronic circuits, it becomes most useful to develop an equivalent circuit for the practical diode. As we shall see, the practical diode may be represented in terms of the ideal diode.

3-1 Diode piecewise models

One way of developing a diode equivalent circuit is to use the piecewise modeling technique.[1] Essentially, this consists of breaking any nonlinear characteristic curve into a number of straight line segments that will approximate the curve to any desired degree. An equivalent circuit based upon ideal diodes may then be evolved. The greater the number of straight line segments used, the more complicated the equivalent circuit becomes.

To illustrate the development of a diode piecewise model, consider the fictitious diode V-I curve of Fig. 3-1*a* which is based upon the diode voltage and current

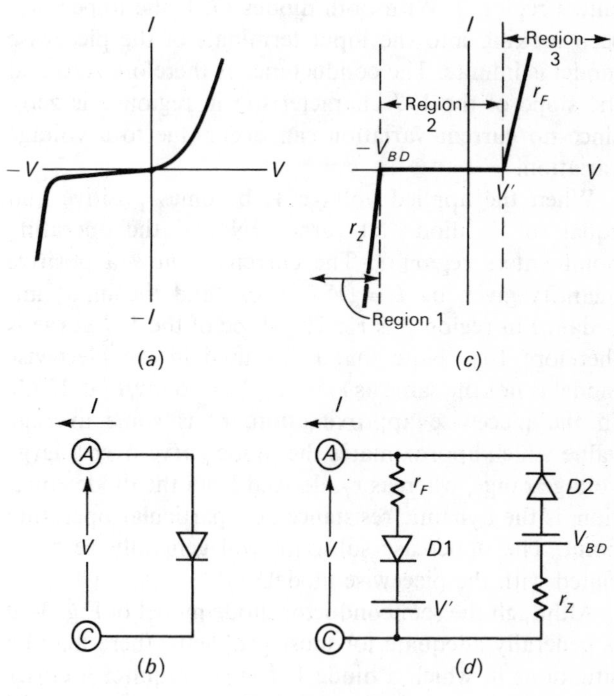

FIGURE 3-1

reference directions shown in Fig. 3-1*b*. Voltage V is referenced as the voltage from the cathode (terminal C) to anode (terminal A) and the current I has a positive reference direction from cathode to anode through the diode. For silicon diodes (which are most commonly used) the V-I curve of Fig. 3-1*a* may be approximated by the straight line (piecewise linear) segments of Fig. 3-1*c*. An equivalent circuit (piecewise model) which

[1] See Phillip Cutler, "Electronic Circuit Analysis," vol. 2, chap. 3, McGraw-Hill Book Company, New York, 1967.

would yield this V-I characteristic is shown in Fig. 3-1d. The diodes in Fig. 3-1d are ideal diodes and this two-ideal-diode model has a V-I characteristic approximating the actual diode characteristic of Fig. 3-1a.

To check the validity of this piecewise model we might reason that for V large and negative (assume a voltage source connected across the input terminals) $D1$ is surely reverse-biased (OFF) and $D2$ is forward-biased (ON). Since these are ideal diodes, they simulate open and closed switches respectively. Therefore, for $V < V_{BD}$ the operating point is in region 1 and I is a negative quantity. The slope of the segment in region 1 is determined by the incremental resistance seen looking into the terminals of the piecewise model in Fig. 3-1d, with $D1$ OFF and $D2$ ON. Clearly this is r_Z. Therefore, in region 1 the slope is $\Delta I/\Delta V = 1/r_Z$.

Next let $V = -V_{BD}$. At this point the V-I characteristic of Fig. 3-1c shows an abrupt discontinuity called a *breakpoint*, and at this breakpoint, $D2$ turns from ON to OFF and I ceases to flow. The operating point now enters region 2. With both diodes OFF the impedance seen looking into the input terminals of the piecewise model is infinite. The conductance is therefore zero and the slope of the V-I characteristic in region 2 is zero, since no current variation can occur due to a voltage variation.

When the applied voltage V becomes positive and equal to V', diode $D1$ turns ON and the operating point enters region 3. The current is now a positive quantity given by $I = (V - V')/r_F$ and the input impedance in region 3 is r_F. The slope of the V-I curve is therefore $1/r_F$. Note that r_F as used in the piecewise model is not the same as $r_f = V_T/I = 26$ mv/I (at 27°C). In the piecewise approximation, r_F is some average value which approximates the diode curve over a large voltage range, whereas r_f, derived from the diode equation, is the dynamic resistance at a particular operating point. The uppercase subscript will generally be associated with the piecewise model.

Although the semiconductor diode model of Fig. 3-1d is generally adequate for most problems, there may be situations in which a diode V-I curve requires a closer approximation, particularly in the reverse characteristic. For example, assume the piecewise V-I curve of Fig. 3-2a is a closer piecewise approximation to a diode curve. The fact that there are three breakpoints (points of discontinuity) in the V-I characteristic indicates that our piecewise model will probably contain at least three ideal diodes. One breakpoint is located at a reverse voltage of 10 volts, another breakpoint is located at the origin, and a third breakpoint is located at a forward voltage of 0.6 volt.

An actual piecewise model utilizing three ideal diodes to simulate this piecewise linear V-I curve is shown in

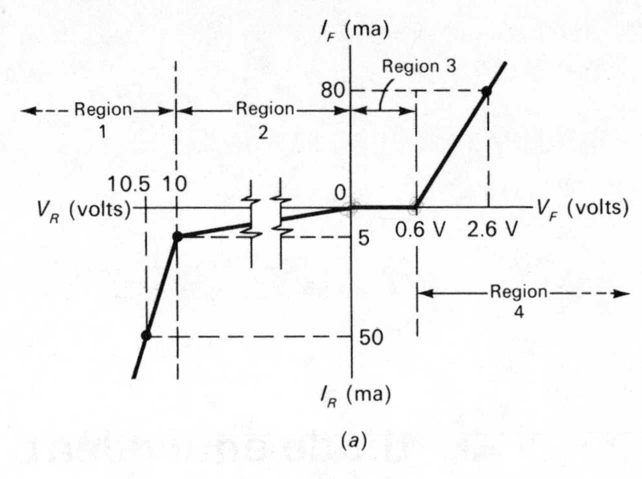

(*a*)

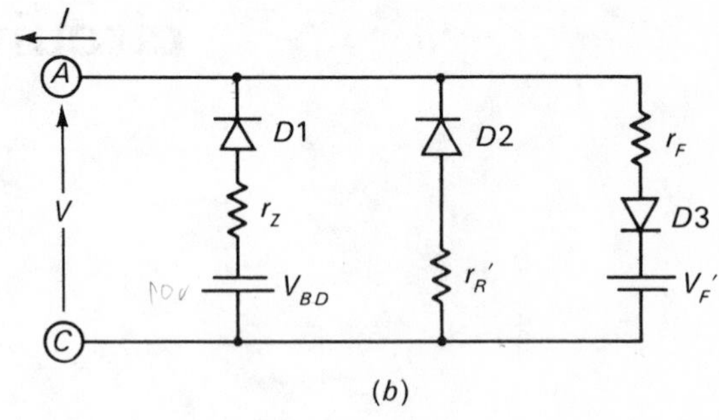

(*b*)

FIGURE 3-2

Fig. 3-2b. It may be verified as follows. For V large and negative, $D3$ is OFF and $D1$ and $D2$ are ON. The slope in region 1 is $g_1 = \Delta I_Z/\Delta V_Z = 45$ ma/0.5 volt $= 90$ millimhos. When $V = V_{BD} = -10$ volts, $D1$ turns OFF and the operating point enters region 2 for all -10 volts $< V < 0$. The slope in region 2 is $g_2 = \Delta I/\Delta V = 5$ *ma*/10 volts $= 0.5$ millimho. Since in region 2 only $D2$ conducts, it follows that $r_R' = 1/g_2 = 2$ kilohms. To determine r_Z we may reason that in region 1, with both $D2$ and $D3$ ON, the slope

$$g_1 = \frac{1}{r_Z} + \frac{1}{r_R'}$$

$$90 \text{ millimhos} = \frac{1}{r_Z} + 0.5 \text{ millimho}$$

$$\frac{1}{r_Z} = 85 \text{ millimhos}$$

$$\therefore r_Z = 1/85 \text{ millimhos} = 11.8 \text{ ohms}$$

At $V = 0$, $D2$ turns OFF, and the operating point enters region 3 where $I = 0$ and the input impedance is infinite since the slope $g_3 = 0$. For $V > V_F'$, $D3$ is ON and $g_3 = 1/r_F$ since all the other branches are open. From the curve we have $r_F = 1/g_3 = \Delta V/\Delta I -$ 2 volts/80 ma $= 25$ ohms.

Another diode piecewise model is presented in Chapter 4 as an exercise in diode circuit analysis.

PROBLEMS WITH SOLUTIONS

PS 3-1 Develop a piecewise model for the diode characteristic of Fig. PS 3-1*a*.

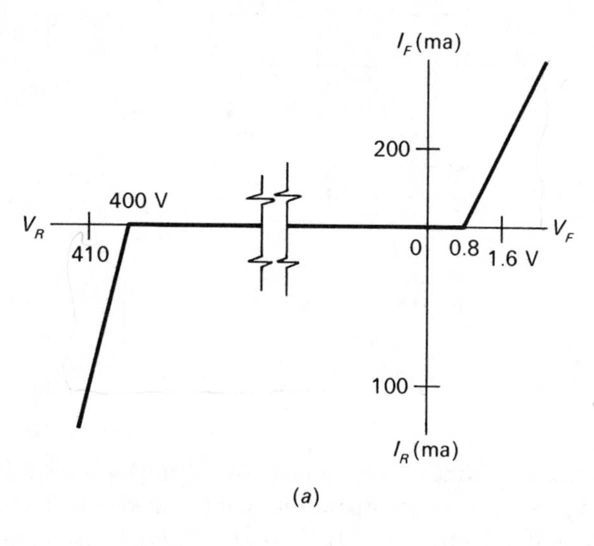

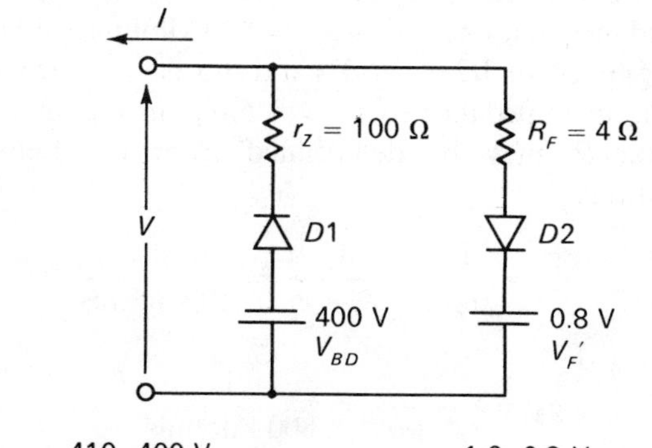

FIGURE PS 3-1

SOLUTION Figure PS 3-1*b* illustrates the piecewise model. It will contain at least two ideal diodes since there are two breakpoints in the *V-I* characteristic. For *V* less (more negative that is) than minus 400 volts the operating point is in the zener region and *D*1 is ON. Diode *D*2 is OFF. Therefore, $V_{BD} = -400$ volts and r_Z, as determined by the slope of the curve, is

$$r_Z = \frac{\Delta V}{\Delta I} = \frac{410 \text{ volts} - 400 \text{ volts}}{100 \text{ ma} - 0 \text{ ma}} = 0.1 \text{ kilohm}$$

For *V* greater than minus 400 volts but less than 0.8 volt (written $-400 \text{ volts} < V < 0.8$ volt) *D*1 and *D*2 are OFF so that the reverse resistance in this region is infinite. For *V* more positive than 0.8 volt, *D*2 turns ON. From the slope of the curve in this region we obtain

$$r_F = \frac{\Delta V}{\Delta I} = \frac{1.6 \text{ volts} - 0.8 \text{ volt}}{200 \text{ ma} - 0 \text{ ma}} = 4 \text{ ohms}$$

PS 3-2 Construct a three-segment, piecewise approximation to the diode characteristic of Fig. PS 3-2*a*. Determine the slopes of the segments.

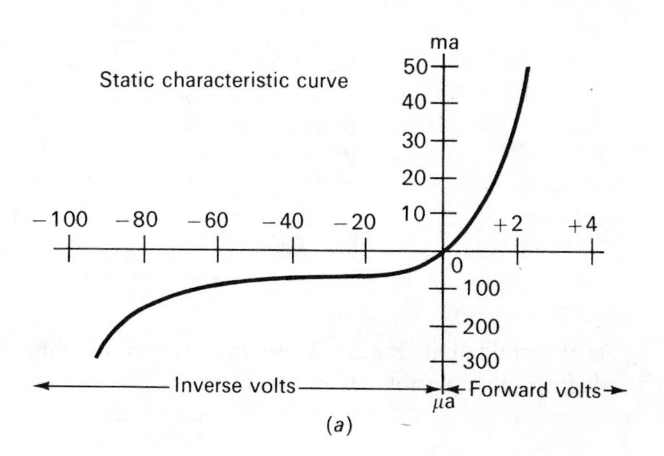

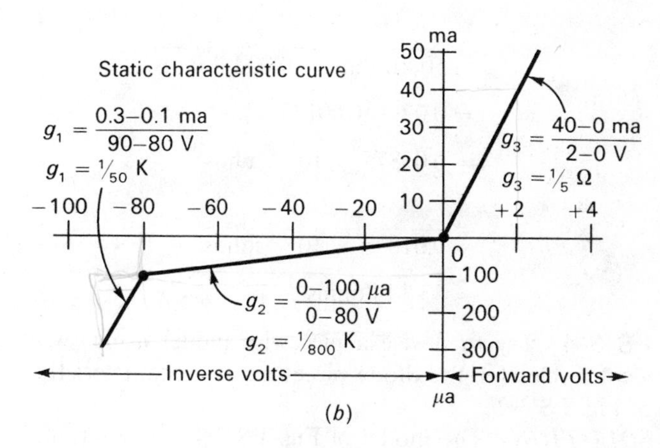

FIGURE PS 3-2

SOLUTION See Fig. PS 3-2*b* for a possible solution.

PS 3-3 Construct a piecewise model to fit the piecewise linear *V-I* characteristic of Fig. PS 3-2*b*.

SOLUTION The piecewise model will contain at least two ideal diodes as there are two breakpoints, one at −80 volts and the other at the origin. The model may, however, have more ideal diodes, although excess diodes should be avoided unless they somehow contribute to better insight. Figure PS 3-3 illustrates a three-diode model based upon the following analysis. Since

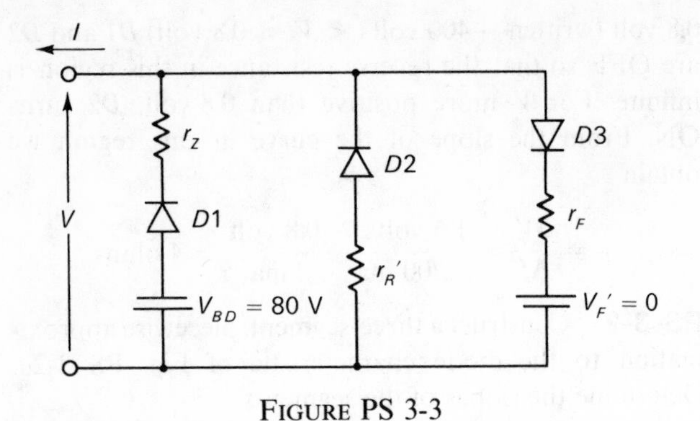

FIGURE PS 3-3

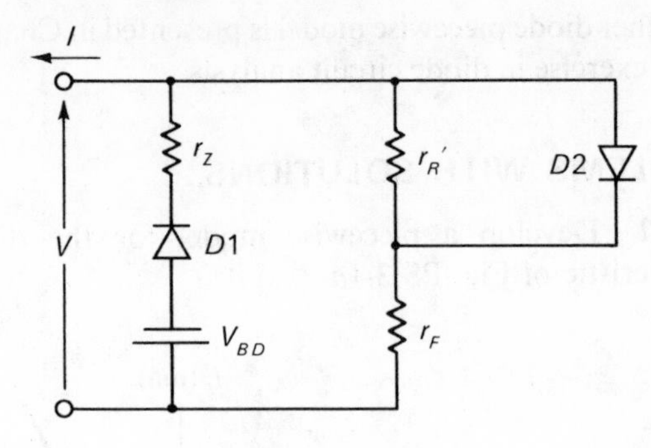

FIGURE PS 3-4

REGION 1: $V < -80$ volts	REGION 2: $-80 < V < 0$	REGION 3: $0 < V$
D1 ON	D1 OFF	D1 OFF
D2 ON	D2 ON	D2 OFF
D3 OFF	D3 OFF	D3 ON
$g_1 = \frac{1}{50\text{K}} = \frac{1}{r_Z} + \frac{1}{r_R'}$	$g_2 = \frac{1}{800\text{K}} = \frac{1}{r_R'}$	$g_3 = \frac{1}{5\Omega} = \frac{1}{r_F}$

r_R' is determined in region 2, we may substitute this result into the equation for g_1 to obtain

$$\frac{1}{r_Z} = g_1 - g_2$$

$$= \frac{1}{50 \text{ kilohms}} - \frac{1}{800 \text{ kilohms}}$$

$$= 0.02 - 0.00125 \text{ mhos}$$

$$= 0.01875 \times 10^{-3} \text{ mhos}$$

$$\therefore r_Z = \frac{1}{0.01875 \times 10^{-3} \text{ mhos}}$$

$$= 53.3 \text{ kilohms}$$

PS 3-4 Try to devise a piecewise model using two ideal diodes for the diode piecewise *V-I* characteristic of Fig. PS 3-2*b*.

SOLUTION The model of Fig. PS 3-4 is a solution. It is based upon the following reasoning. With V large and more negative than V_{BD}, D1 is ON, D2 is OFF, and the operating point is in region 1. The input impedance is then determined by $r_Z \| (r_R' + r_F)$. For $-V_{BD} < V < 0$, the operating point is in region 2 and D1 and D2 are both OFF. With D1 OFF the input impedance rises to $r_R' + r_F = 800$ kilohms. When V swings positive, D2 turns ON and shorts r_R', which drops the input impedance to $r_F = 5$ ohms in region 1. The resistances may be determined from the following equations:

1) $$g_1 = \frac{1}{r_Z} + \frac{1}{r_R' + r_F} = \frac{1}{50 \text{ kilohms}}$$

2) $$g_2 = \frac{1}{r_R' + r_F} = \frac{1}{800 \text{ kilohms}}$$

3) $$g_3 = \frac{1}{r_F} = \frac{1}{5 \text{ ohms}}$$

Solving these equations simultaneously yields

4) $$r_R' = 800 \text{ kilohms} - 5 \text{ ohms} \approx 800 \text{ kilohms}$$

5) $$r_Z = 53.3 \text{ kilohms}$$

PA 3-1 Construct a simple two-segment approximation to the forward characteristics of the silicon diode of Fig. PA 3-1*a* at each of the three temperatures indicated. Determine r_F and V_F' for each temperature.

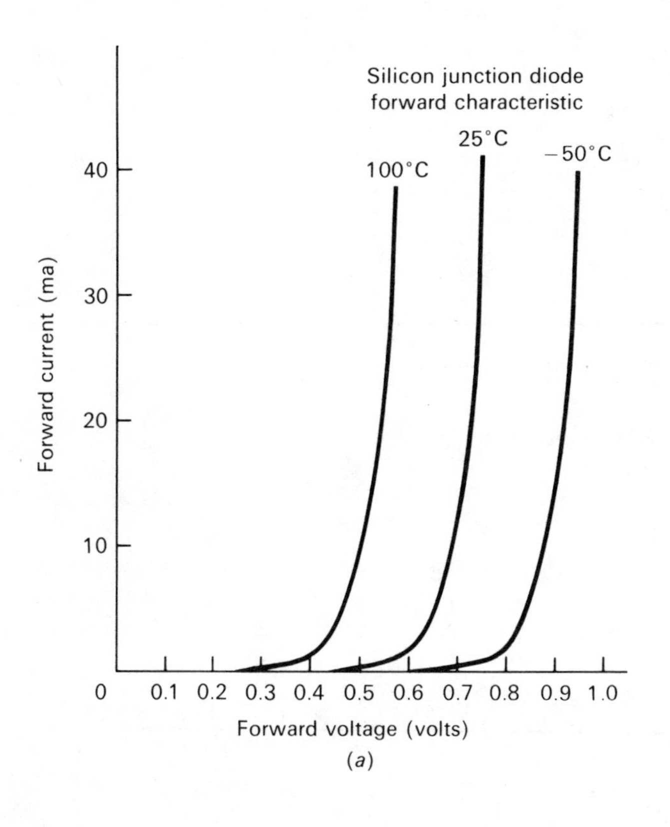

(*a*)

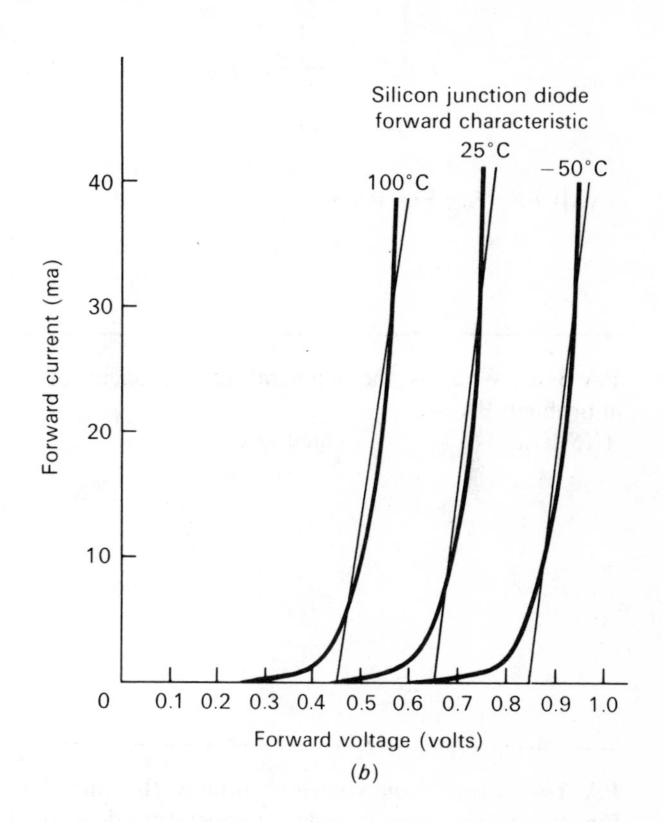

(*b*)

ANSWER See Fig. PA 3-1*b*, which represents a possible solution. For the segments chosen we have, at −50°C,

$$V_F' = 0.85 \text{ volt}, \; r_F \cong 0.1 \text{ volt}/34 \text{ ma} = 2.9 \text{ ohms}$$

at 25°C,

$$V_F' = 0.65 \text{ volt}, \; r_F \cong 0.1 \text{ volt}/34 \text{ ma} = 2.9 \text{ ohms}$$

at 100°C,

$$V_F' = 0.45 \text{ volt}, \; r_F \cong 0.1 \text{ volt}/34 \text{ ma} = 2.9 \text{ ohms}$$

PA 3-2 Construct a piecewise model to fit the piecewise linear approximation of the 25°C *V-I* curve of Fig. PA 3-1*b*.

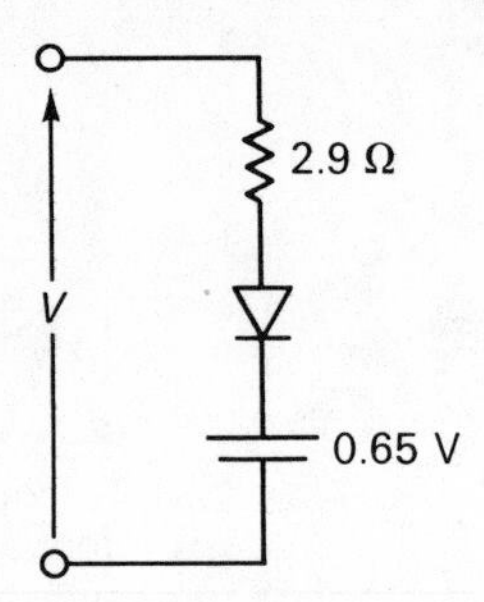

ANSWER See Fig. PA 3-2.

PA 3-3 What is the temperature coefficient of V_F in problem PA 3-1?

ANSWER −2.66 mv/°C

PA 3-4 Could you somehow modify the model of Fig. PA 3-2 so as to include a temperature-dependent voltage source related to the temperature coefficient (T.C.) determined in Fig. PA 3-3?

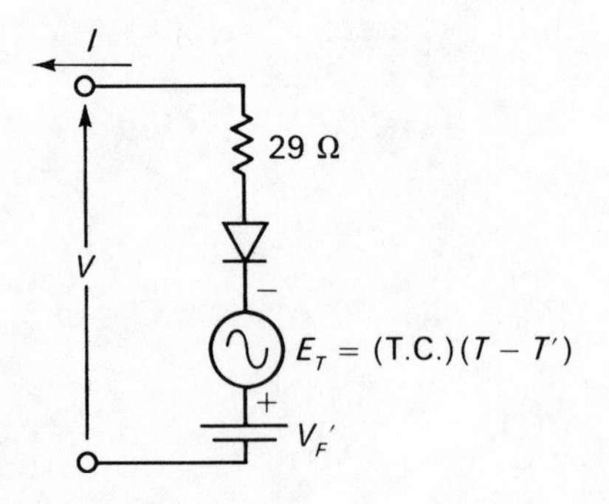

ANSWER See Fig. PA 3-4.

PROBLEMS WITHOUT ANSWERS

P 3-1 What is the minimum number of diodes required in a piecewise model to fit the *V-I* curve of Fig. P 3-1?

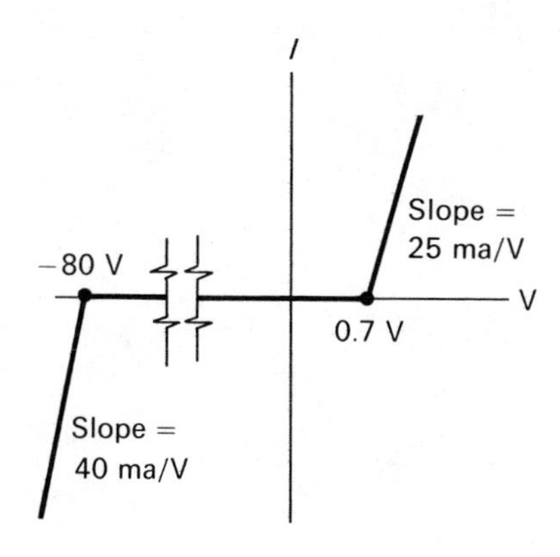

P 3-2 Construct a piecewise model to fit the *V-I* curve of Fig. P 3-1.

P 3-3 Construct a three-diode piecewise model to fit the piecewise linear *V-I* curve of Fig. P 3-3.

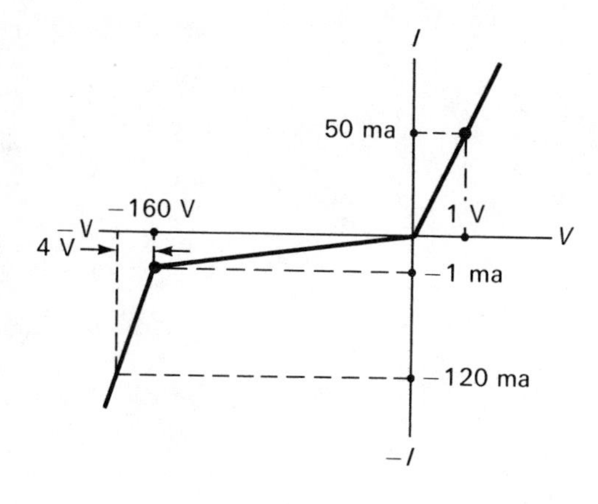

4 diode circuit analysis

The usual problem encountered in analyzing diode circuits is that the state of the diodes involved is unknown. If the circuit contains diodes that may be approximated in behavior by ideal diodes, then each diode may have either one of two states, that is, ON or OFF. Should it turn out that the range of voltages encountered might cause some diode to turn ON in the reverse direction due to operation in the breakdown region, we would have three states to consider.

4-1 Single-diode circuits

Circuits containing only one diode may readily be analyzed by imagining the diode to be momentarily removed from the circuit and then determining the Thévenin's equivalent circuit the diode looks back into. If V_{OC} is of such a polarity that the anode of the diode would tend to be biased positive relative to the cathode, the diode will be turned ON. This situation is illustrated in Fig. 4-1. If it turns out that the anode of the diode

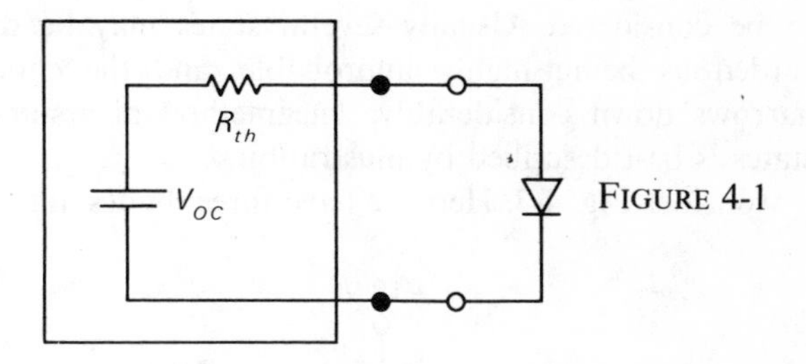

FIGURE 4-1

tends to be biased negative with respect to the cathode, the diode is turned OFF. If it should turn out that V_{OC} in the reverse direction exceeds the breakdown voltage of the diode, then the diode is turned ON in the reverse direction and, to a first approximation, simulates a voltage source equal to the reverse breakdown voltage.

As an illustrative problem, consider Fig. 4-2*a*. It is desired to determine the current I. If we imagine the

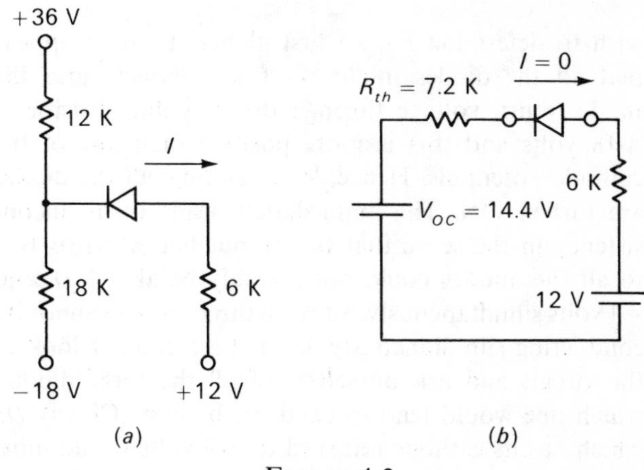

FIGURE 4-2

diode removed from the circuit, we obtain the Thévenin's equivalent circuit of Fig. 4-2*b*. Clearly the diode will be reverse-biased, since the voltage on the cathode of the diode is 2.4 volts positive with respect to the anode. Hence, the diode is reverse-biased and $I = 0$.

4-2 Assumed-state analysis

When a circuit contains several diodes and/or fixed voltage and current sources, the method of assumed states is useful. In assumed-state analysis we simply make an educated guess as to the state of the diodes. Some may be chosen OFF and some may be chosen ON. The circuit is then analyzed on the basis of the diode states assumed. The resultant currents and voltages in the circuit are then examined to see if they are compatible with the diode states assumed. If not, some new assumptions must be made. Since each diode may have two states, an n diode circuit will have 2^n possible states. This means in a three-diode circuit there would eight possible combinations of ON and OFF to be considered. Usually several states may be discarded as being highly improbable, and the choice narrows down considerably. The method of assumed states is best described by illustration.

Consider Fig. 4-3. Here we have three diodes and we

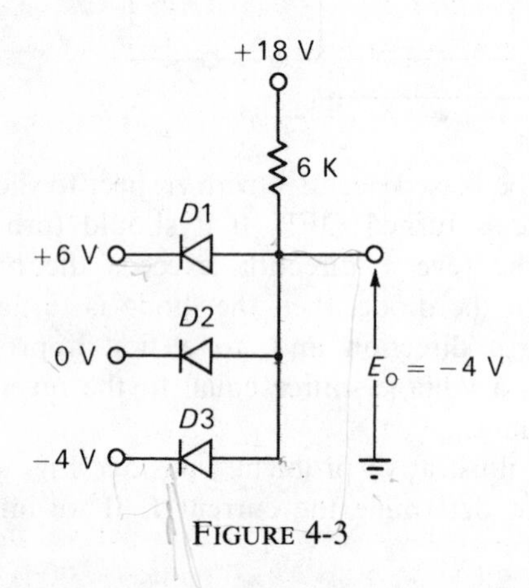

FIGURE 4-3

wish to determine E_O. At first glance it would appear that all the diodes might want to conduct since the anode return voltage through the 6-kilohm resistor is +18 volts and this is more positive than any of the cathode potentials. Hence, let us assume all the diodes are turned ON. This immediately leads to an inconsistency in the sense that the output line which is tied to all the anodes could not possibly be at +6, 0, and −4 volts simultaneously. Since all three diodes cannot be conducting simultaneously, let us take another look at the circuit and ask ourselves, of all the three diodes, which one would tend to conduct the best? Clearly $D3$ which has its cathode returned to −4 volts would most tend to conduct. Hence, let us assume $D3$ is ON and $D1$ and $D2$ are OFF. If $D3$ is ON, its anode is also at −4 volts. Consequently, $D2$ must be reverse-biased by 4 volts and $D1$ is reverse-biased by 10 volts. Hence with $D3$ ON, $D1$ and $D2$ would truly be OFF. Suppose we had assumed that $D2$ was ON and $D1$ and $D3$ were OFF. If $D2$ was ON, its anode and also the anodes of $D1$ and $D3$ would be tied to 0 volts. $D1$ under these circumstances would indeed be reverse-biased by 6 volts. However, with the anode of $D3$ at 0 volts, $D3$ would have to turn ON since the anode is 4 volts positive with respect to the cathode. Consequently, the assumption of $D2$ ON and $D1$ and $D3$ OFF must be an error. The only valid case is with $D3$ ON and $D1$ and $D2$ OFF so that $E_O = -4$ volts.

This particular circuit belongs to a class of circuits called logic gates which are commonly found in digital computers. The trick in simplifying the analysis of these logic gates is to pose the question, of all the diodes connected to the common output line, which one will conduct the best? The state of the diodes may then be verified from this initial assumption.

As another illustration of assumed-state analysis, consider Fig. 4-4*a* in which we desire to determine the currents I_1 and I_2. Let us assume as a first guess that both $D1$ and $D2$ are conducting. Assuming $D1$ and $D2$ short circuits, we may then calculate the voltage V_J by Millman's theorem to obtain

$$V_J = \frac{\dfrac{-6\text{ volts}}{2\text{ kilohms}} + \dfrac{10\text{ volts}}{4\text{ kilohms}} + \dfrac{12\text{ volts}}{8\text{ kilohms}}}{\dfrac{1}{2\text{ kilohms}} + \dfrac{1}{4\text{ kilohms}} + \dfrac{1}{8\text{ kilohms}}} \tag{1}$$

$$= 1.143\text{ volts}$$

Now, referring to Fig. 4-4*b*, with V_J taken at 1.143 volts, it follows that $D1$ would surely be turned ON, but $D2$ would have to be reverse-biased since V_J is negative with respect to the +10-volt cathode return of $D2$. Hence the assumption that $D1$ and $D2$ were ON is in error.

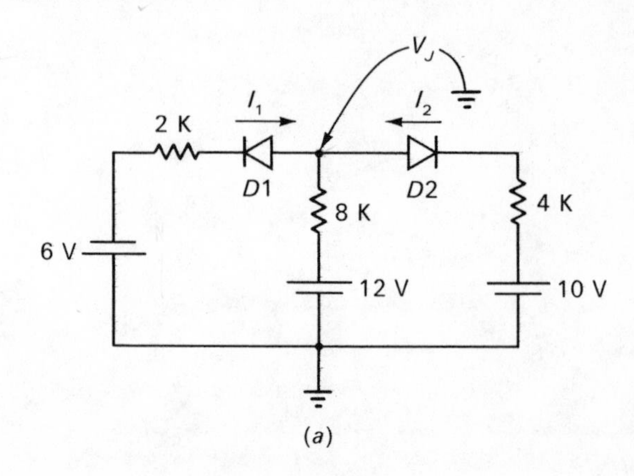

FIGURE 4-4

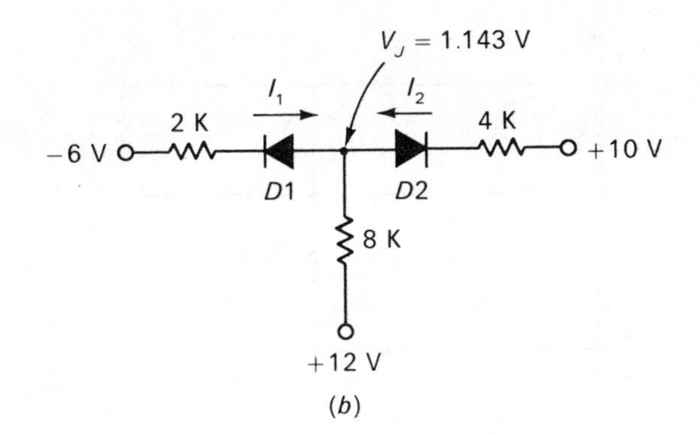

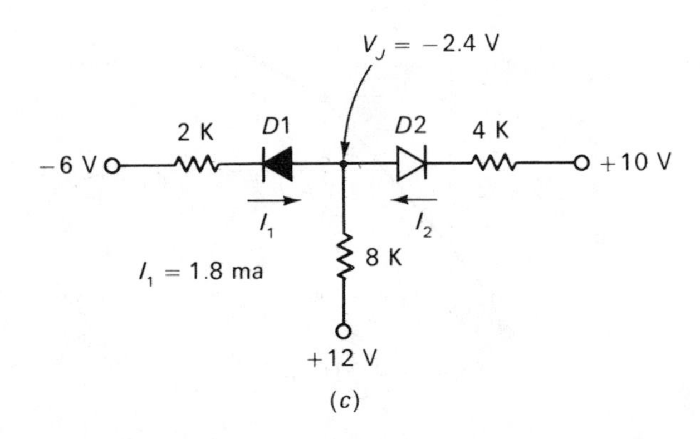

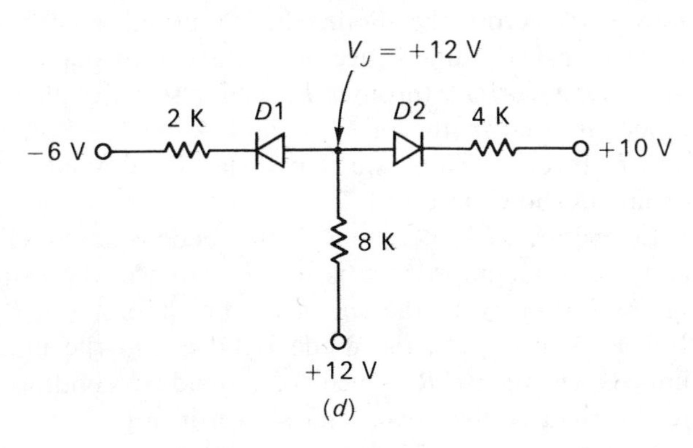

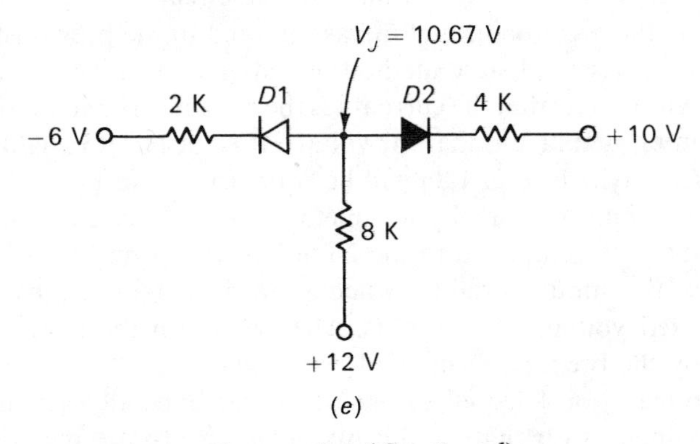

FIGURE 4-4 (*Continued*)

We might then reconsider our assumptions, and this time assume $D1$ ON and $D2$ OFF as shown in Fig. 4-4*c*. With $D2$ OFF and $D1$ ON, the voltage $V_J = -2.4$ volts. Under these conditions $D2$ would indeed be reverse-biased, $D1$ would be forward-biased, and the current I_1 would be given by

$$2) \quad I_1 = \frac{-2.4 \text{ volts} - (-6 \text{ volts})}{2 \text{ kilohms}} = 1.8 \text{ ma}$$

Note that I_1 turned out to be a positive quantity in the direction from cathode to anode through $D1$, which agrees with the only direction in which $D1$ can conduct. I_2, of course, $= 0$. This, then, is the correct solution.

By way of illustration, if we had considered $D1$ and $D2$ OFF as shown in Fig. 4-4*d*, V_J would equal $+12$ volts and this would surely cause $D1$ and $D2$ to want to turn on. Hence $D1$ and $D2$ could not be OFF. Had we assumed $D2$ ON and $D1$ OFF, as shown in Fig. 4-4*e*, V_J would be calculated as 10.67 volts, and this would surely cause $D1$ to turn ON. Hence an inconsistency appears in this case.

4-3 Breakpoint analysis

If it is desired to study the response of a multiple-diode circuit to large variations in some applied voltage or current, we can usually use the breakpoint method of analysis. It is based upon the fact that at the very point that a diode makes a transition from OFF to ON it has zero voltage drop across it and zero current through it. This breakpoint corresponds to the origin of the ideal diode V-I curve illustrated in Fig. 4.5. Usually, but not

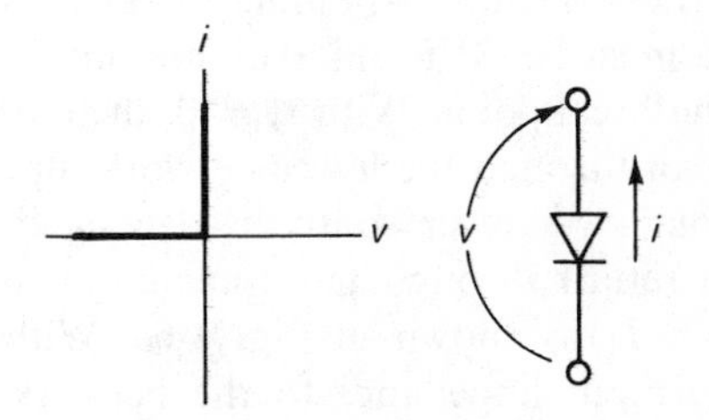

FIGURE 4-5

necessarily, a circuit containing n diodes will exhibit n breakpoints. As with assumed-state analysis, there may exist some question as to which diodes are ON or OFF for different ranges of the input variable. This question may be resolved by varying the input variable over the range of interest and making some educated guesses to determine the order in which the diodes change states. The validity of our guess should be checked by looking for inconsistencies just as we did in assumed-state analysis.

The basic difference between assumed-state analysis, as we shall employ it, and breakpoint analysis is that the former is primarily employed for some fixed value of the variable, whereas the latter is most useful in determining circuit response to a range of the input variable. In many cases a combination of both methods may be successfully utilized.

The breakpoint method of analysis is best mastered by carefully studying the following examples.

Sketch the volt-ampere characteristic (V-I curve) seen looking into the input terminals of the circuit of Fig. 4-6*a*.

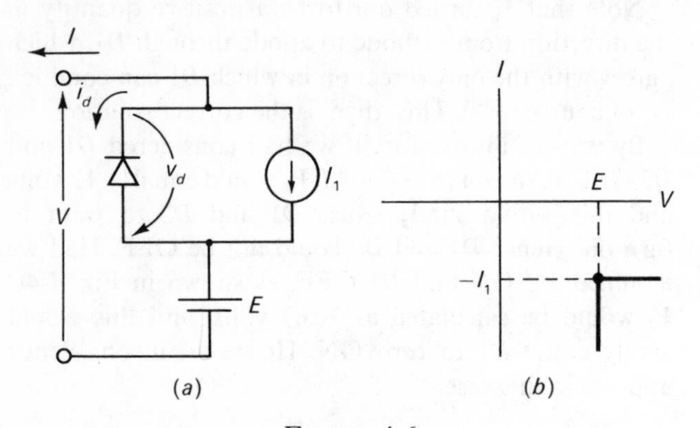

FIGURE 4-6

To solve this problem easily, we might note that since there is one diode there will be one breakpoint. The breakpoint conditions at the input terminals may be determined by setting $v_d = 0$ and $i_d = 0$. If $v_d = 0$, $V = E$, and if $i_d = 0$, $I = -I_1$. Thus we have the breakpoint coordinates shown in Fig. 4-6*b*. For values of V tending to be less than E the diode is surely ON and the input impedance into the network is zero. Hence, the slope of the V-I curve is infinite (vertical). For $V > E$ the diode must be OFF, since we are now on the other side of the breakpoint. With $i_d = 0$, the current source I_1 must exit through the lower network input terminal, flow through whatever source is driving the network, and then return through the upper input terminal, so that $I = -I_1$ as shown in Fig. 4-6*b*. With the diode OFF, the input impedance to the network is infinite. Therefore I cannot vary with E and the slope of the V-I curve is flat to the right of the breakpoint.

As another example of breakpoint analysis consider Fig. 4-7*a*. To develop the V-I curve we again note there is one diode and, consequently, one breakpoint. The breakpoint conditions are shown in Fig. 4-7*b*. With

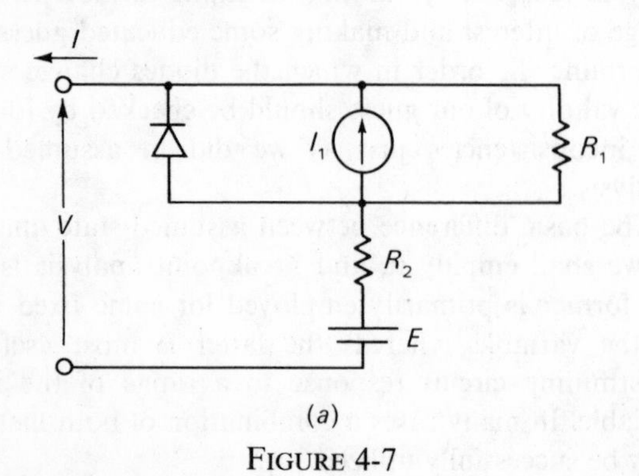

FIGURE 4-7

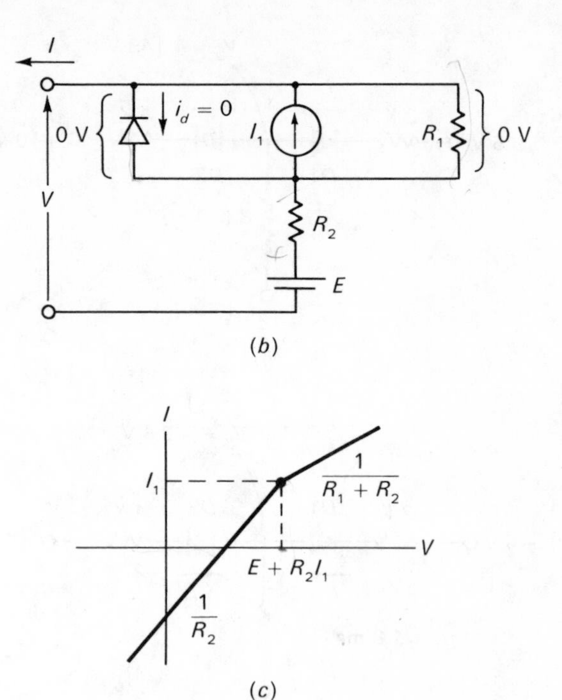

FIGURE 4-7 (*Continued*)

zero volts across the diode, there must also be zero volts across R_1 and hence zero current through R_1. With zero current through R_1 and also through the diode, it follows that $I = I_1$, and $V = E + R_2I_1 + 0 = E + R_2I_1$. Thus we have the breakpoint coordinates shown in Fig. 4-7*c*.

For values of $V < E + R_2I_1$ the diode is surely ON and the input impedance is just R_2. Hence, the slope of the V-I curve to the left of the breakpoint is $1/R_2$. For $V > E + R_2I_1$, the diode is OFF and the input impedance is $R_1 + R_2$ which corresponds to conditions to the right of the breakpoint shown in Fig. 4-7*c*.

In Fig. 4-8*a* we will analyze the circuit of Fig. 4-4*a* by the breakpoint method as opposed to the previously used assumed-state method of analysis. In Fig. 4-8*a*, we will assume that the current I is the output variable we are interested in and that the voltage V is the input variable. Clearly, I in Fig. 4-8*a* will be equal to $-I_1$ in Fig. 4-4*a*. To apply the breakpoint method, we will assume that the voltage applied to the 2-kilohm resistor in Fig. 4-8*a* is the input variable, whereas in Fig. 4-4*a* it was a fixed voltage of -6 volts. After we finish the analysis by the breakpoint method, we should be able to determine I for V equal to -6 volts and hopefully get the same answer that we did by the assumed-state method of analysis.

To begin the analysis, let us conceptually vary V from values which are large and negative to values which are large and positive. For V large and negative as assumed in Fig. 4-8*a*, it seems most reasonable that $D1$ would be ON and $D2$ would be OFF. The ON diode is

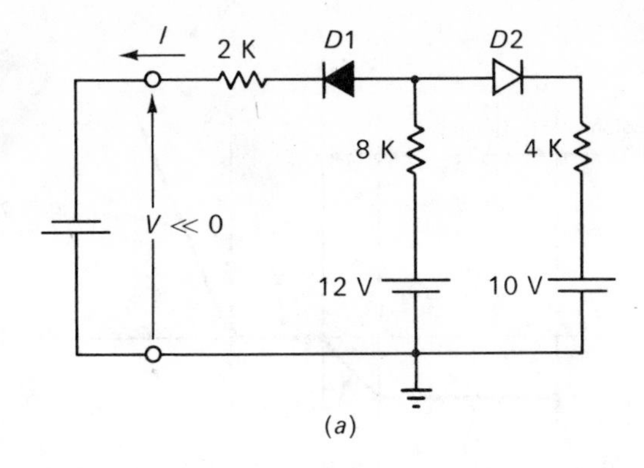

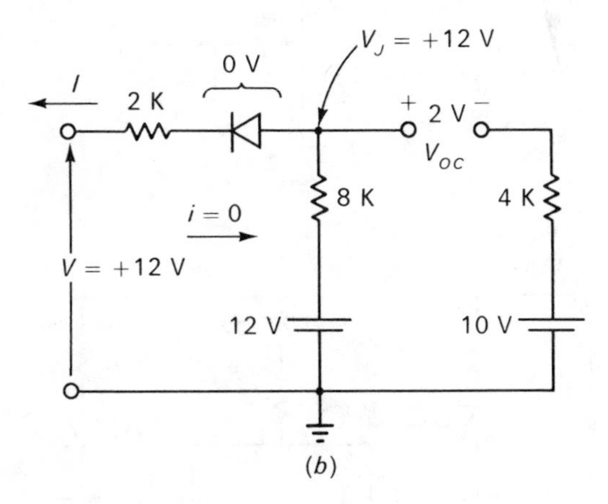

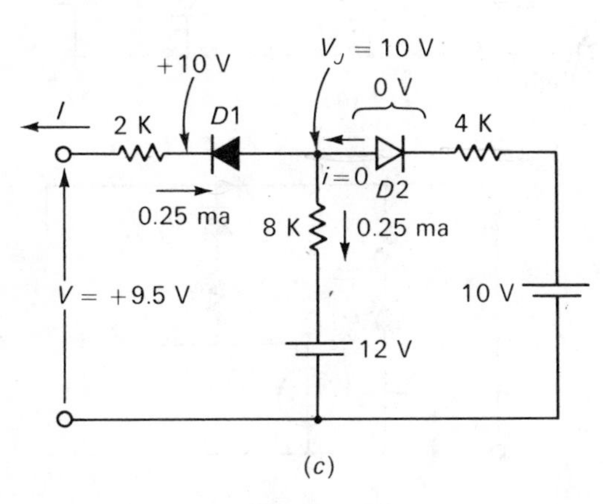

FIGURE 4-8

drawn shaded to distinguish it from the OFF diode. The input impedance Z_{in} in this region of operation with $D1$ ON and $D2$ OFF is 10 kilohms.

Now let us assume that V is increasing and that as V increases the very first breakpoint we encounter is $D1$ turning OFF. Circuit conditions at $D1$'s breakpoint will appear as shown in Fig. 4-8b. The voltage across $D1$ is zero, and simultaneously, the current through $D1$ is also zero. If the current through $D1$ is zero and $D2$ still is assumed OFF as before, it follows that the anode of $D2$ would be held at +12 volts and the cathode at +10 volts, which leaves 2 volts of open-circuit voltage to forward-bias $D2$. This inconsistency forces us to conclude that we made an error in assuming that as V becomes more positive, the first thing to occur is that $D1$ turns from ON to OFF.

Let us then assume that as V becomes more positive the next change to occur is that $D2$ turns from OFF to ON. At the breakpoint of $D2$, we have the circuit conditions illustrated in Fig. 4-8c. At the $D2$ breakpoint, the current through the diode and the voltage across it are simultaneously zero. With no current flowing through the 4-kilohm resistor and zero volts across $D2$, it follows that the junction voltage V_J must be equal to 10 volts. The current through the 8-kilohm resistor is then 0.25 ma. With $D1$ ON, it follows that the current through $D1$ must also be 0.25 ma in the forward direction. With $D1$ ON, there will also be zero volts across $D1$ so that we have +10 volts on the right-hand side of the 2-kilohm resistor and, with 0.25 ma through it, the voltage $V = +9.5$ volts. The terminal current $I = -0.25$ ma. There are no inconsistencies evident in this circuit. Hence, we may assume that when V exceeds +9.5 volts, $D1$ is ON, $D2$ is ON, and the input impedance is 2 kilohms in series with the parallel combination of 8 kilohms and 4 kilohms which yields a $Z_{\text{in}} = 2.67$ kilohms.

Now the operating point is in a region with $D1$ ON and $D2$ ON. If V becomes still more positive, it seems plausible that $D1$ must turn OFF since its cathode is

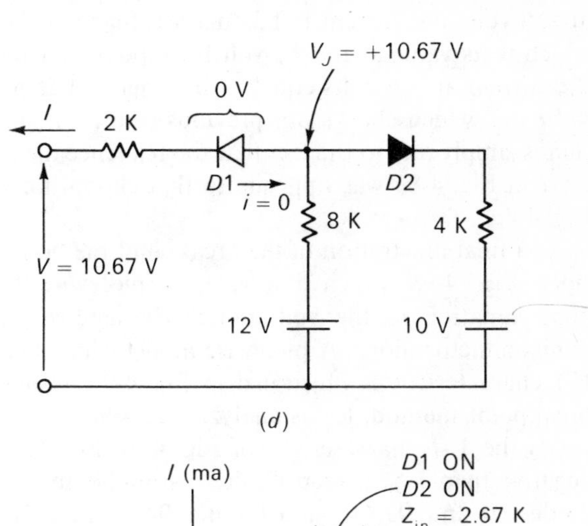

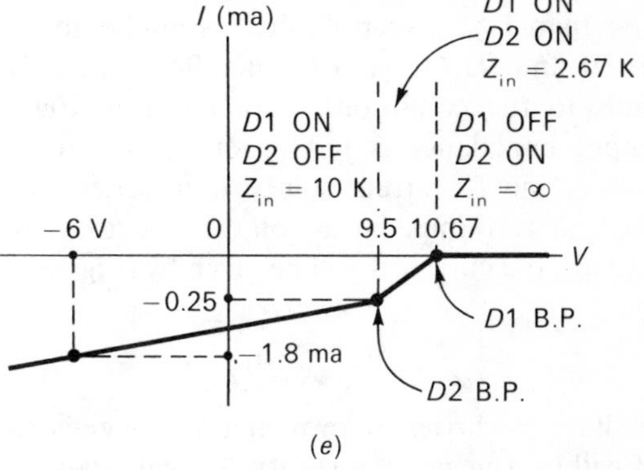

FIGURE 4-8 (*Continued*)

being pulled increasingly positive. The breakpoint for $D1$ with $D2$ ON is illustrated in Fig. 4-8d. The junction voltage V_J, with $D2$ considered a short circuit, is 10.67 volts, and with zero voltage drop across $D1$ and across the 2-kilohm resistor, it follows that the applied input $V = 10.67$ volts. The current $I = 0$, since the current through $D1$ is zero at its breakpoint. For values of voltage more positive than 10.67 volts, $D1$ remains OFF, $D2$ remains ON, and Z_{in} is infinite since $D1$ is an open circuit. The complete volt-ampere characteristic is illustrated in Fig. 4-8e. Notice there are two breakpoints (points of discontinuity). The slope of the V-I characteristic in each region is given by the slope $\Delta I/\Delta V = 1/Z_{in} = g_{in}$, where g is a conductance. Notice we have the complete V-I characteristic before us, and if this were plotted on graph paper the value of I for any given V would be determined by inspection. Even without graph paper, the coordinates of any point on the V-I characteristic may be computed, since the slopes and the coordinates of the breakpoints are known. For example, to determine I at $V = -6$ volts and in order to correlate our solution using the breakpoint method with the same problem when analyzed by the assumed-state method, we may reason as follows. At $+9.5$ volts the current is -0.25 ma. In going from $+9.5$ volts down to -6 volts, or for a total voltage decrease of 15.5 volts, the current decrease will be given by $\Delta I = \Delta V/Z_{in} = -15.5$ volts/10 kilohms $= -1.55$ ma. This means that at -6 volts the current is 1.55 ma less than -0.25 ma, which is its value at the 9.5-volt breakpoint. Therefore, the current at -6 volts equals -0.25 ma $-$ 1.55 ma $=$ -1.8 ma, which checks our previous result. The minus sign is simply due to the fact that the reference direction for I in Fig. 4-8a was opposite to that chosen for I_1 in Fig. 4-4a.

As a final illustration of the breakpoint method, consider Fig. 4-9a which illustrates a piecewise linear approximation to the volt-ampere characteristic of a semiconductor diode. A piecewise model which fits this V-I characteristic is illustrated in Fig. 4-9b. Using the breakpoint method, let us analyze Fig. 4-9b to see if it yields the V-I characteristic of Fig. 4-9a. For V more negative than V_{BD}, it seems quite reasonable to assume diodes $D1$ and $D2$ ON and $D3$ and $D4$ OFF. This corresponds to the conditions of region 1 in Fig. 4-9a. The input impedance is just r_Z since the infinite impedance of the I_S current source is in series with r_R'. Assuming a zero-impedance voltage source across the V terminals, the current I will be given by (Fig. 4-9c)

$$I = -I_S + \frac{-V_{BD} - V}{r_Z}$$

The voltage V carries its own sign and in regions 1, 2, and 3 will be a negative quantity. The fact that V_{BD} is a negative quantity is already taken care of in this equa-

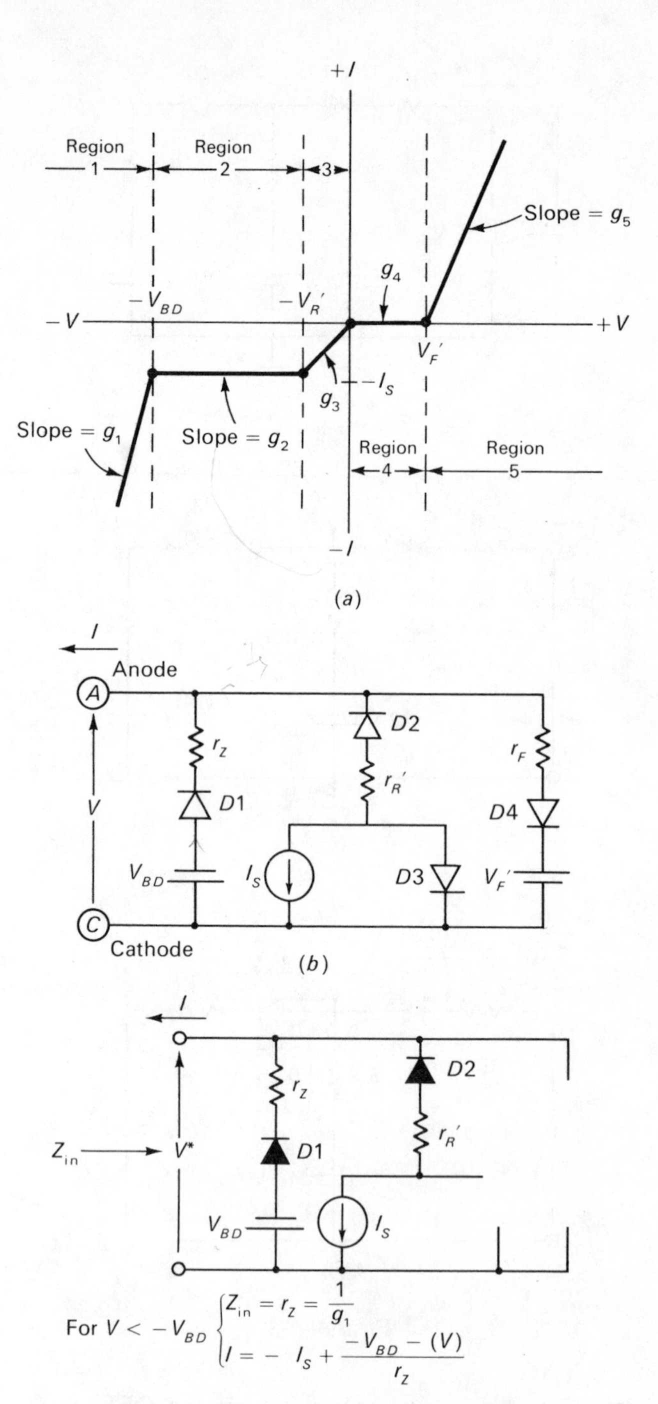

FIGURE 4-9

tion and only the magnitude need be substituted. The first breakpoint occurs at $V = -V_{BD}$. This is readily determined by setting the voltage across $D1$ equal to zero and simultaneously the current through it also equal to zero. Therefore, there is no voltage drop across r_Z and $V = -V_{BD}$. At this breakpoint the current source I_S is finding a path down and out through the bottom terminal, up through the external voltage source, which is not illustrated, back into the top terminal, and then down through $D2$ and r_R' to complete the path for I_S. Clearly $I = -I_S$ at this breakpoint.

For V greater than $-V_{BD}$ but less than V_R', the operating point is in region 2. Diode $D1$ has turned OFF and $D2$ is still ON. The input impedance is infinite due to the infinite impedance of the current source as shown in Fig. 4-9*d*. The external current is forced to $I = -I_S$.

The next breakpoint occurs when $D3$ turns from OFF to ON as shown in Fig. 4-9*e*. We may safely assume that $D3$ turns ON before $D2$ turns OFF, by virtue of the fact that Figure 4-9*a* indicates that in region 3 the impedance has become less than in region 2. This is evidenced by the steeper slope of the V-I characteristic. If $D2$ had turned OFF first, the input impedance would still be infinite in region 3 as it was in region 2. This, however, is not the case. Of course, in this particular problem we are cheating somewhat, in the sense that we already have the V-I characteristic to guide us. From Fig. 4-9*e* it is evident that $I = -I_S$ and $V = -r_R'I_S = -V_R'$. The equivalent circuit in region 3 is shown in Fig. 4-9*f*. The input impedance is simply r_R', and the current with both diodes ON is given by $I = V/r_R'$. The I_S current source in no way affects the external

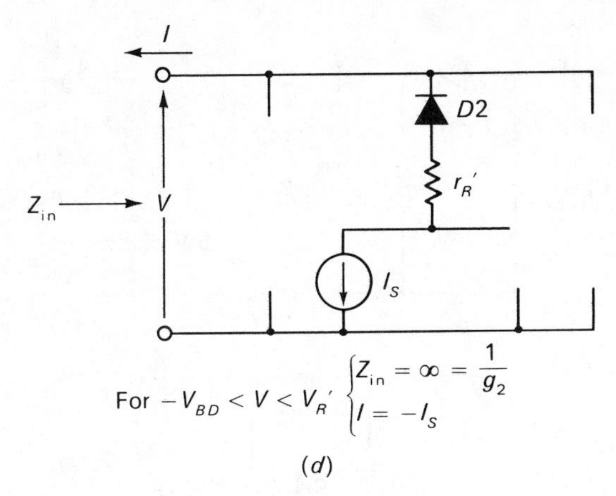

(*d*)

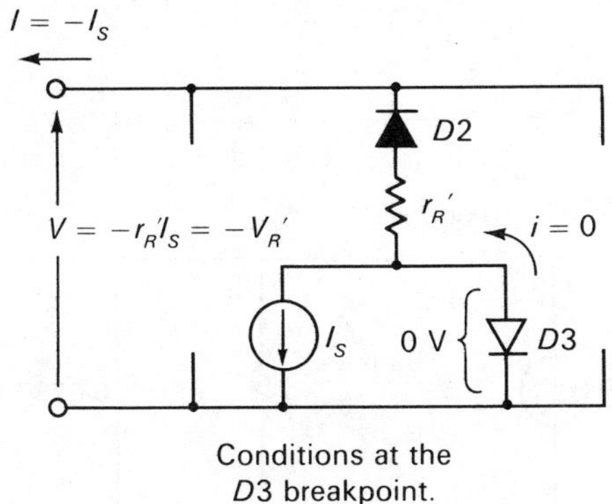

(*e*)

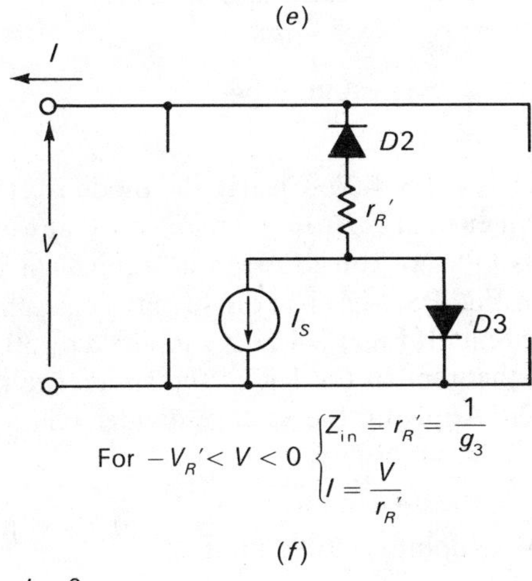

(*f*)

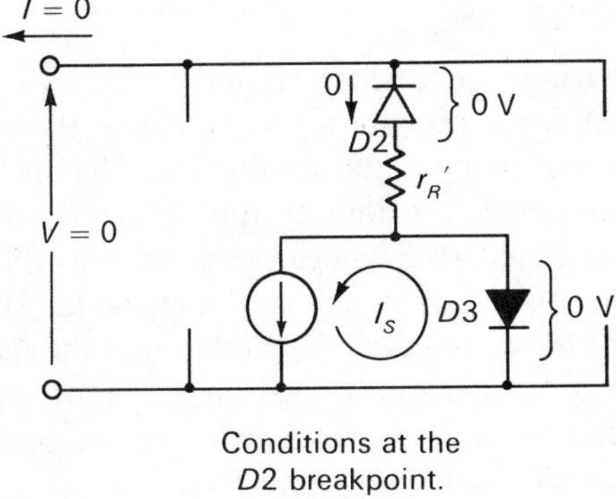

(*g*)

FIGURE 4-9 (*Continued*)

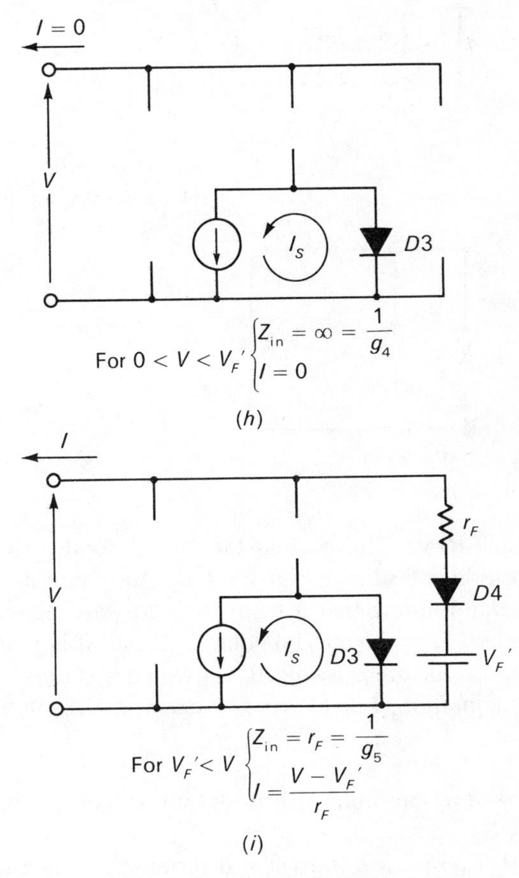

FIGURE 4-9 (*Continued*)

current I in this region, since it is diverted through the zero resistance of $D3$.

As V becomes increasingly less negative, we encounter another breakpoint at the origin, beyond which the curve again becomes flat. This leads us to conclude that the next breakpoint we encounter is due to $D2$ turning from ON to OFF as shown in Fig. 4-9*g*. At the $D2$ breakpoint, with $D3$ ON and all the other diodes OFF, $I = 0$ and $V = 0$. The I_S current source sinks itself in the ON diode $D3$. The operating point now enters region 4 and the circuit conditions are as shown in Fig. 4-9*h*. Clearly $Z_{in} = \infty$ and $I = 0$ for all values of V greater than zero but less than V_F'.

The next breakpoint occurs at $V = V_F'$, at which point diode $D4$ turns ON. For values of V greater than V_F', the input impedance is r_F and the current I is given by $I = (V - V_F')/r_F$. This situation is shown in Fig. 4-9*i*.

PROBLEMS WITH SOLUTIONS

PS 4-1 Determine the current I in Fig. PS 4-1*a*.

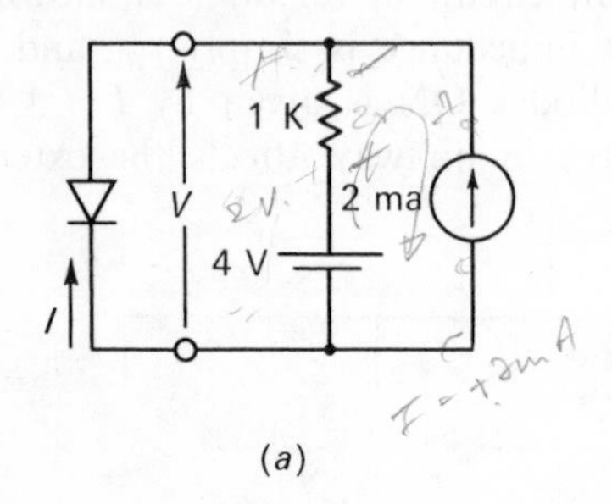

FIGURE PS 4-1

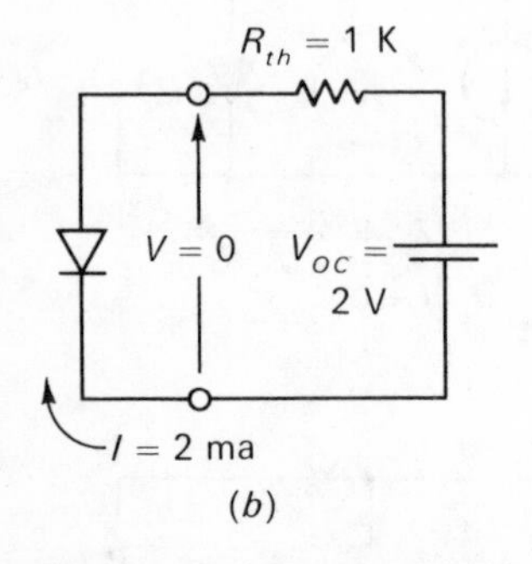

SOLUTION Thevenizing the circuit to the right of the diode, we obtain Fig. PS 4-1*b*. Assuming an ideal diode and noting that it tends to be forward-biased, we may write I = 2 volts/1 kilohm = 2 ma. If a practical silicon diode were assumed, we would write as a first approximation I = (2 volts − 0.6 volts)/1 kilohm = 1.4 ma.

PS 4-2 Determine the ac output voltage e_O in Fig. PS 4-2*a*.

SOLUTION We must first determine the state of the diode. The dc conditions that the diode is exposed to are indicated in Fig. PS 4-2*b*. Clearly the diode is ON and the quiescent current is I_{dc} = 2.5 ma. Now that we know the diode is ON, we can draw an ac equivalent circuit as shown in Fig. PS 4-2*c*. The diode may be assumed to be a short circuit. If Fig. PS 4-2*c* is simplified by applying Thévenin's theorem to the left of the anode, we obtain Fig. PS 4-2*d*. Applying the voltage-divider relationship to Fig. PS 4-2*d* we obtain

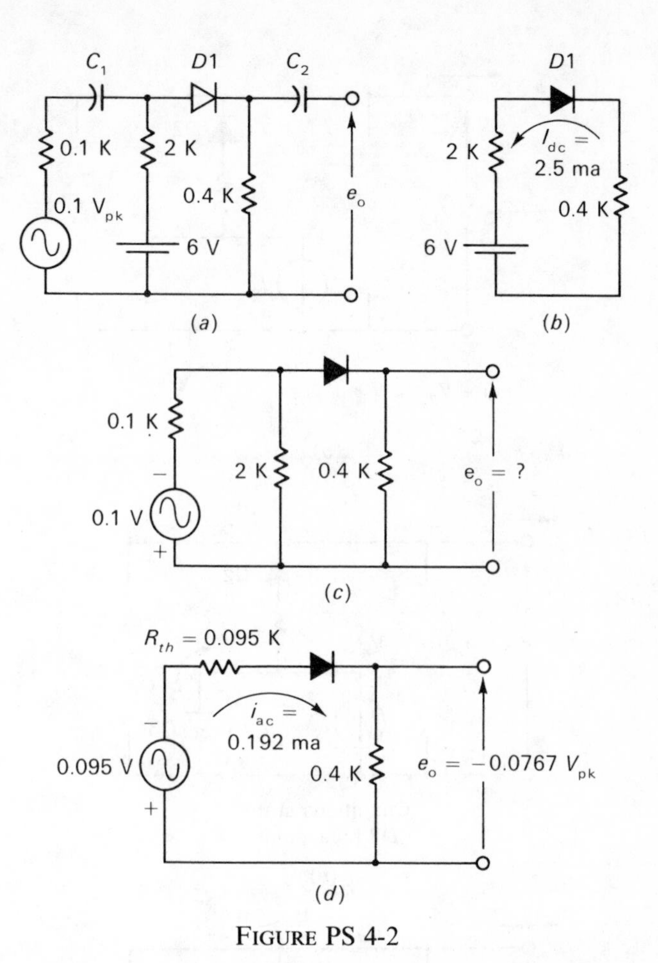

FIGURE PS 4-2

$$e_O = \frac{0.4 \text{ kilohm}}{0.4 \text{ kilohm} + 0.095 \text{ kilohm}} (0.095 \text{ volts})$$

$$= 0.0767 \text{ volt}_{pk}$$

One important point which should be checked is that the ac signal under no condition overcomes the dc bias which is trying to hold the diode ON. That is, if the superposition theorem is loosely applied and the driving voltage is assumed to be negative as in Fig. PS 4-2*d*, we might be tempted to say that the diode will not conduct. We must, however, remember that the diode is conducting a dc current of 2.5 ma in the opposite direction. Since this reversed-current ac component is less than the dc component, the diode will always remain forward-biased, although at this particular instant the net current through the diode is 2.5 ma − 0.192 ma, which still leaves a net forward current.

PS 4-3 Determine E_O in Fig. PS 4-3.

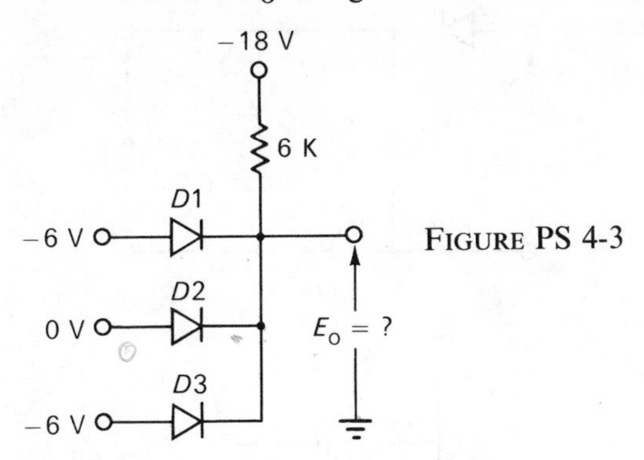

FIGURE PS 4-3

SOLUTION This circuit is one of the logic-gate types referred to in the text. The solution for E_O is most easily determined by posing the question, which diode of all the diodes connected to the common node point (the output line, that is) is trying to conduct the hardest? Clearly, $D2$ would tend to conduct the hardest as its anode is returned to the most positive potential. With $D2$ ON, $E_O = 0$ volts and $D1$ and $D3$ are reverse-biased.

PS 4-4 Determine E_O in Fig. PS 4-4.

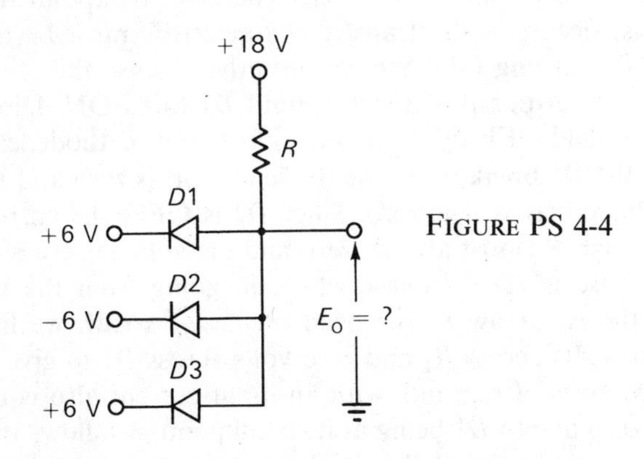

FIGURE PS 4-4

SOLUTION In this logic gate circuit the diode trying to conduct the hardest is $D2$. With $D2$ ON, $E_O = -6$ volts and $D1$ and $D3$ are reverse-biased.

PS 4-5 Determine E_O in Fig. PS 4-5.

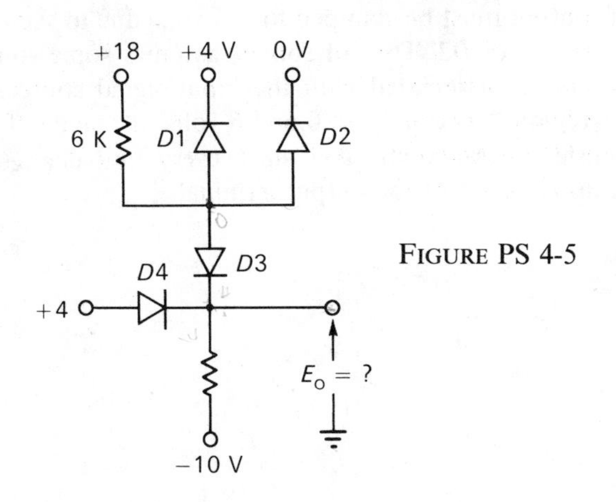

FIGURE PS 4-5

SOLUTION There are several diodes in this circuit and although it could be solved by applying the method of assumed states we would have $2^4 = 16$ possibilities to consider. A much better way of analyzing this circuit is to break it down into individual logic gates. For example, let us momentarily assume the anode of $D3$ is disconnected from the anodes of $D1$ and $D2$. Hence $D1$, $D2$, and the 6-kilohm resistor form one gate. Of the two diodes $D1$ and $D2$, it follows that $D2$ is trying to conduct the hardest. Hence, the common line connecting the anodes of $D2$ and $D1$ to the 6-kilohm resistor will be tied to zero volts. Now let us reconnect the anode of $D3$ to this zero-volt common line and pose the question, of the two diodes $D3$ and $D4$, which is trying to conduct the hardest? Clearly, $D4$ is trying to conduct the hardest, which fixes the output line at $+4$ volts and the cathode of $D3$ also at $+4$ volts. Since the anode of $D3$ was returned to zero volts, $D3$ is reverse-biased.

As an exercise, the reader may consider other possibilities and explore them for inconsistencies to verify the solution just obtained.

PS 4-6 Determine the currents I_1 and I_2 in Fig. PS 4-6*a*.

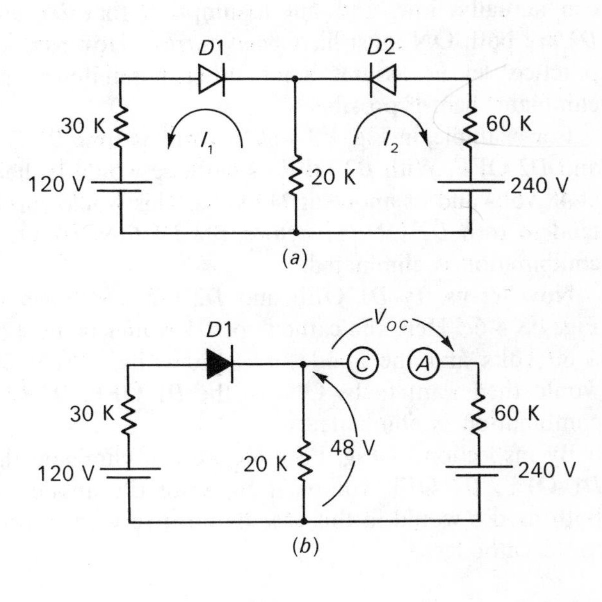

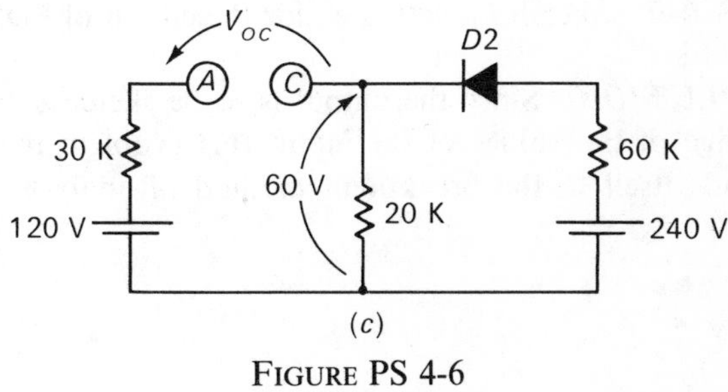

FIGURE PS 4-6

SOLUTION Applying the method of assumed states we might, as a first guess, choose $D1$ and $D2$ ON. With zero volts across $D1$ and $D2$, we may then solve for I_1 and I_2 by determinants, as follows:

1) 120 volts = 50 kilohms I_1 + 20 kilohms I_2

2) 240 volts = 20 kilohms I_1 + 80 kilohms I_2

3) $$I_1 = \frac{\begin{vmatrix} 120 \text{ volts} & 20 \text{ kilohms} \\ 240 \text{ volts} & 80 \text{ kilohms} \end{vmatrix}}{\begin{vmatrix} 50 \text{ kilohms} & 20 \text{ kilohms} \\ 20 \text{ kilohms} & 80 \text{ kilohms} \end{vmatrix}} = \frac{4.8 \times 10^6}{3600 \times 10^6}$$

$$= 1.33 \text{ ma}$$

4) $$I_2 = \frac{\begin{vmatrix} 50 \text{ kilohms} & 120 \text{ volts} \\ 20 \text{ kilohms} & 240 \text{ volts} \end{vmatrix}}{3600 \times 10^6} = \frac{9.6 \times 10^6}{3600 \times 10^6}$$

$$= 2.66 \text{ ma}$$

Notice that neither I_1 or I_2 turned out to be negative quantities. This means that the currents actually flow in the directions assumed in Fig. PS 4-6*a*; since this is in the forward direction for the diodes, these currents can actually flow, and our assumption that $D1$ and $D2$ are both ON must have been correct. However, for practice, let us assume some other possibilities and eliminate them if possible.

For example, in Fig. PS 4-6*b*, we will assume $D1$ ON and $D2$ OFF. With $D2$ OFF its cathode would be held at 48 volts and its anode at 240 volts. This would surely tend to turn $D2$ ON. Therefore, the $D1$ ON, $D2$ OFF combination is eliminated.

Now let us try $D1$ OFF and $D2$ ON as shown in Fig. PS 4-6*c*. Here the cathode of $D1$ would be held at +60 volts and the anode at +120 volts. Surely $D1$ would then want to be ON, so the $D1$ OFF, $D2$ ON combination is eliminated.

By inspection of Fig. PS 4-6*a*, we can eliminate the $D1$ OFF, $D2$ OFF combination, since the anodes of both diodes would in this case be positive with respect to the cathodes.

PS 4-7 Sketch E_o versus e_{in} for the circuit of Fig. PS 4-7*a*.

SOLUTION Since the output is to be sketched for a range of the values of the input, this problem readily lends itself to the breakpoint method of analysis. We might begin by assuming e_{in} large and negative. Under these circumstances $D1$ would surely be OFF and e_o is then zero for all values of e_{in}. The first breakpoint that must occur on the transfer characteristic must be due to $D1$ turning ON. We can intuitively sense this, since e_{in} cannot possibly affect e_o until $D1$ turns ON. Diode $D2$ is held OFF by the 6-volt source on its cathode lead. At the $D1$ breakpoint, the diode current is zero and the voltage across it is zero. Since $D2$ is OFF, the current through R_1 must also be zero and the voltage across R_1 likewise is zero. Consequently, in going from the tail of the e_{in} arrow to the head of the e_{in} arrow, we find zero volts across R_1 and zero volts across $D1$ to give us zero volts of e_{in}; and, with an input current also equal to zero due to $D1$ being at its breakpoint, it follows that the coordinates of the $D1$ breakpoint on the transfer characteristic lie on the origin. This is shown in Fig. PS 4-7*b*. As e_{in} becomes increasingly positive, the next change which occurs must be due to $D2$ turning ON. $D2$ turns ON when e_{in} = +6 volts. Once $D2$ turns ON the output must be clamped to +6 volts due to the zero resistance of $D2$. This, of course, assumes some source impedance associated with the input signal source. In the region between e_{in} = 0 and 6 volts, the slope of the transfer characteristic is 1 since every unit change in e_{in} appears across the output terminal.

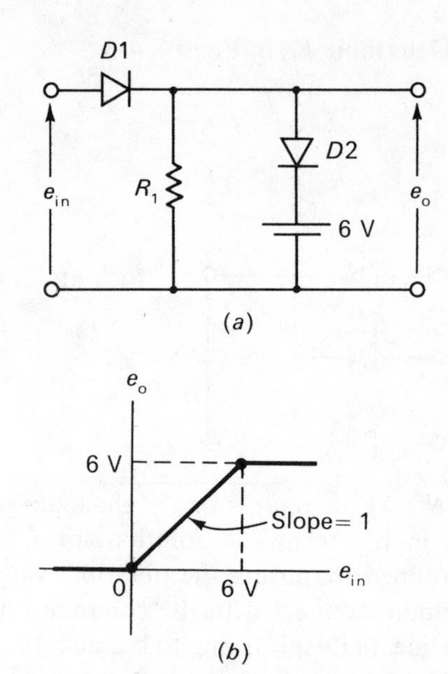

FIGURE PS 4-7

PROBLEMS WITH ANSWERS

PA 4-1 Determine E_O in Fig. PA 4-1.

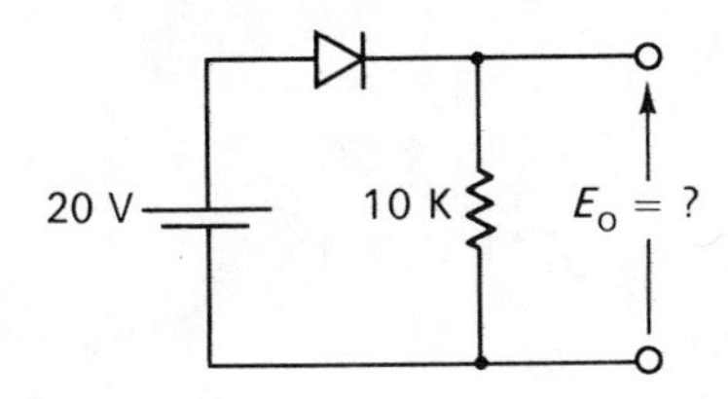

ANSWER $E_O = 20$ volts

PA 4-2 Determine E_O in Fig. PA 4-2.

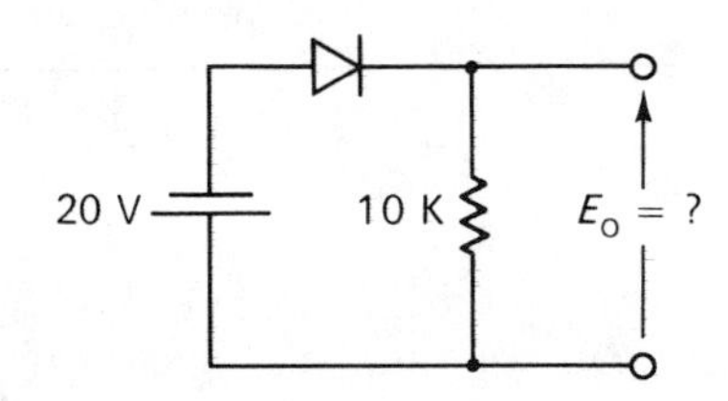

ANSWER $E_O = 0$

PA 4-3 Determine I in Fig. PA 4-3.

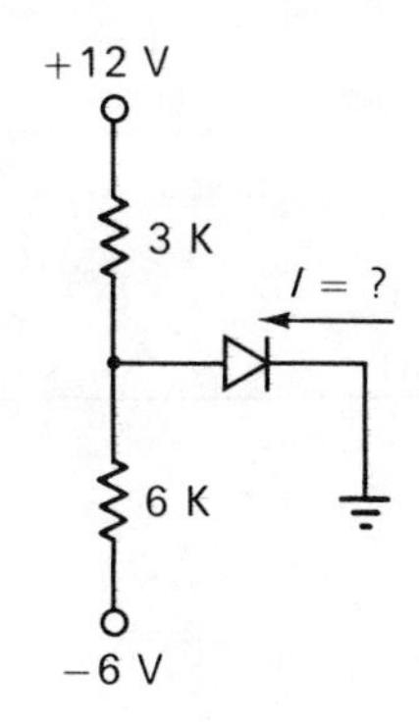

ANSWER $I = 3$ ma

PA 4-4 Determine V in Fig. PA 4-4.

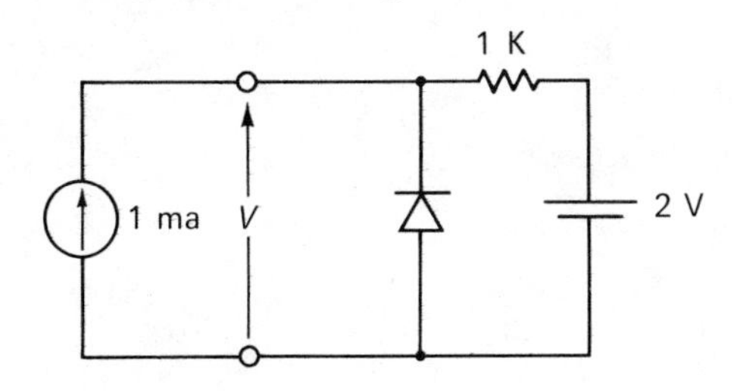

ANSWER $V = 1$ volt

PA 4-5 Determine I_1 and I_2 in Fig. PA 4-5.

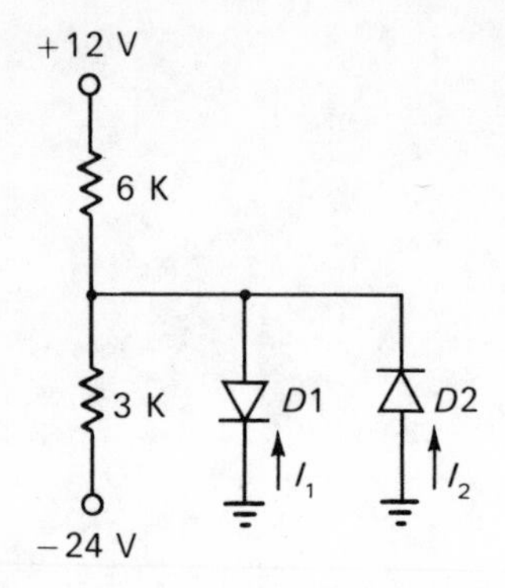

ANSWER

$I_1 = 0$
$I_2 = 6$ ma

PA 4-6 Determine E_O in Fig. PA 4-6.

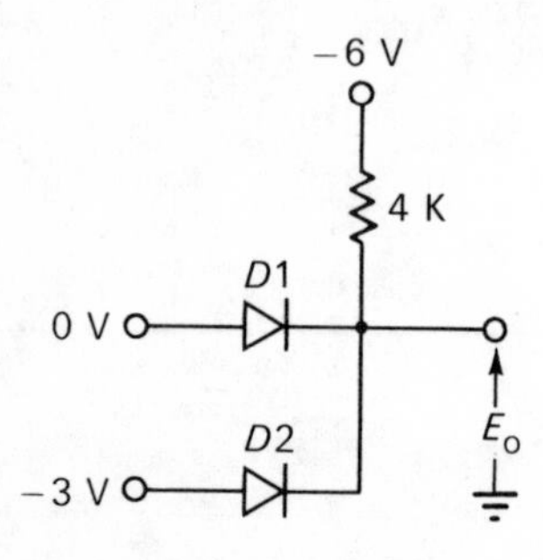

ANSWER $E_O = 0$

PA 4-7 Determine E_O in Fig. PA 4-7.

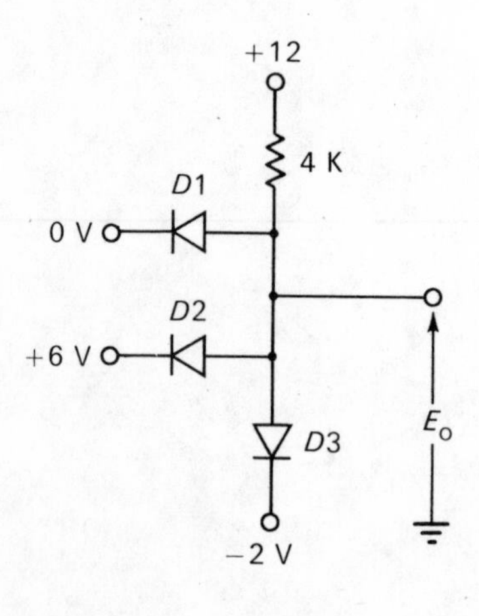

ANSWER $E_O = -2$ volts

PA 4-8 Determine E_O in Fig. PA 4-8.

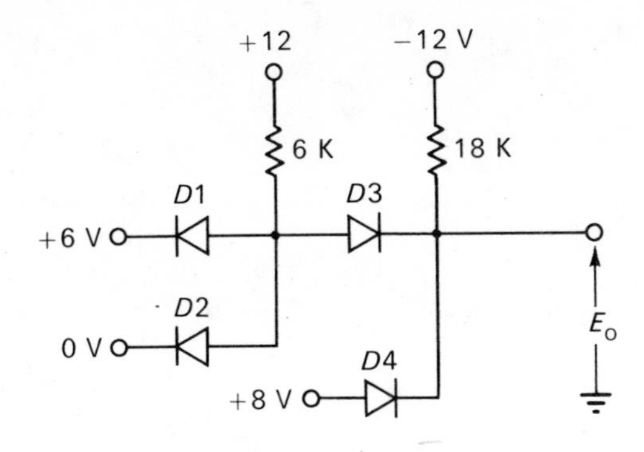

ANSWER $E_O = +8$ volts

PA 4-9 Plot the transfer characteristic E_O versus E_{in} for the circuit of Fig. PA 4-9*a*.

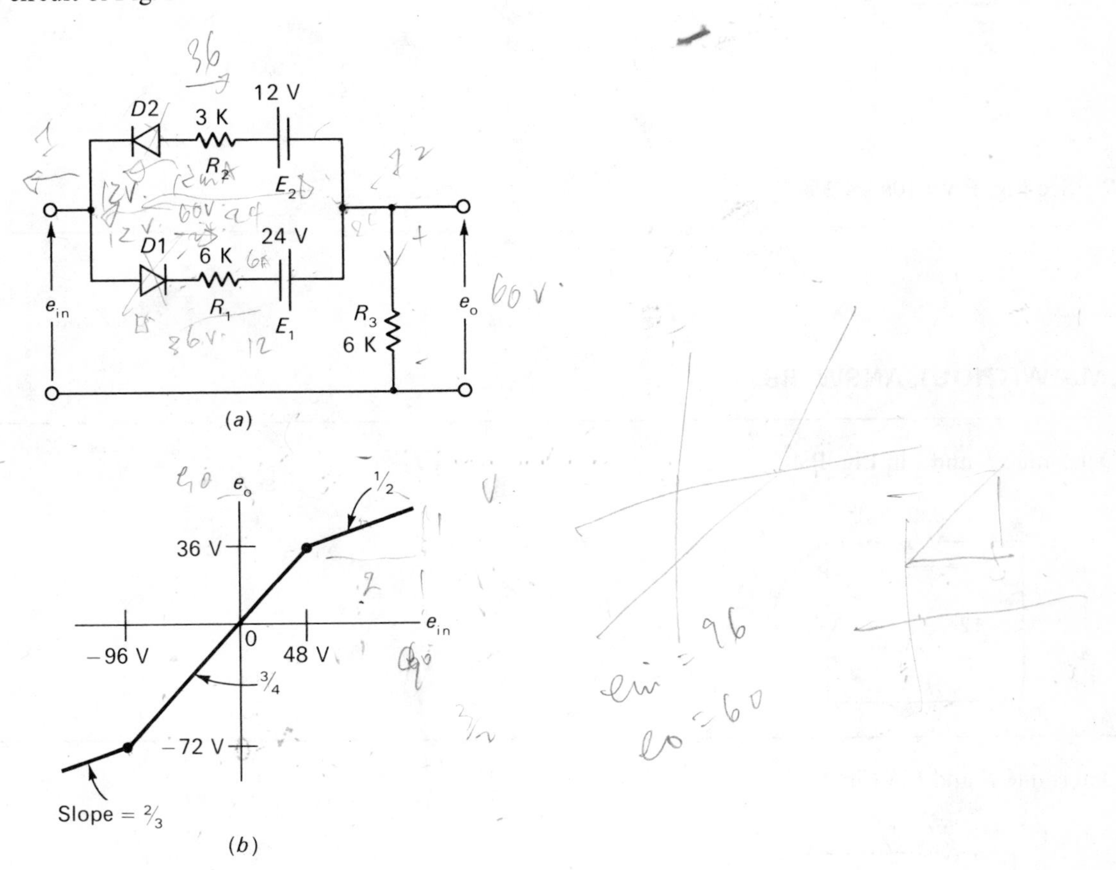

ANSWER See Fig. PA 4-9*b*.

PA 4-10 Sketch the *V-I* characteristic for the circuit of Fig. PA 4-10*a*.

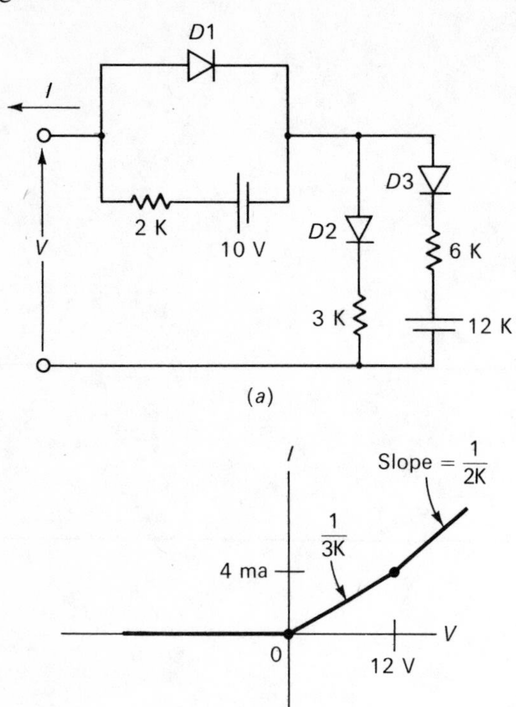

(*b*)

ANSWER See Fig. PA 4-10*b*.

PROBLEMS WITHOUT ANSWERS

P 4-1 Determine *E* and *I* in Fig. P 4-1.

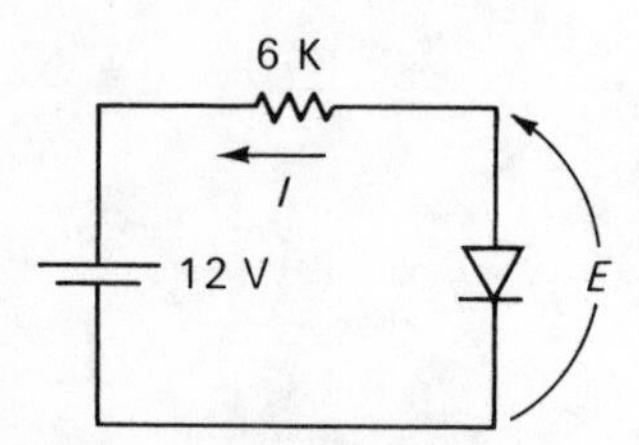

P 4-2 Determine *E* and *I* in Fig. P 4-2.

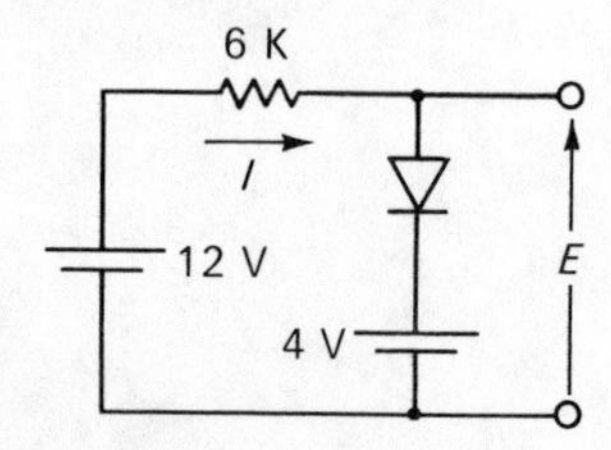

P 4-3 Determine E in Fig. P 4-3.

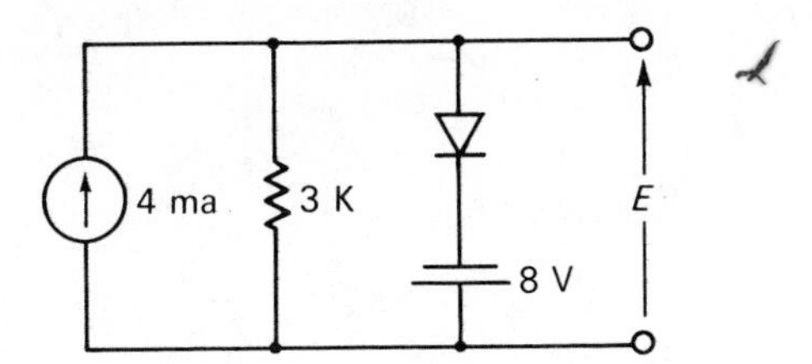

P 4-4 Determine E in Fig. P 4-4.

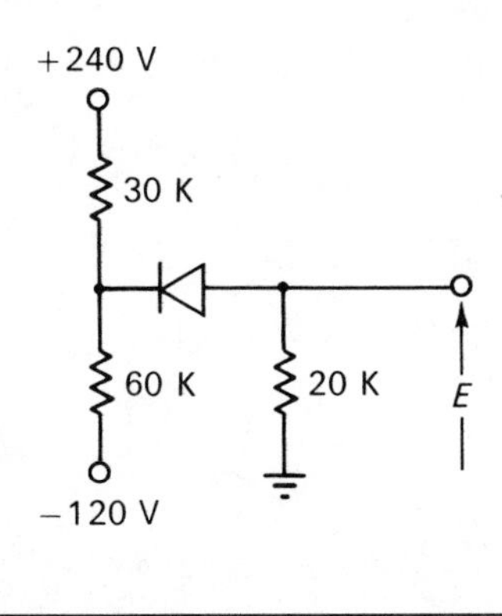

P 4-5 Determine e_o in Fig. P 4-5.

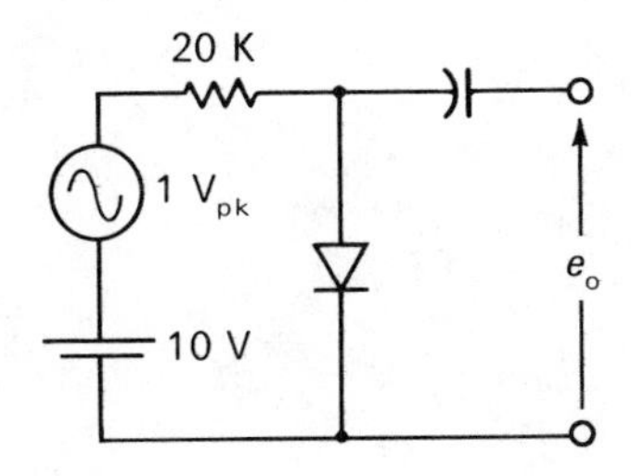

P 4-6 Sketch the V-I characteristic for Fig. P 4-6.

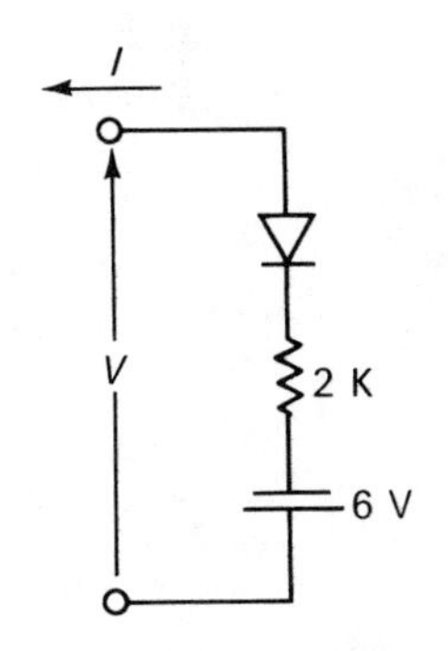

P 4-7 Sketch the V-I characteristic for the circuit of Fig. P 4-7.

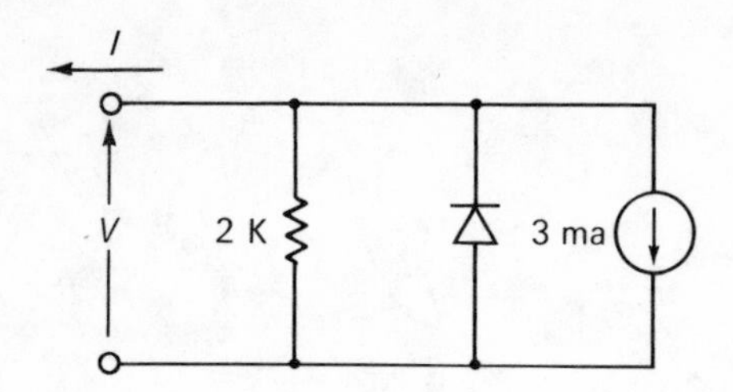

P 4-8 Sketch the V-I characteristic for the circuit of Fig. P 4-8.

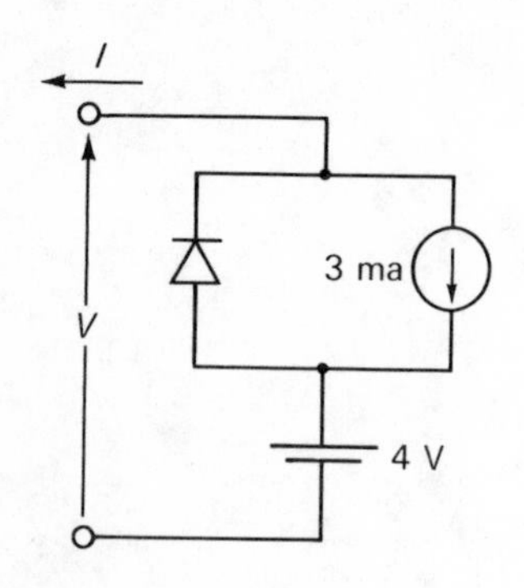

P 4-9 Devise a piecewise model to fit the V-I characteristic of Fig. P 4-9.

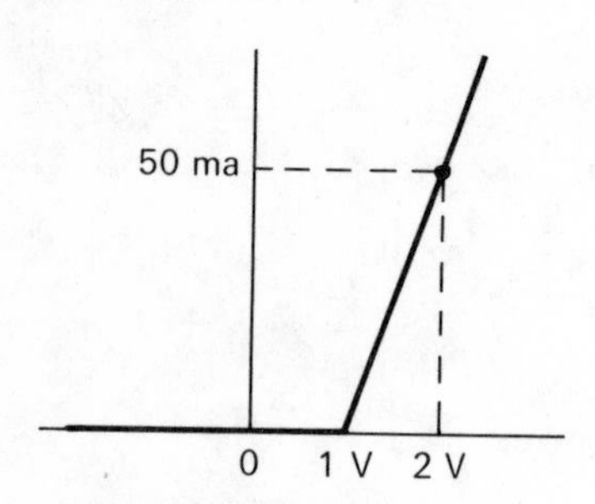

P 4-10 Determine E in Fig. P 4-10.

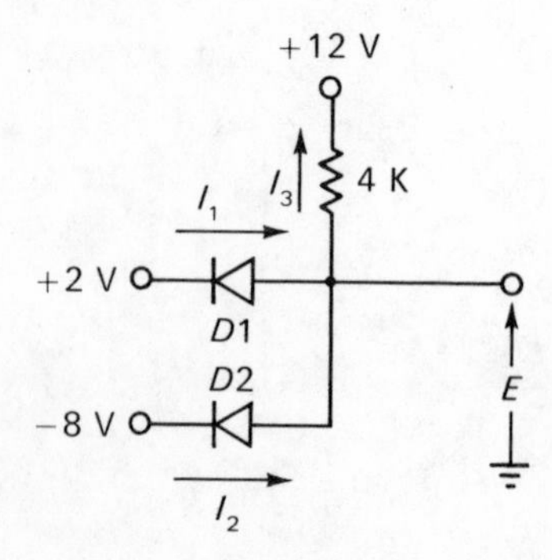

P 4-11 Determine E in Fig. P 4-11.

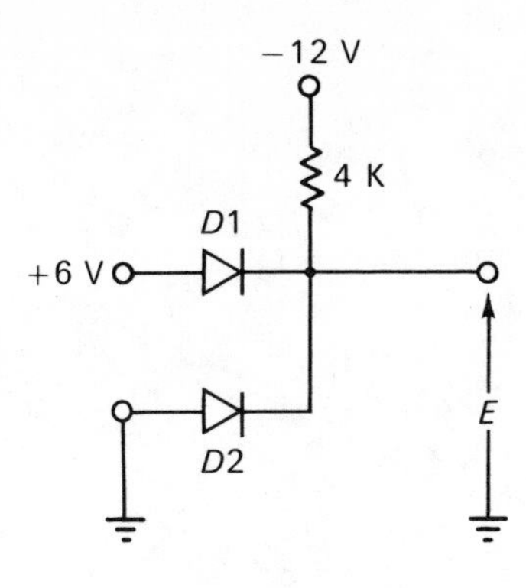

P 4-12 Determine the transfer characteristic (e_o/e_{in}) for the circuit of Fig. P 4-12.

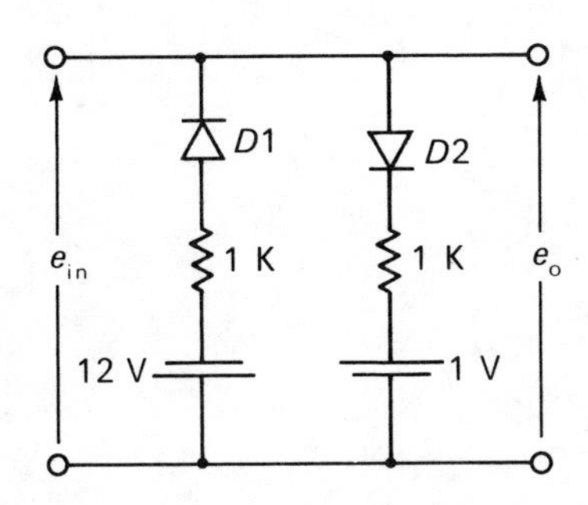

P 4-13 Sketch the V-I characteristic for the circuit of Fig. P 4-13.

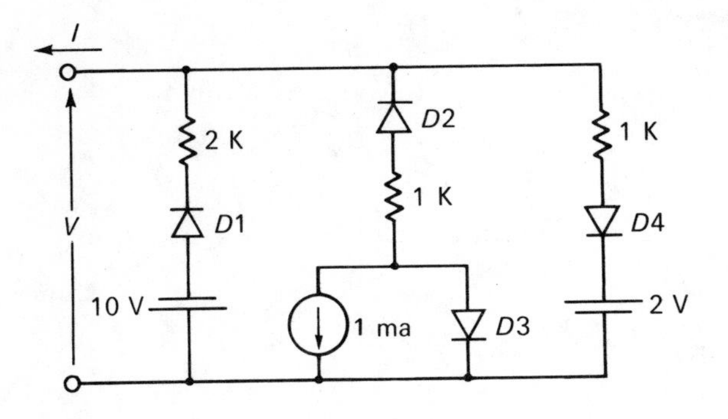

5 the junction transistor (transistor action)

It is most helpful if, prior to a detailed study of transistors, we first obtain a broad picture of how amplification of an input signal may be achieved. This is conveniently accomplished by considering the circuit of Fig. 5-1, which, as we shall see later, is somewhat analogous to the mechanism by which a transistor amplifies.

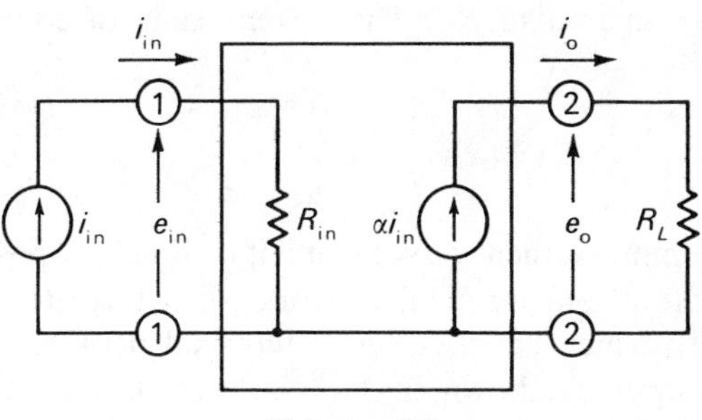

FIGURE 5-1

5-1 The controlled current source

Fig. 5-1 illustrates a box, commonly called a *four-terminal* or *two-port network* because it has a pair of input terminals (1-1) and output terminals (2-2). Assume this box is characterized by a certain input resistance R_{in} seen looking into the input terminals and by an ideal current source αi_{in} seen looking into the output terminals. This αi_{in} current source is called a *dependent current source*, and its value is a function of a remote current i_{in} which flows into the input terminals. The quantity α is a dimensionless constant (numeric) of proportionality which simply expresses how much bigger or smaller this dependent current source is than i_{in} itself. For this reason, α may also be called a *current gain*, since it is the ratio

$$\alpha = \frac{i_o}{i_{in}} = \frac{\alpha i_{in}}{i_{in}}$$

To explore the possibilities of power gain in this circuit, we may write

$$1) \quad P_G = \frac{P_o}{P_{in}} = \frac{i_o{}^2 R_L}{i_{in}{}^2 R_{in}} = \frac{(\alpha i_{in})^2 R_L}{i_{in}{}^2 R_{in}} = \alpha^2 \frac{R_L}{R_{in}}$$

Equation 1 indicates that the power gain can be greater than one if α is a number greater than one and if R_L is greater than R_{in}. However, greater than unity power gain may be achieved if α is less than one just so long as the ratio R_L/R_{in} is sufficiently greater than one. Physically, what this equation tells us is that if we have an ideal current source on the output side, we can make the load resistor, R_L, just as large as we please and thereby keep increasing the power developed in it. And if the input resistance, R_{in}, is very small, then only a small amount of power will be developed at the input terminals. Hence power gain should be achieved. An expression for voltage gain may be similarly derived as follows:

$$A_v = \frac{e_o}{e_{in}} = \frac{-R_L i_o}{-R_{in} i_{in}} = \frac{R_L \alpha i_{in}}{R_{in} i_{in}} = \alpha \frac{R_L}{R_{in}} \tag{2}$$

Note in equation 2 that even if α is a number less than one, considerable voltage gain is still possible if R_L is large compared to R_{in}. The current gain, of course, is given by

$$A_i = \frac{i_o}{i_{in}} = \frac{\alpha i_{in}}{i_{in}} = \alpha \tag{3}$$

In summary, then, we see that if it is somehow possible to create a four-terminal network (or a three-terminal network which has a common terminal between input and output as shown in Fig. 5-1) which simulates a dependent current source on the output side, is a function of the current into the input side, and which also exhibits a low input impedance, it is possible to achieve power gain or voltage gain. Should α be a number greater than one, then it would even be possible to achieve current gain. As we shall see, the transistor essentially simulates a network of this type.

5-2 The junction transistor

The *junction transistor* is a solid-state amplifying device made of suitably doped semiconductor materials. Transistors are available with power-handling capabilities ranging from microwatts to over hundreds of watts and in a frequency range extending from direct current to thousands of megacycles. Although the power-handling ability of transistors decreases as the frequency increases, the picture is continuously improving as technology advances.

Transistors are manufactured by a variety of methods, which also change as the technology advances. The details of these manufacturing methods are beyond the scope of this text and of no particular concern with respect to our immediate goal of analyzing circuits. The important thing is that all these techniques have one common goal, and that is the formation of a single semiconductor crystal containing two *PN* junctions in close proximity.

Basically transistors are of two types, *PNP* and *NPN*. Fig. 5-2*a* illustrates the structure of a *PNP* transistor. It is divided into three regions called emitter (*E*), base (*B*), and collector (*C*). The emitter and base regions are separated by a *PN* junction called the *emitter junction*, and the collector and base regions are separated by another *PN* junction called the *collector junction*. The emitter and base regions with the emitter junction may be considered as a diode called the *emitter diode* (D_E) as shown in Fig. 5-2*b*. Similarly the base and collector regions with the collector junction form a diode called

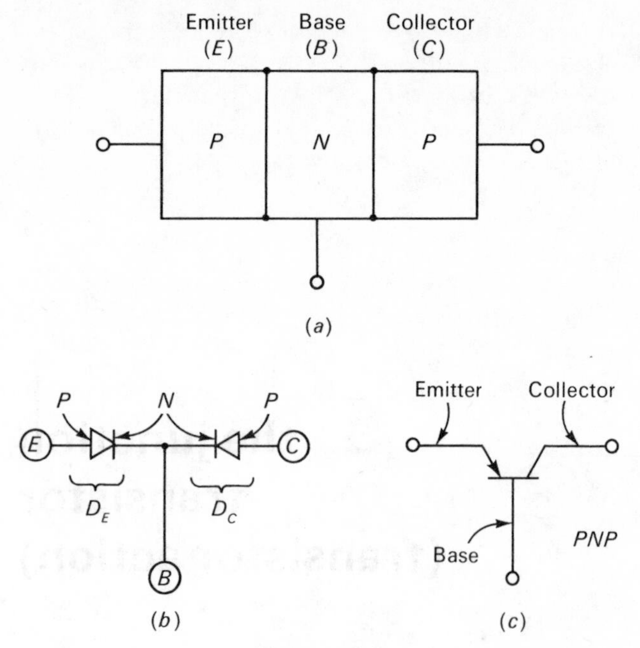

FIGURE 5-2

the *collector diode* (D_C). Fig. 5-2*b* should not be construed as being an equivalent circuit of the transistor, however. It turns out that in a transistor these diodes are so close together that the current in one diode affects the current in the other diode, and hence the transistor cannot just be considered as two isolated diodes as shown in Fig. 5-2*b*. Fig. 5-2*b* is presented solely as an aid in remembering the polarity of the diodes involved in a *PNP* transistor. The schematic symbol of a *PNP* transistor is shown in Fig. 5-2*c*.

Fig. 5-3*a* illustrates the structure of an *NPN* transistor.

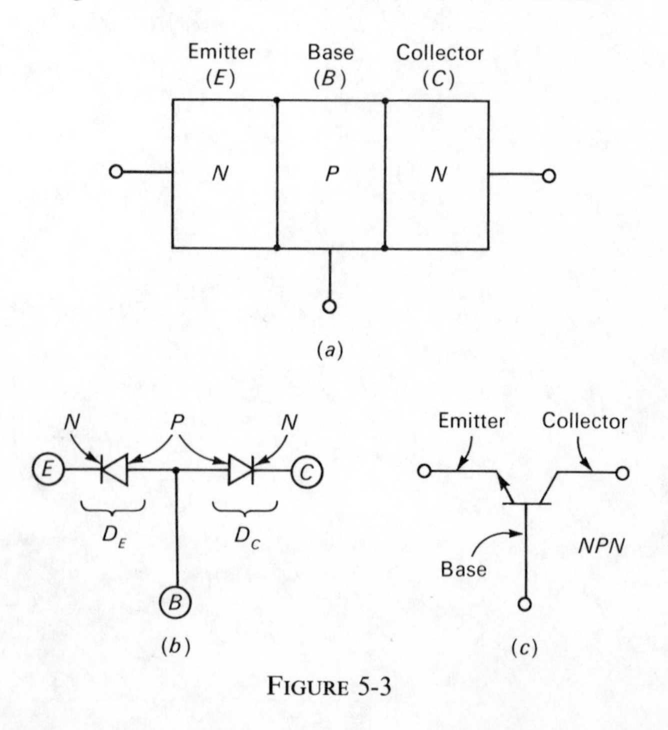

FIGURE 5-3

It is identical to a *PNP* transistor, except the emitter and collector are composed of *N*-type semiconductor and the base is composed of *P*-type material. Fig. 5-3*b* illustrates the polarity of the diodes in an *NPN* transistor and Fig. 5-3*c* is the schematic representation.

5-3 Transistor current components

Figure 5-4 illustrates one method of biasing a transistor so that it may be capable of amplification. If we choose to

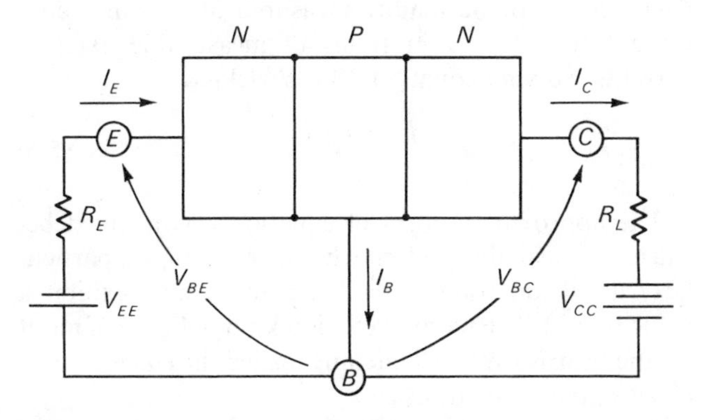

FIGURE 5-4

think of the transistor as merely being two diodes connected back to back as shown in Fig. 5-5, it would seem that the emitter diode would be forward-biased due to

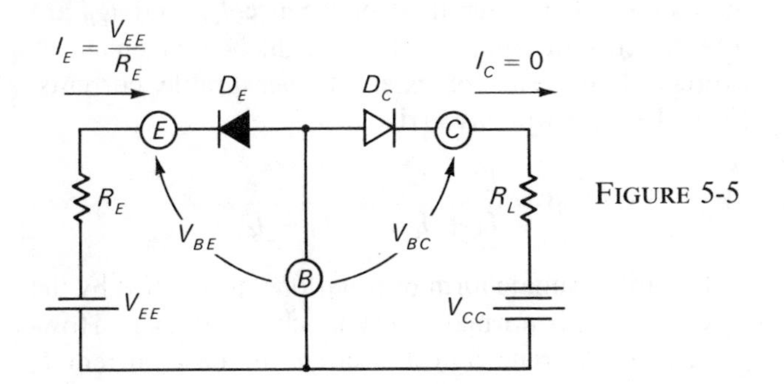

FIGURE 5-5

the V_{EE} supply, and the collector diode would be reverse-biased due to the V_{CC} supply. Assuming these are ideal diodes, we could write $I_E = V_{EE}/R_E$ since there would be zero voltage drop across D_E. With D_C reverse-biased, its current should be equal to zero and hence $V_{BC} = V_{CC}$. This situation would be true if Fig. 5-5 truly represented an *NPN* transistor. In actuality we would find that I_C is not equal to zero and, in fact, is almost identical with I_E in a high-quality transistor.

To understand how the transistor is capable of performing amplification, consider Fig. 5-6. Voltage source V_{CC}, commonly called the *collector supply*, is intended to reverse-bias the collector junction. Since this is an *NPN* transistor, it follows that the *N*-type collector must be held positive with respect to the base. Resistor R_L is a load resistor across which the useful output is normally developed. If the emitter lead is assumed to be open, the emitter current I_E must equal zero. That collector current which flows when the emitter current is zero is called I_{CO} or I_{CBO}. This current, for our immediate purposes, may be considered as the leakage current of the collector junction when it is reverse-biased. It consists of thermally generated electrons in the *P* region of the collector junction which are swept across the junction to the *N* region and thermally generated holes on the collector side which are swept across the collector-junction depletion region into the *P*-type base. Just as long as the collector junction is reverse-biased (but not in its breakdown region) by a few hundred or more millivolts, this thermally generated leakage current will be theoretically unaffected by changes in V_{BC}. In effect, then, the load resistor, R_L, is fed from a constant-current source. With the emitter lead open, the collector current is just equal to I_{CBO}. Now if the emitter lead is connected up through R_E and V_{EE} as shown in Fig. 5-6, the emitter junction becomes forward-biased. If we think of the forward-biased emitter junction as being an ordinary forward-biased diode, we may reason that the current across the emitter junction will consist of electrons going from

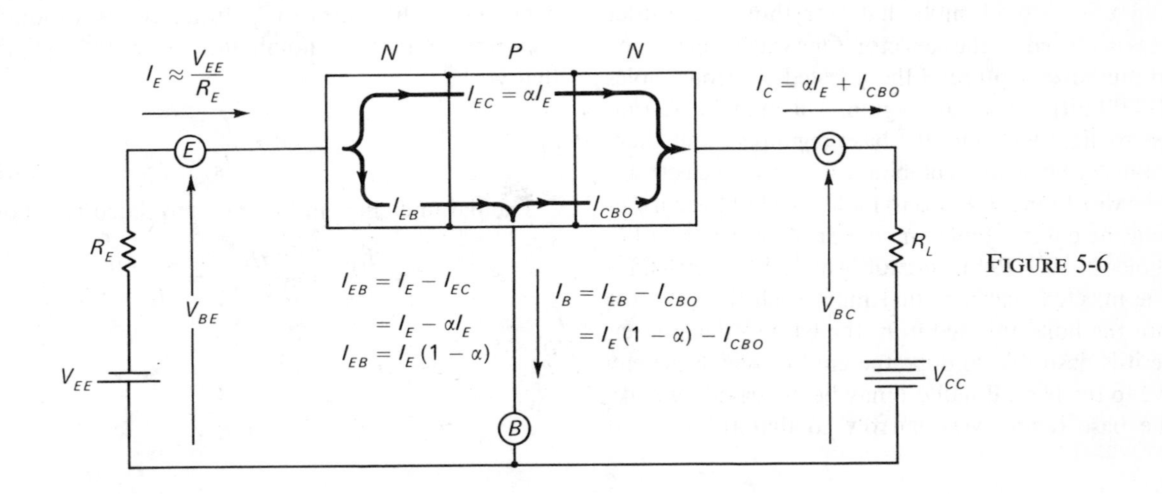

FIGURE 5-6

the emitter to the base and holes going from the base to the emitter. Some of the electrons which cross from emitter to base will combine with holes in the base region. For each electron which is so captured in the base region, there will be an electron forced out of the base lead in the direction of I_B. Likewise, each hole that leaves the base to enter the emitter causes the base region to acquire a negative charge. To maintain charge neutrality, an electron will be forced out of the base lead also in the I_B direction. Both the holes going from base to emitter and the electrons going from emitter to base which combine with holes in the base give rise to a component of base current called I_{EB}. In a well-designed transistor most of the electrons which enter the base from the emitter will not immediately recombine with base region holes. If these injected electrons should drift to the collector junction, they will encounter an electric field across the collector-junction depletion region which will sweep them across the junction into the collector region where they contribute to the collector current. It is these current carriers which are injected from the emitter to the base and subsequently collected by the collector that are indicated by I_{EC}. In a *PNP* transistor, I_{EC} would consist of holes that survived the journey from emitter to collector.

Insofar as the collector junction in Fig. 5-6 is concerned, the electrons that make up I_{EC} are no different than the thermally generated electrons that make up part of I_{CBO}. Hence these injected carriers will also appear to emanate from a constant-current source, and the net effect will be to cause an apparent increase in the collector leakage current. Thus the collector current consists of the thermally generated leakage current, I_{CBO}, plus I_{EC}, the component of injected emitter current. The more we forward-bias the emitter diode, the greater I_{EC} becomes. We may define a parameter alpha as

1) $$\alpha = \frac{I_{EC}}{I_E} \qquad (5\text{-}1)$$

It is desirable to make α as close to unity as possible, since an $\alpha = 1$ would imply that everything the emitter injects is captured by the collector. One way of increasing α is to minimize capture of the injected electrons (holes in a *PNP* unit) in the base region. This may be accomplished by lightly doping the base region, so that there are not many holes to recombine with injected electrons. Another way to increase α is to make most of the current crossing the emitter junction consist of electrons rather than holes in an *NPN* transistor. This is beneficial since it is the injected electrons that may reach the collector and not the holes injected from the base to the emitter. Hence it is desirable to dope the emitter region heavily relative to the base. Finally, α may be increased by making the base region very narrow so that the injected carriers will have a better chance of drifting through to the collector without recombining in the base region.

From Fig. 5-6 the following relations become apparent.

2) $$I_C = \alpha I_E + I_{CBO} \qquad (5\text{-}2)$$

3) $$I_B = I_E(1 - \alpha) - I_{CBO} \qquad (5\text{-}3)$$

4) $$I_E = \frac{V_{EE} - V_{BE}}{R_E} \approx \frac{V_{EE}}{R_E} \qquad (5\text{-}4)$$

Since I_{EC} in equation 1 is an internal transistor current which may not be readily measured, it becomes convenient to define α in terms of measurable external currents. Solving equation 2 for α yields

5) $$\alpha = \frac{I_C - I_{CBO}}{I_E} \approx \frac{I_C}{I_E} \qquad (5\text{-}5)$$

The approximate form of equation 5 is justified because I_C is usually much much larger than I_{CBO}, particularly in silicon transistors. Typically, α may exhibit a value of 0.98. The parameter α is a kind of figure of merit for the transistor. It is sometimes called the *common-base current gain* of the transistor.

Another figure of merit for the transistor is the parameter β (beta), defined as the current gain

6) $$\beta = \frac{I_{EC}}{I_{EB}}$$

Since I_{EB} is a much smaller current than I_{EC}, β is a number much greater than one. Since I_{EC} and I_{EB} are internal current components, it might be convenient to express β in terms of external, measurable currents. From Fig. 5-6 we may arrive at

7) $$\beta = \frac{I_C - I_{CBO}}{I_B + I_{CBO}} \approx \frac{I_C}{I_B + I_{CBO}} \qquad (5\text{-}6)$$

The approximate form of equation 7 is justified by the fact that I_{CBO} is normally very much less than I_C. However, if the current gain β is large, the base current I_B may be relatively small, and hence I_{CBO} may not be negligible with respect to I_B. In many cases though, the I_{CBO} term and the denominator are both negligible so that we have

8) $$\beta \approx \frac{I_C}{I_B} \qquad (5\text{-}7)$$

The parameters α and β are interrelated as follows:

9) $$\beta = \frac{I_{EC}}{I_{EB}} = \frac{\alpha I_E}{I_E - I_{EC}} = \frac{\alpha I_E}{I_E - \alpha I_E}$$

or

10) $$\beta = \frac{\alpha}{1 - \alpha} \qquad (5\text{-}8)$$

or

$$\alpha = \frac{\beta}{\beta + 1} \tag{5-9}$$

11)

Thus if a transistor has an α equal to 0.98, the corresponding $\beta = 49$. Typically, β may vary by a factor of two to three in any given transistor type.

To summarize our discussion in this chapter, we wish to first point out that the transistor-current components we have discussed are valid for the so-called *active* or *amplifying region* of the transistor. In the active region, the emitter diode is forward-biased and it injects carriers into the base region. The collector diode is reverse-biased and it collects the injected carriers. The collected carriers which contribute to the collector current appear to emanate from a constant-current source. Hence, any load connected in the collector lead is driven by an apparent constant-current source. Thus we have succeeded in mechanizing a box which simulates the four-terminal network of Fig. 5-1.

For a *PNP* transistor it is only necessary to consider holes rather than electrons as being the injected carriers, and the polarity of all the diodes and external bias voltages must be reversed.

PROBLEMS WITH SOLUTIONS

PS 5-1 The input side of the four-terminal network shown in Fig. PS 5-1 is driven by a 1-ma input-signal

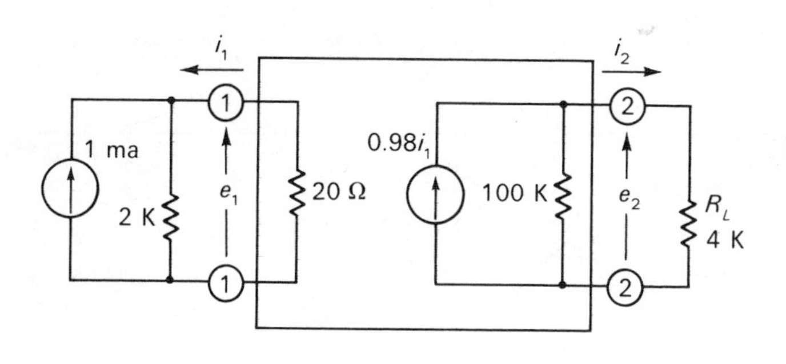

FIGURE PS 5-1

current source of internal impedance $Z_{th} = 2$ kilohms. The dependent current source on the output side has 100 kilohms internal impedance. Estimate the voltage gain A_v, current gain A_i, power gain A_p, and the actual output voltage e_2.

SOLUTION

1) $$A_v = \frac{e_2}{e_1} = \frac{-4 \text{ kilohms } i_2}{0.02 \text{ kilohm } i_1} \approx \frac{-4 \text{ kilohms}(0.98 i_1)}{0.02 \text{ kilohm } i_1}$$

$$= 196$$

The approximate form of the above equation is permissible because, taken as a current divider, 20 ohms $<$ 2 kilohms and 4 kilohms $<$ 100 kilohms.

2) $$A_i = \frac{i_2}{i_1} \approx \frac{0.98 i_1}{i_1} = 0.98$$

3) $$A_p = A_v A_i = 196(0.98) = 192$$

4) $$e_2 = -4 \text{ kilohms } i_2 \approx -4 \text{ kilohms}(0.98 i_1)$$

$$\approx -4 \text{ kilohms}(0.98)(-1 \text{ ma})$$

$$= 3.92 \text{ volts}$$

Note from the figure that $i_1 \approx -1$ ma.

PS 5-2 Determine the polarity of V_{BE} and V_{BC} in Fig. PS 5-2 in order to bias the transistor into its active operating region. Assuming ideal diodes, estimate V_{EC}.

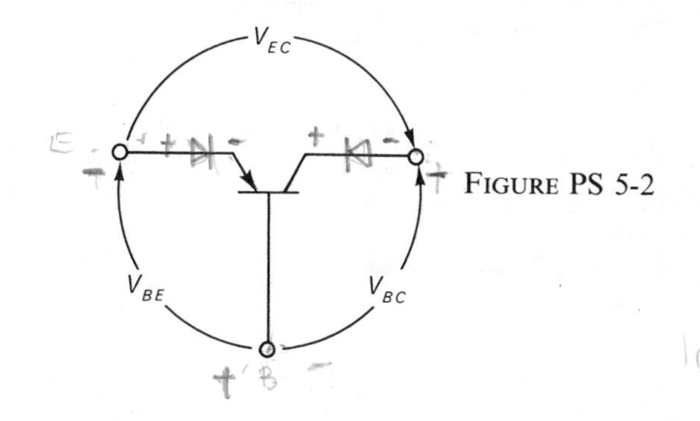

FIGURE PS 5-2

SOLUTION For active region operation, the emitter diode must be forward-biased and the collector diode must be reverse-biased. Since we have a *PNP* transistor, it follows that V_{BE} must be a positive quantity, and V_{BC} must be a negative quantity. Remember the first subscript represents the reference point to which we are comparing the voltage of the second subscript. Likewise, graphically, the tail of the arrow is the reference point. To determine V_{EC} we might reason that if the emitter diode is ideal and forward-biased, $V_{BE} = 0$. Hence $V_{EC} = -V_{BE} + V_{BC} = 0 + V_{BC} = V_{BC}$.

Having a rigorous means of denoting voltage and current is exceedingly important in analyzing electronic circuits. If the subject of voltage or current notation is hazy to you, review it in the first text of this outline series ("Outline for DC Circuit Analysis–With Illustrative Problems").

PS 5-3 Determine I_E, I_B, I_C, and V_{EC} in the circuit of Fig. PS 5-3. Assume a silicon transistor with the following parameters: $\alpha = 0.975$, $I_{CBO} = 1$ μa. Also determine the transistor's β.

SOLUTION The 2-volt supply in the emitter lead will tend to forward-bias the emitter diode. Since this is a silicon transistor, the forward voltage drop across the emitter diode will be approximately 0.6 volt as in

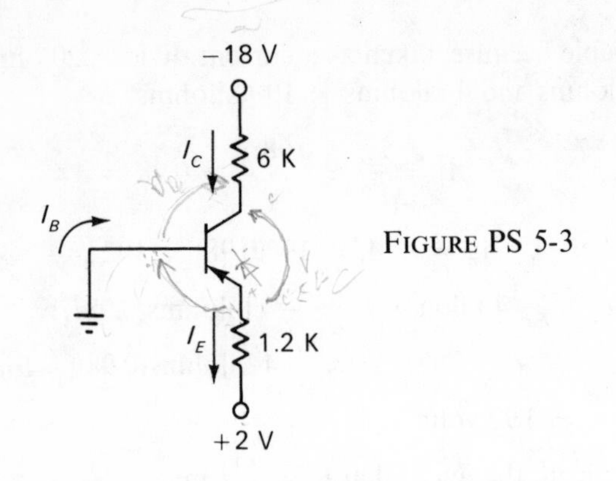

FIGURE PS 5-3

an ordinary silicon diode. Hence $V_{BE} \approx 0.6$ volt. Note that the emitter must be positive relative to the base. To determine I_E we need only write an expression for the voltage across the 1.2-kilohm resistor and divide this by the value of the resistor. Thus

$$I_E = \frac{2 \text{ volts} - 0.6 \text{ volt}}{1.2 \text{ kilohms}} = 1.17 \text{ ma}$$

The collector current is given by

$$\begin{aligned} I_C &= \alpha I_E + I_{CBO} = 0.975(1.17 \text{ ma}) + 0.001 \text{ ma} \\ &= 1.14 \text{ ma} \end{aligned}$$

$$\begin{aligned} I_B &= I_E(1 - \alpha) - I_{CBO} \\ &= 1.17 \text{ ma}(1 - 0.975) - 0.001 \text{ ma} \\ &= 0.024 \text{ ma} \end{aligned}$$

$$V_{EC} = V_{EB} + V_{BC}$$

Note that V_{EB} will be the negative of V_{BE}. Therefore,

$$\begin{aligned} V_{EC} &= -0.6 \text{ volt} + (-18 \text{ volts} + 6 \text{ kilohms } I_C) \\ &= -0.6 \text{ volt} - 18 \text{ volts} + 6 \text{ kilohms}(1.14 \text{ ma}) \\ &= -11.8 \text{ volts} \end{aligned}$$

Since V_{EC} is a negative quantity, it implies that the collector terminal is 11.8 volts negative with respect to the emitter terminal.

$$\beta = \frac{\alpha}{1 - \alpha} = \frac{0.975}{1 - 0.975} = 39$$

By equation 5-6 we may determine β in terms of the calculated external currents. Thus

$$\beta \approx \frac{I_C}{I_B + I_{CBO}} = \frac{1.14 \text{ ma}}{0.0282 + 0.001 \text{ ma}} = 38.9$$

which, allowing for round-off errors, is in agreement with the previously calculated value of β.

PROBLEMS WITH ANSWERS

PA 5-1 Determine e_2 and the power dissipated in R_L in the circuit of Fig. PA 5-1.

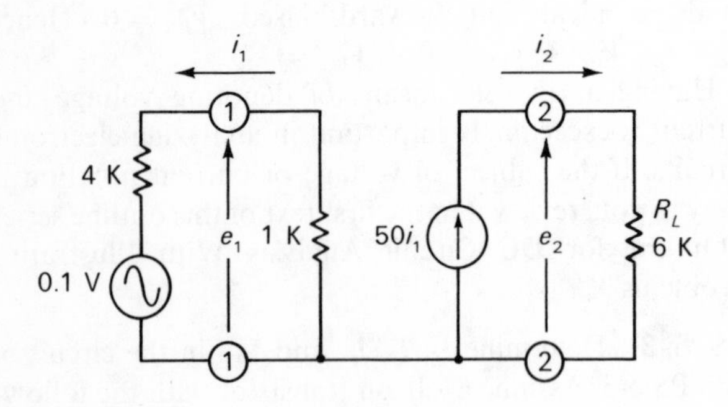

ANSWER $\quad e_2 = 6$ volts
$P_L = 0.6$ mw

PA 5-2 A milliammeter in the collector lead of the transistor in the circuit of Fig. PA 5-2 reads 0.5 ma for values of V_{CC} ranging from 0.5 volt to 60 volts. Above 60 volts the collector current starts to increase markedly. What can we say about this transistor?

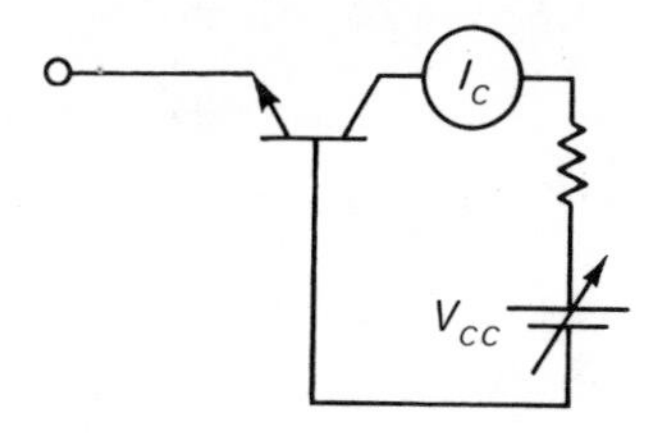

ANSWER $I_{CBO} = 0.5$ ma and collector diode exhibits reverse breakdown at 60 volts.

PA 5-3 The transistor used in Fig. PA 5-2 is now connected in the circuit of Fig. PA 5-3. If $I_B = 0$, what can we say about the transistor? What is I_C equal to?

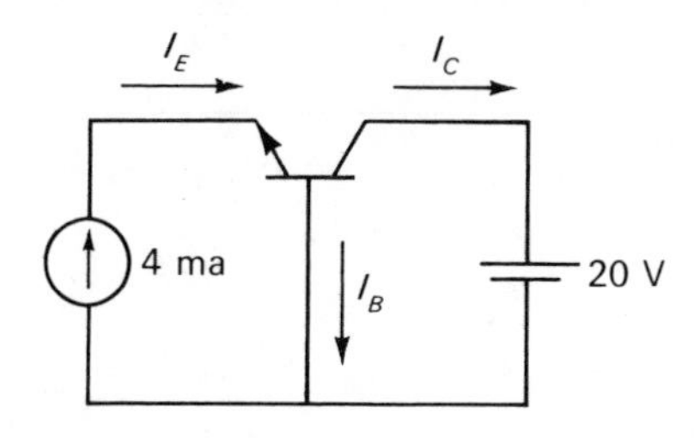

ANSWER

$$\alpha = 0.875$$
$$\beta = 7$$
$$I_C = 4 \text{ ma}$$

PA 5-4 The transistor of Fig. PA 5-3 is connected in the circuit of Fig. PA 5-4. Determine I_C.

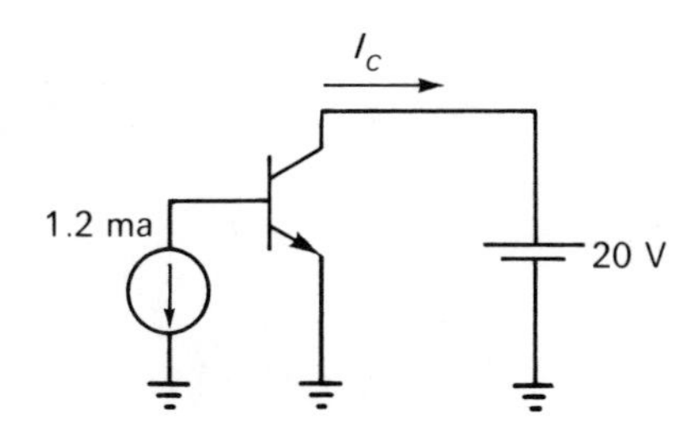

ANSWER $I_C = 12.4$ ma

PA 5-5 The transistor of Fig. PA 5-3 is connected in the circuit of Fig. PA 5-5. Assuming negligible V_{BE} voltage, determine the maximum permissible value of R_L which will permit the transistor to remain in its active region.

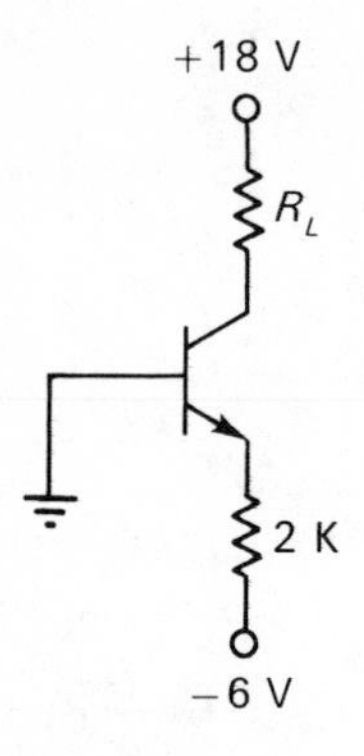

ANSWER $R_L = 6$ kilohms

PROBLEMS WITHOUT ANSWERS

P 5-1 How must the emitter and collector junctions of a transistor be biased so that operation in the active region is maintained?

P 5-2 Explain how a transistor might be capable of yielding power gain to an input signal.

P 5-3 The function of the emitter is to ________ and the collector is to ________.

P 5-4 In the active region the collector current looks as if it is emanating from a ________ ________ source.

P 5-5 In the active region, what must the polarity of V_{BE} and V_{BC} for a *PNP* transistor be?

P 5-6 What must be the polarity of V_{BE} and V_{BC} for an *NPN* transistor biased in its active region?

P 5-7 If the collector junction is reverse-biased and the emitter lead is left floating, the resultant collector current is termed __________.

P 5-8 The temperature of a transistor biased into its active region is gradually increased. The base current is observed to correspondingly decrease, reach zero, and then go negative. Explain this phenomenon.

P 5-9 The α of a transistor equals 0.997. Determine the corresponding β.

P 5-10 If $\beta = 80$, determine the corresponding α.

P 5-11 Determine I_E and I_B if $I_C = 6$ ma and $\beta = 40$. Assume the transistor is biased in its active region and that I_{CO} is negligible.

P 5-12 Determine V_{EC} in Fig. P 5-12. Assume negligible V_{BE} voltage drop.

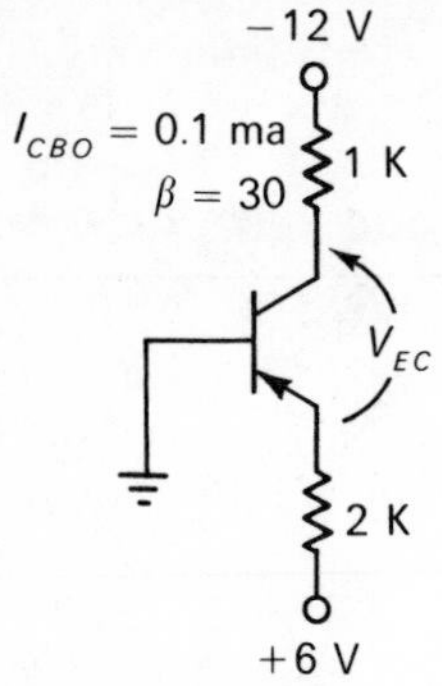

−12 V
$I_{CBO} = 0.1$ ma
$\beta = 30$
1 K
V_{EC}
2 K
+6 V

P 5-13 Determine V_{EC} in Fig. P 5-13. Assume an ideal emitter diode.

−12 V
$I_{CBO} = 0.1$ ma
$\beta = 30$
1 K
V_{EC}
60 K
−6 V

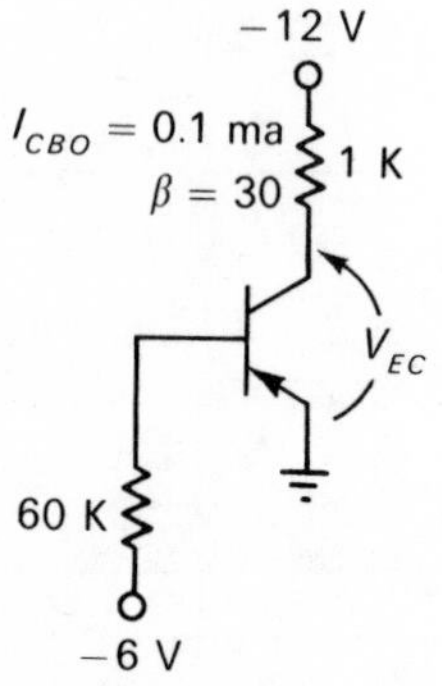

P 5-14 Determine V_{BC} in Fig. P 5-14. Assume an ideal emitter diode.

−12 V
1 K
−2 V

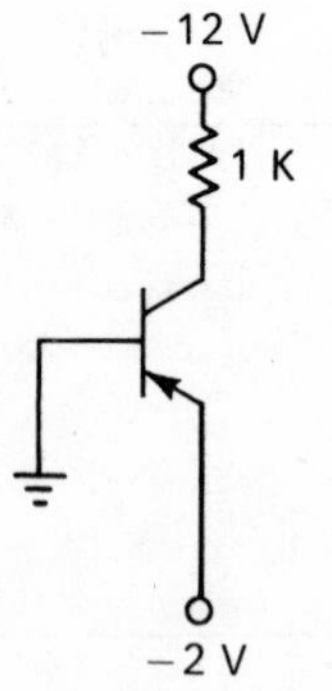

6 the common-base model

Now that we have some insight into the physical behavior of the transistor, it becomes convenient to synthesize an equivalent circuit (model) which may be substituted for the actual transistor. The use of these models often proves advantageous in circuit analysis.

6-1 The common-base dc model

One approach to modeling the transistor is to develop an equivalent circuit which approximates the physical behavior of the device. For example, consider Fig. 6-1*a* which represents an *NPN* transistor. In this model,

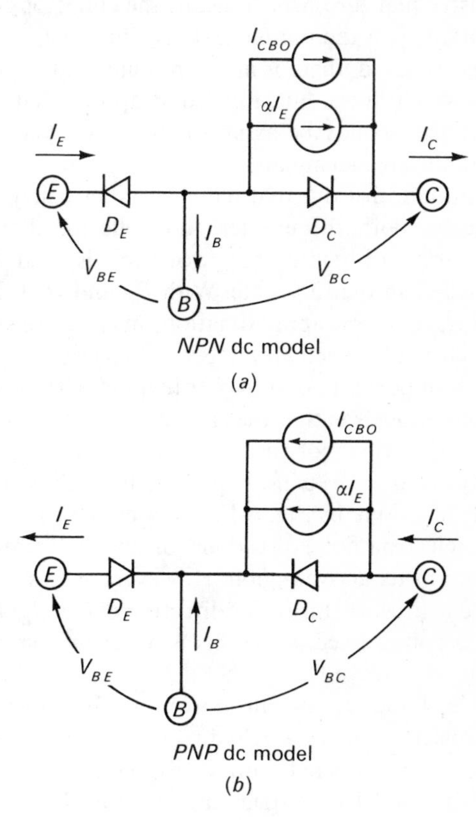

FIGURE 6-1

diodes D_E and D_C are the emitter and collector diodes respectively. You will recall that in the active region the collector diode is reverse-biased and the emitter diode is forward-biased. If a current I_E is made to flow in the emitter lead, a fractional part of this current αI_E flows out of the collector lead. We also know that the reverse-biased collector diode has a leakage current I_{CBO} associated with it. The net collector current I_C, which appears to emanate from a constant-current source, is given by $I_C = \alpha I_E + I_{CBO}$. This equation tells us that the collector current consists of two components which may be represented by two parallel current sources in shunt with the collector diode shown in Fig. 6-1*a*. As long as the collector diode is held reverse-biased these current sources cannot flow through D_C, but must instead exit out the collector lead. To guarantee that the collector diode is reverse-biased, it is only necessary that the external circuitry associated with the transistor maintain V_{BC} positive. The emitter diode, which in the active region is held forward-biased by the external circuitry, may, as a first approximation, be assumed a short circuit.

The model of Fig. 6-1*a* is adequate for most of our applications and will be refined as required.[1]

If both the emitter and collector diodes are somehow reverse-biased, the transistor is said to be *cut-off* or in the OFF state. This seems reasonable in the light of our model since if $I_E = 0$, the dependent αI_E current source must also equal zero, which means the collector current is essentially just the minute leakage current I_{CBO}. With D_E reverse-biased, there is also a minute emitter reverse current which flows, but, to a good approximation, an OFF transistor may be assumed to be an open circuit between all three terminals.

If it should turn out that the external circuitry somehow causes both the emitter and collector diodes to become forward-biased, the transistor is said to be *saturated*, or in the ON state. With D_E and D_C ON, the transistor, to a first approximation, may be considered a short circuit between all three terminals.

A most important point to bear in mind when working with these models is that the choice of assumed positive reference directions for the terminal currents, I_E, I_B, I_C, is purely arbitrary. We just happen to have chosen these current directions in Fig. 6-1*a* to agree with the actual current (electron flow) directions for an *NPN* transistor operating in its active region. However, once you have assumed a direction for I_E, the direction of the αI_E current source becomes fixed, since this is a dependent source. The rule to observe here is that if I_E is directed into the emitter lead, the αI_E current source must be directed out of the collector lead; or, if I_E is directed out of the emitter, αI_E should be directed toward the emitter. In other words, the model must guarantee that as I_E varies, I_C varies accordingly. If one chooses the αI_E current source direction first, then the emitter current direction must be chosen accordingly. This precaution must be observed whenever dependent sources of any nature are involved.

To avoid problems, the I_{CBO} current source should be directed out of the collector lead in an *NPN* transistor, and this is irrespective of the assumed I_C direction since this is not a dependent current source.

In general, we shall use the current and voltage reference directions shown in Fig. 6-1*a* when dealing with *NPN* transistors.

Figure 6-1*b* shows a *PNP* transistor model. Note that the I_E and αI_E current source directions are in agreement. If either one is reversed, the other must also be reversed. The choice of all the other current directions is arbitrary, but it is advisable to always choose the I_{CBO} current source in this direction when dealing with a *PNP* transistor. The fact that in a *PNP* transistor conduction is primarily due to hole flow is of no significance in this model. All currents may be considered electron flow.

In general, the model of Fig. 6-1*b* with its current and voltage reference directions will be used when dealing with a *PNP* transistor. As you can see, the assumed direction of I_E will surely be the actual direction I_E flows if the emitter diode is forward-biased. Also since $I_C = \alpha I_E + I_{CBO}$ for any type transistor in its active region, the choice of I_C direction seems quite natural. Furthermore, $I_E = I_B + I_C$ in the active region for either type transistor. Thus it seems natural to choose the I_B and I_C current directions into the transistor and the I_E direction out of the emitter.

The models of Fig. 6-1*a* and 6-1*b* are sometimes called common-base models. The common-base model is recognizable by the αI_E current source that it contains.

6-2 The common-base piecewise model

Figure 6-2*a* shows a plot of I_C versus V_{BC} for different values of injected I_E. This plot is called the common-base collector characteristic. The flatness of the curve is indicative of the fact that I_C is barely effected by V_{BC}; rather it depends primarily upon the injected emitter current I_E. Physically this ties in with the fact that, looking into the collector of the transistor, we see something that approximates a constant-current source whose magnitude is highly dependent upon the injected emitter current. Note that $I_C \approx I_E$ which is indicative of the parameter α being ≈ 1. The $I_E = 0$ curve is actually the I_{CBO} characteristic of the transistor, since I_{CBO} was defined as the collector current with $I_E = 0$. Note also that even with $V_{BC} = 0$, I_C is still approximately equal to

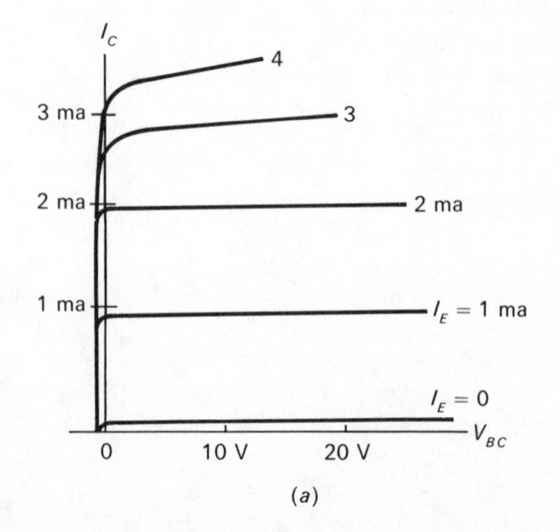

(*a*)

FIGURE 6-2

[1]A more accurate model (the Ebers and Moll) would allow for the role of emitter and collector being interchanged so that the collector can also inject and the emitter can also collect.

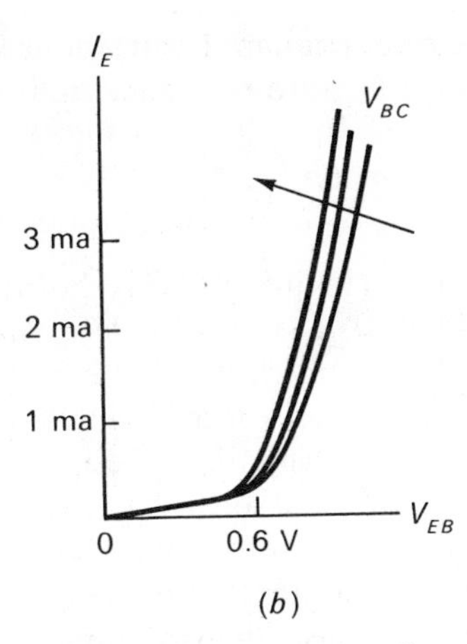

(b)

FIGURE 6-2 (*Continued*)

I_E. This is due to the internal barrier potential across the collector-junction depletion region being of such a polarity that the collector collects that which the emitter injects. To actually reduce the collector current to zero it is necessary to slightly forward-bias the collector diode.

The region to the left of the I_C axis corresponds to the saturated or ON state of the transistor, and the region below the $I_E = 0$ curve is the cutoff region. In between lies the active or amplifying region.

Figure 6-2*b* illustrates the input characteristics of the transistor. As might be expected, the I_E versus V_{EB} curve in the active region strikingly resembles that of a forward-biased diode. For a silicon transistor, the break in the forward characteristic is about 0.6 volt, whereas in germanium it is more rounded and about 0.2 volt. Note that V_{BC} has only a slight effect on the input curves, which indicates that, although there is some apparent internal feedback within the transistor, it may usually be neglected, at least at frequencies low enough to neglect reactive effects due to internal and stray capacities.

The models of Fig. 6-1*a* and 6-1*b* are sufficiently accurate for most purposes. In some instances it may, however, be desirable to use a more accurate model. One approach is to make some piecewise linear approximations to the transistor characteristics and to then develop a model which fits the approximated characteristics over the range of interest.

Figure 6-3*a* illustrates a piecewise approximation to Fig. 6-2*a*. The point on the V_{BC} axis labeled BV_{CBO} is the breakdown voltage of the reverse-biased collector diode with $I_E = 0$. Normally the operating point of the transistor will not extend into this region.

In the piecewise approximation of Fig. 6-3*a*, the I_E curves are equally spaced, parallel lines. This means the parameter alpha (α), which is given by

1) $$\alpha = \left.\frac{\Delta I_C}{\Delta I_E}\right|_{\Delta V_{BC}=0} \tag{6-1}$$

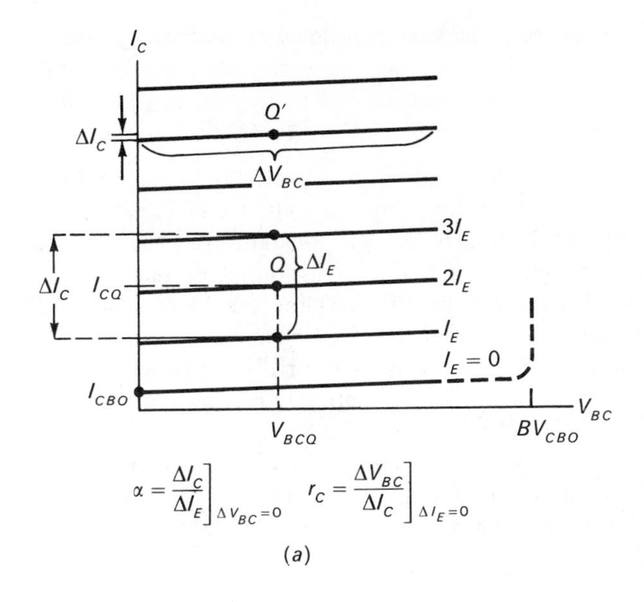

(a)

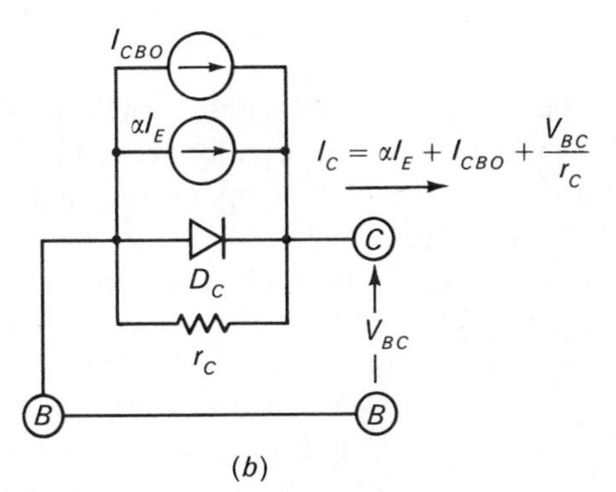

(b)

FIGURE 6-3

is a constant value throughout the illustrated active region. Inspection of Fig. 6-2*a* illustrates that in reality α is not a constant, but actually tends to decrease as I_C increases. We deduce this because the curves are compressed at the higher I_C values which indicates a smaller change in I_C for a given change in I_E. Setting $\Delta V_{BC} = 0$ in equation 6-1 simply means that alpha should be evaluated at some constant collector voltage V_{BC}. This stipulation is necessary since in the practical characteristics shown in Fig. 6-2*a* alpha is somewhat sensitive to V_{BC}.

Alpha may be graphically evaluated by assuming some quiescent operating point (Q) as shown in Fig. 6-3*a*, and then considering an increment in emitter current, ΔI_E, while noting the corresponding increment in collector current, ΔI_C, with the collector voltage V_{BC} held constant.

We might also note that there is some slight tilt or slope to the curves of Fig. 6-3*a*. Physically, this implies that there must be some equivalent leakage resistance across the collector diode in order to permit I_C to be affected by V_{BC}. In other words, the collector diode is not truly a constant-current source in the active region.

This collector leakage resistance r_C is given by

2) $$r_C = \left.\frac{\Delta V_{BC}}{\Delta I_C}\right|_{\Delta I_E = 0} \qquad (6\text{-}2)$$

To evaluate r_C graphically, we hold I_E constant and note the change in I_C due to a change in V_{BC} about the Q point. In Fig. 6-3*a*, r_C is constant everywhere in the active region because the curves consist of equally spaced, parallel lines. Since this isn't the case with a practical transistor, it is advisable that r_C and α be evaluated at the same Q point. It is only for illustrative clarity that the construction for evaluating r_C is shown at the quiescent operating point Q' instead of Q.

An *NPN* active-region piecewise model to fit the characteristics of Fig. 6-3*a* is shown in Fig. 6-3*b*. The current emanating from the collector lead will be given by

3) $$I_C = \alpha I_E + I_{CBO} + \frac{V_{BC}}{r_C} \qquad (6\text{-}3)$$

Usually the V_{BC}/r_C term is negligibly small in equation 6-3. Also the I_{CBO} term is normally negligible in silicon transistors except at high temperatures, and in germanium transistors up to moderate temperatures. Manufacturers data sheets should be consulted whenever possible, but, in lieu of this, we might use 160° and 70° as junction temperature limits for silicon and germanium respectively. Case temperature refers to the actual temperature at the housing or case of the transistor, which is actually higher than the ambient temperature due to power dissipation within the transistor. For the same reason the actual junction temperature within the transistor is higher than the case temperature. We will learn more about this later. The point is that, particularly in silicon transistors, equation 6-3 is approximated by

4) $$I_C = \alpha I_E \qquad (6\text{-}4)$$

A piecewise approximation to the common-base input characteristic of Fig. 6-2*b* is shown in Fig. 6-4*a*. In silicon transistors V_{EB}' is about 0.6 to 0.7 volt, and it is 0.15 to 0.3 volt in germanium. The reciprocal of the slope of the curve is the forward resistance, and it is given by

5) $$R_{ib} = \left.\frac{\Delta V_{EB}}{\Delta I_E}\right|_{\Delta V_{BC} = 0} \qquad (6\text{-}5)$$

A piecewise model to fit Fig. 6-4*a* is shown in Fig. 6-4*b*. The emitter diode, D_E, in this model is now an ideal diode since all its imperfections are allowed for by V_{EB}' and R_{ib}. The R_{ib} resistor may be referred to as the large-signal, common-base input resistance. Since all the curves of Fig. 6-2*b* are so close together for most all V_{BC} values, it is convenient to lump them all together in Fig. 6-4*a*. The physical implication here is that varying V_{BC} causes negligible effect upon the input characteristics. This is generally true, as long as the operating point is not close to the saturated region, for it is here that the collector diode starts turning ON, which means conditions at the collector will start being coupled back to the emitter.

Figures 6-3*b* and 6-4*b* may now be combined to form the *NPN* common-base, active-region model shown in Fig. 6-5. The emitter diode is shaded to indicate it is forward-biased, and r_C is shown in dashed lines to indicate it is usually negligible. Figure 6-6 illustrates the *PNP* equivalent model.

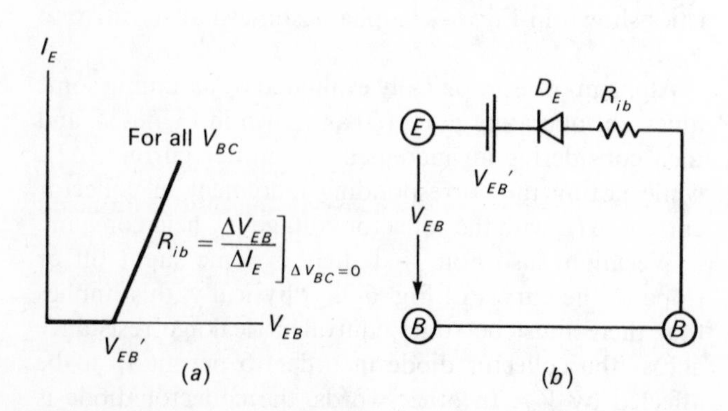

FIGURE 6-4

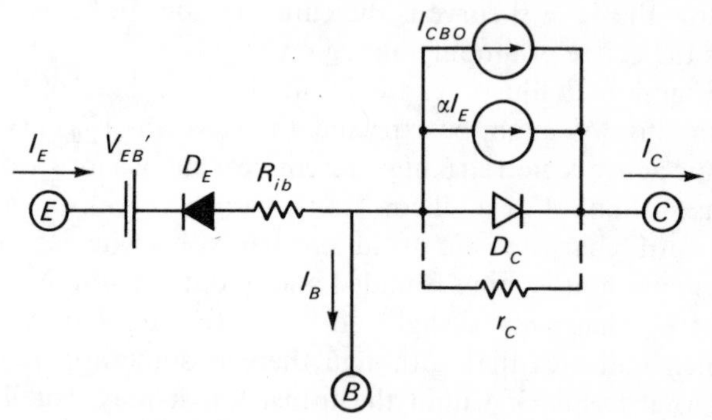

FIGURE 6-5

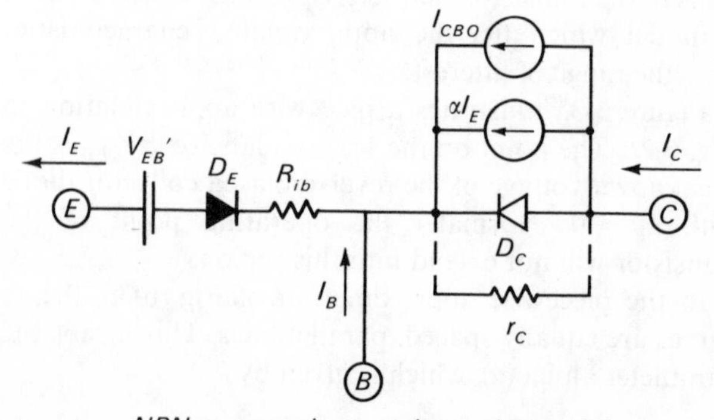

FIGURE 6-6

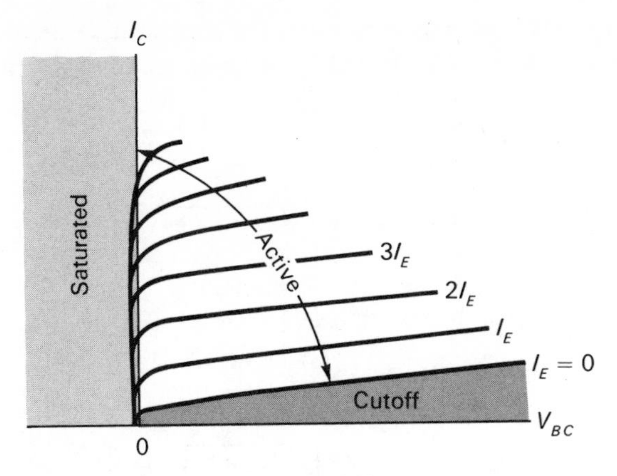

FIGURE 6-7

Figure 6-7 reviews the three operating regions of the transistor. Remember, in the cut-off region both D_E and D_C are reverse-biased, and in the saturated region both are forward-biased. In the active (amplifying) region D_E is forward-biased and D_C is reverse-biased.

PROBLEMS WITH SOLUTIONS

PS 6-1 Determine I_E, I_B, I_C, and V_{EC} in the circuit of Fig. PS 6-1*a*. Assume $r_c = \infty$, $\beta = 79$, $I_{CBO} = 0.01\ \mu a$, $V_{EB}' = 0.6$ volt, and $R_{ib} = 50$ ohms.

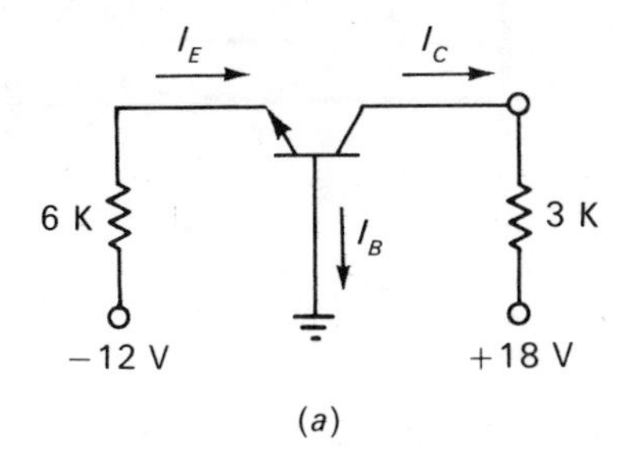

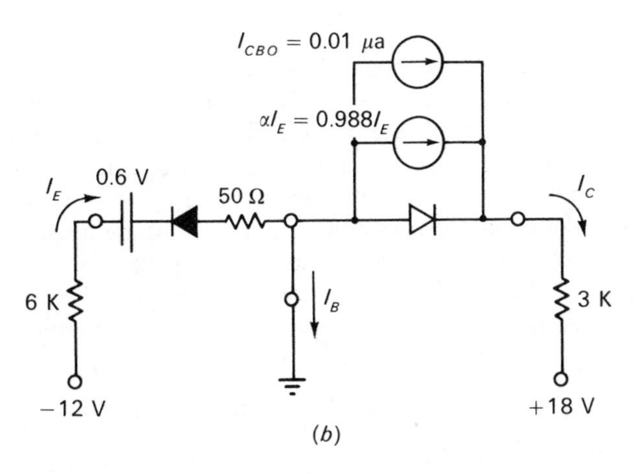

FIGURE PS 6-1

SOLUTION A piecewise common-base model is used to construct the equivalent circuit of Fig. PS 6-1*b*. From equation 5-9 we obtain

1) $$\alpha = \frac{\beta}{\beta + 1} = \frac{79}{80} = 0.988$$

Assuming D_E ON and D_C OFF, we may write

2) $$I_E = \frac{12 \text{ volts} - 0.6 \text{ volt}}{6 \text{ kilohms} + 0.05 \text{ kilohm}} = 1.88 \text{ ma}$$

From equation 5-2

3) $I_C = \alpha I_E + I_{CBO} \approx 0.988\,(1.88 \text{ ma}) + 0 = 1.86 \text{ ma}$

From equation 5.3

4) $$\begin{aligned} I_B &= I_E(1 - \alpha) - I_{CBO} \\ &= 1880\ \mu a\,(1 - 0.988) - 0.01\ \mu a \\ &= 23.5\ \mu a \end{aligned}$$

5) $$\begin{aligned} V_{BC} &= 18 \text{ volts} - 3 \text{ kilohms } I_C \\ &= 18 \text{ volts} - 5.58 \text{ volts} = 12.3 \text{ volts} \end{aligned}$$

6) $$\begin{aligned} V_{EC} &= V_{EB} + V_{BC} \\ &= 0.6 \text{ volt} + 0.05 \text{ kilohm}\,(1.88 \text{ ma}) \\ &\qquad + 12.4 \text{ volts} = 13 \text{ volts} \end{aligned}$$

This problem could, of course, have been solved with-without drawing the model of Fig. PS 6-1*b*. However, we will use the model initially, as it enhances comprehension of transistor action and in more complex circuits is almost a necessity.

PS 6-2 Estimate the power dissipated in the transistor in the circuit of Fig. PS 6-2.

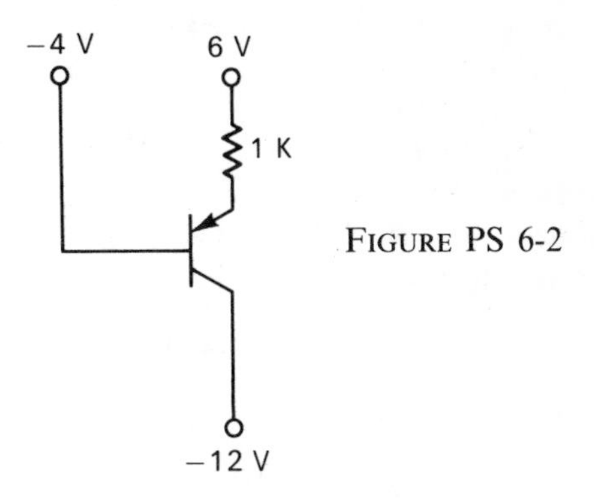

FIGURE PS 6-2

SOLUTION The power dissipated in the transistor is the sum of the power dissipated in the emitter diode and the collector diode. Usually the emitter-diode power dissipation is much less than the collector-diode dissipation because the emitter-junction voltage is much less than the collector-junction voltage. Thus the dissipation is given by

1) $$P_{EC} = V_{EB} I_E + V_{BC} I_C \approx (V_{EB} + V_{BC})\, I_E$$

Assuming a silicon transistor with about 0.65 volt emitter-junction drop we have

2) $$I_E = \frac{6 + 4 - 0.65 \text{ volt}}{1 \text{ kilohm}} = 9.35 \text{ ma}$$

Noting that V_{BC} = up 4 volts and down 12 volts = −8 volts, we may substitute in equation 1 to obtain

3) $$P_{EC} = (0.65 \text{ volt} + 8 \text{ volts})\, 9.35 \text{ ma} = 80.9 \text{ mw}$$

PS 6-3 Determine R_E and R_L in Fig. PS 6-3 for a Q point at $I_C = 1.5$ ma, $V_{EC} = 6$ volts. Since I_{CBO} is negligible compared to I_C, it may be neglected in this problem.

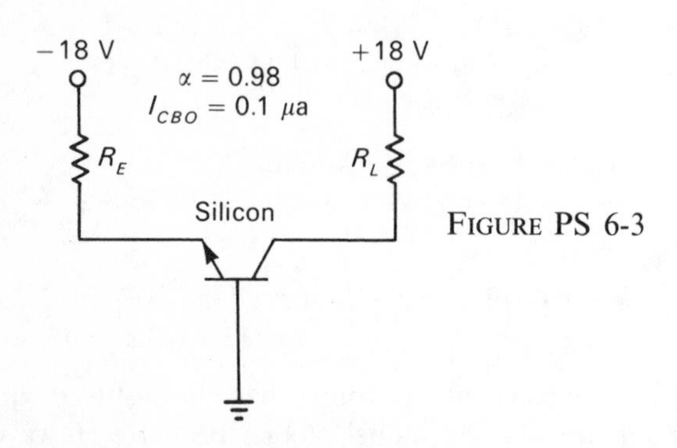

FIGURE PS 6-3

SOLUTION

1) $$I_C = \alpha I_E + I_{CBO} \approx \alpha I_E$$

2) $$1.5 \text{ ma} = 0.98\, I_E$$

Solving for I_E in equation 2

3) $$I_E = \frac{1.5 \text{ ma}}{0.98} \approx 1.53 \text{ ma}$$

4) $$R_E = \frac{18 \text{ volts} - 0.6 \text{ volt}}{1.53 \text{ ma}} = 11.4 \text{ kilohms}$$

5) $$V_{EC} = V_{EB} + V_{BC}$$

Substituting values

6) $$6 \text{ volts} = 0.6 \text{ volt} + V_{BC}$$

Solving for V_{BC}, we obtain

7) $$V_{BC} = 5.4 \text{ volts}$$

8) $$V_{BC} = 18 \text{ volts} - R_L I_C$$

Substituting equations 3, 1, and 7 into 8, we obtain

9) $$5.4 \text{ volts} = 18 \text{ volts} - R_L\,[(0.98)\, 1.53 \text{ ma}]$$

Solving for R_L,

10) $$R_L = 8.4 \text{ kilohms}$$

PS 6-4 Estimate I_E and I_C in Fig. PS 6-4*a*.
SOLUTION At first glance it would seem

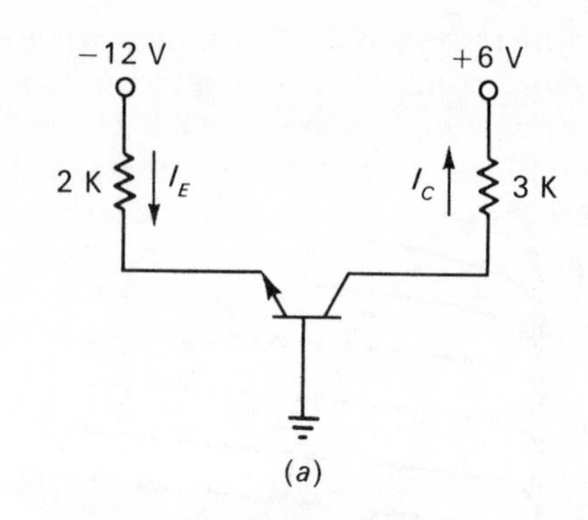

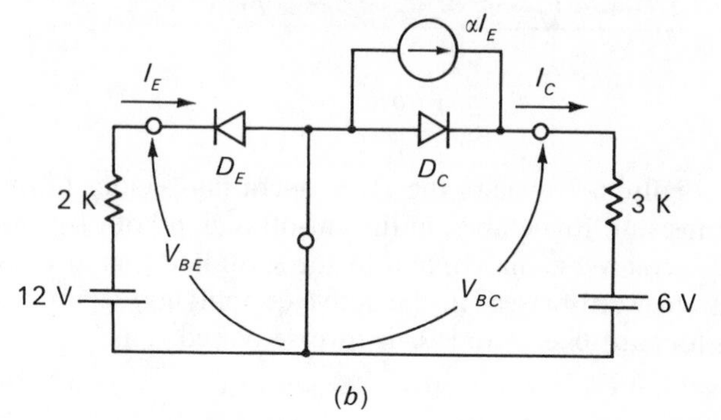

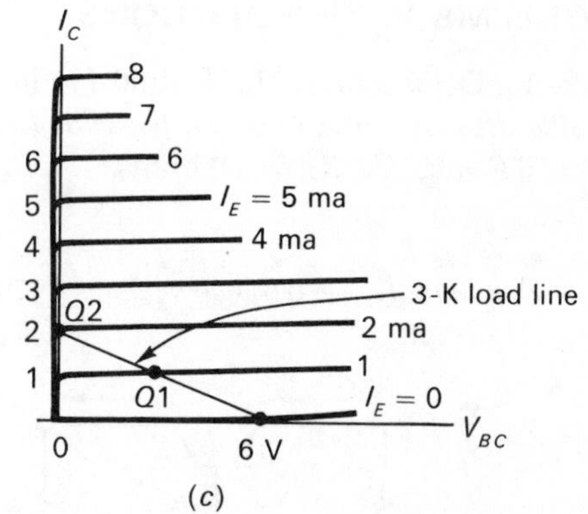

FIGURE PS 6-4

1) $$I_E \approx 12 \text{ volts}/2 \text{ kilohms} = 6 \text{ ma}$$

2) $$I_C \approx I_E = 6 \text{ ma}$$

Just in case you agree with this solution and think the problem was a cinch—you are wrong! You haven't really checked the state of the emitter and the collector diodes. Usually if you are certain that the operation is in the active region, this procedure can be ignored. However, there may be occasions such as this where it's wise to check the diode states. Inspecting Fig. PS 6-4*b*, it is quite evident that D_E will be ON. To check if D_C is OFF as is required for an active-region operation, we compute V_{BC}

3) $$V_{BC} = 6 \text{ volts} - 3 \text{ kilohms } (6 \text{ ma}) = -12 \text{ volts}$$

Since V_{BC} turned out to be a negative quantity, it implies that the collector terminal (the second subscript) is negative with respect to the first subscript (base) which is our reference point. This means the collector diode wants to turn ON and, indeed, is ON. The αI_E current source is therefore shorted through the ON diode D_C and the collector is shorted (clamped) to the base. Thus $V_{BC} = 0$ and

4) $$I_C = \frac{6 \text{ volts} - 0 \text{ volts}}{3 \text{ kilohms}} = 2 \text{ ma}$$

With both diodes ON the transistor is in the saturated rather than the active region.

To graphically illustrate what is happening, consider Fig. PS 6-4*c* which plots I_C versus V_{BC}. From Fig. PS 6-4*c* we may write the general expression for V_{BC}

5) $$V_{BC} = V_{CC} - R_L I_C$$

where V_{CC} is the general description of the voltage source in the collector lead. Equation 5 is known as the *load line equation.*[1] Equation 5 is a linear equation and may be plotted as a straight line called the load line on the collector characteristics shown in Fig. PS 6-4*c*. Since the load line equation is linear, we need only two points to plot it. One point which may be conveniently chosen is found letting $I_C = 0$. If $I_C = 0$, $V_{BC} = V_{CC}$. This is the 6-volt intercept on the V_{BC} axis. The next convenient point to choose is where $V_{BC} = 0$. Then solving for the corresponding I_C yields $I_C = V_{CC}/R_L = 6$ volts/3 kilohms $= 2$ ma. This is the intercept on the I_C axis. It is also the point marked $Q2$. Fig. PS 6-4*c* indicates that when $I_E = 0$, I_C is approximately 0, since the emitter is not injecting anything for the collector to collect. Hence there is no drop across the 3-kilohm load resistor which means $V_{BC} = 6$ volts. If we inject some emitter current, say 1 ma, the operating point will move to $Q1$. There will be a drop across the 3-kilohm resistor, and hence V_{BC} becomes less. If we increase the injected emitter current to 2 ma, the collector current is essentially 2 ma and we note that V_{BC} is equal to 0 since the drop across the load resistor uses up all the reverse bias the V_{CC} supply was furnishing to the collector diode. The operating point is now at $Q2$; hence we see that with 2 ma of injected emitter current the transistor will go into saturation, since at this point $V_{BC} = 0$. However, in this problem, I_E is actually 6 ma, which drives us deep into saturation since the $I_E = 6$ ma curve intersects the load line after it has bent into the saturated region.

A quick way of determining whether or not the transistor is saturated is to estimate the maximum possible I_C that can flow before $V_{BC} = 0$. In this case, with $V_{BC} = 0$, the maximum I_C that can flow is 2 ma. If our calculations indicate that I_C is greater than the saturation value of collector current, it follows that the transistor will be driven into saturation and hence the collector diode must be ON and $V_{BC} = 0$. The collector current should then be calculated on the basis of assuming the collector diode is ON.

[1] See Phillip Cutler, "Outline for DC Circuit Analysis with Illustrative Problems," chap. 16, McGraw-Hill Book Company, New York, 1968.

PS 6-5 Sketch E_O as a function of I_E in Fig. PS 6-5*a*.

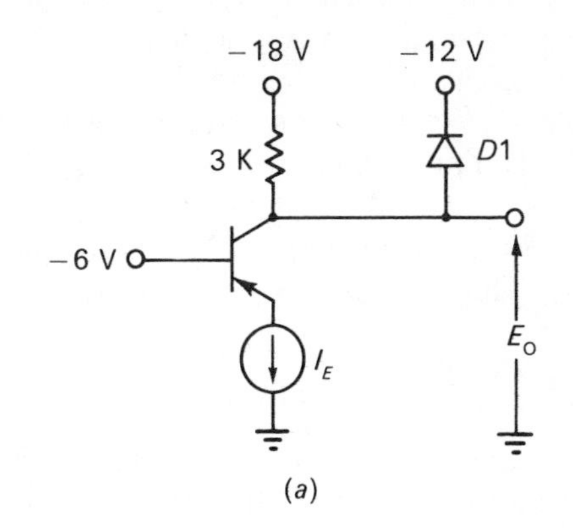

(*a*)

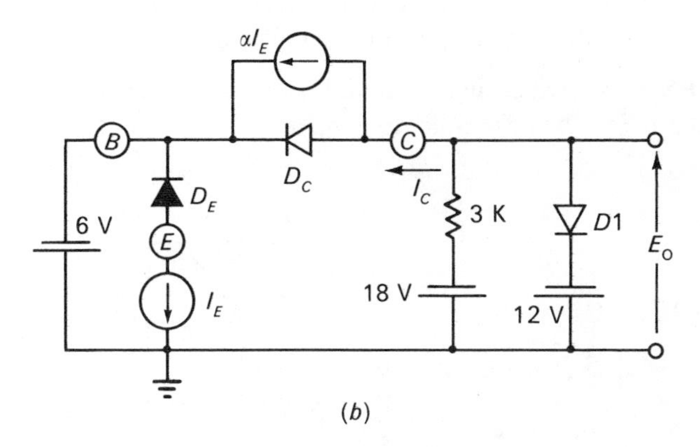

(*b*)

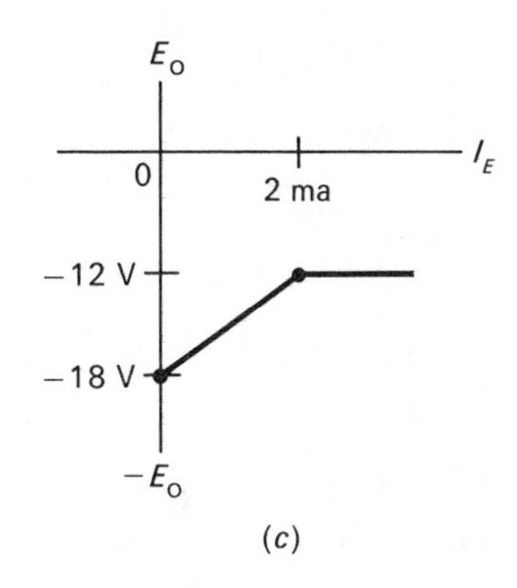

(*c*)

FIGURE PS 6-5

SOLUTION An equivalent circuit using the common-base, piecewise model is shown in Fig. PS 6-5*b*. Diode D_E is surely ON due to the current source I_E. Let us assume D_C OFF, which places the transistor in the active region. Let us also assume D_1 OFF. Also, for simplicity, assume $\alpha = 1$. Therefore, $I_C = I_E$. With these assumptions we may write

1) $$E_O = -18 \text{ volts} + 3 \text{ kilohms } I_E$$

For $I_E = 0$, $E_O = -18$ volts, which is plotted in Fig. PS 6-5*c*. As I_E increases, I_C increases which causes E_O to rise in the positive direction. As E_O becomes more positive, we might expect D_C or $D1$ to turn ON. However, since $D1$ has its cathode returned to -12 volts, whereas D_C has its cathode returned to -6 volts through the base lead, it seems reasonable to assume that $D1$ will turn ON first. Hence let us assume $D1$ breaks next. At the $D1$ breakpoint, $E_O = -12$ volts, and since the current through $D1$ is 0 at its breakpoint, it follows that $I_C = [-12 \text{ volts} - (-18 \text{ volts})]/3 \text{ kilohms} = 2$ ma. Once $D1$ turns ON, E_O is clamped to -12 volts for all further values of I_E. The collector diode remains reverse-biased by

2) $$V_{BC} = +6 \text{ volts} - 12 \text{ volts} = -6 \text{ volts}$$

Once $D1$ turns ON, the current through the 3-kilohm resistor remains fixed at 2 ma and any additional collector current that is required flows from the -12 volt supply through $D1$ into the collector lead.

PS 6-6 Determine I_C and V_{EC} in the circuit of Fig. PS 6-6*a*. The base is floating (open).

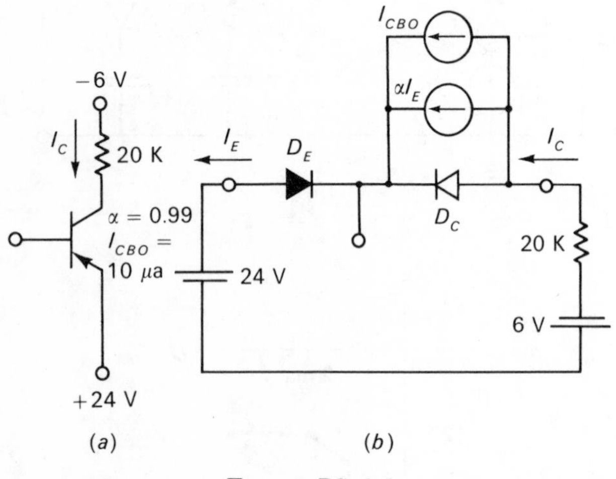

FIGURE PS 6-6

SOLUTION At first glance it would seem that the reverse-biased collector diode is effectively in series with the emitter diode, as shown in Fig. PS 6-6*b*. This would seem to indicate that I_C and I_E equal I_{CBO}, since the leakage resistance of the reverse-biased collector diode is in series with the forward-biased emitter diode. Thus, the emitter diode is actually forward-biased by the leakage current I_{CBO}, and hence we might conclude $I_C = I_{CBO}$. However, this is in error because it neglects the fact that when the leakage current crosses the emitter junction it gives rise to a component of emitter current, I_E. This, in turn, produces a current, αI_E, in the dependent current source in shunt with the collector diode. Analytically we might write

1) $$I_C = \alpha I_E + I_{CBO}$$

But with the base open it follows that

2) $$I_E = I_C$$

Substituting equation 2 into 1 we have

3) $$I_C = \alpha I_C + I_{CBO}$$

Solving equation 3 for I_C yields

4) $$I_C = \frac{I_{CBO}}{1 - \alpha}$$

Since the quantity $1 - \alpha$ is a number much less than 1, it follows that a collector current with the base lead open can be much larger than the leakage current, I_{CBO}. In fact, we may define a leakage current called I_{CEO}, which is the collector leakage current with the base open. Equation 4 may be rewritten in a slightly different form by manipulating equation 5-8 algebraically to obtain

5) $$\frac{1}{1 - \alpha} = \beta + 1$$

Thus equation 4 may be rewritten as

6) $$I_{CEO} = I_{CBO}(\beta + 1) \tag{6-6}$$

Since β is a number which may typically range from 20 to 200, it is evident that the leakage current may be enormously multiplied when the base lead of the transistor is left open. Since I_{CBO} is essentially thermally generated leakage current, it follows that the quantity I_{CEO} is exceedingly temperature-sensitive. It is for this reason that it is poor practice to design transistor circuits in which the base lead may be left open.

The essence of this problem lies in the fact that when the base lead is left open, all of I_{CBO} is forced to flow through the emitter diode in a direction which forward-biases it. This causes the emitter to inject a component of current into the dependent αI_E current source which is in shunt with the I_{CBO} current source. The net effect, then, is to cause a collector current, I_C, which is actually greater than I_{CBO}.

PROBLEMS WITH ANSWERS

PA 6-1 Estimate I_C and V_{EC} in Fig. PA 6-1.

ANSWER $I_C \approx 1$ ma
$V_{EC} \approx -12$ volts

PA 6-2 Estimate in Fig. PA 6-2 the maximum permissible value of R_L, so that R_L is driven by a constant-current source.

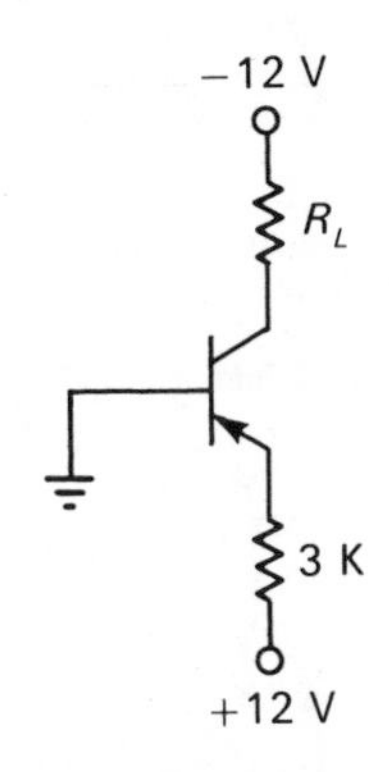

ANSWER $R_{L\max} \approx 3$ kilohms

PA 6-3 Assuming a silicon transistor, determine I_C, I_B, V_{EB}, and V_{EC} in Fig. PA 6-3.

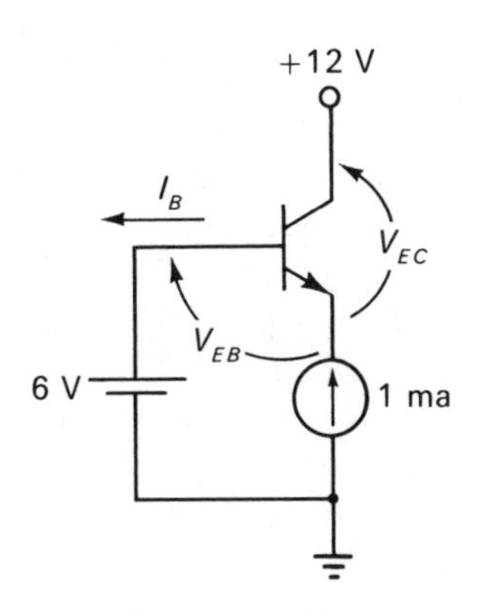

ANSWER $I_C = 1.08$ ma
$I_B = -0.08$ ma
$V_{EB} \approx 0.6$ volt
$V_{EC} = 5.4$ volts

PA 6-4 Assuming a silicon transistor, determine I_E and V_{EC} in Fig. PA 6-4.

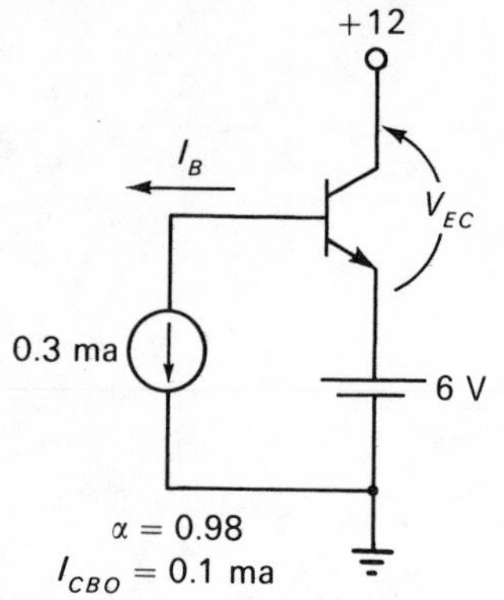

+12

I_B

V_{EC}

0.3 ma

6 V

$\alpha = 0.98$

$I_{CBO} = 0.1$ ma

ANSWER $V_{EC} = 6$ volts
$I_E = 20$ ma

PA 6-5 Determine I_E and V_{EC} in Fig. PA 6-5.

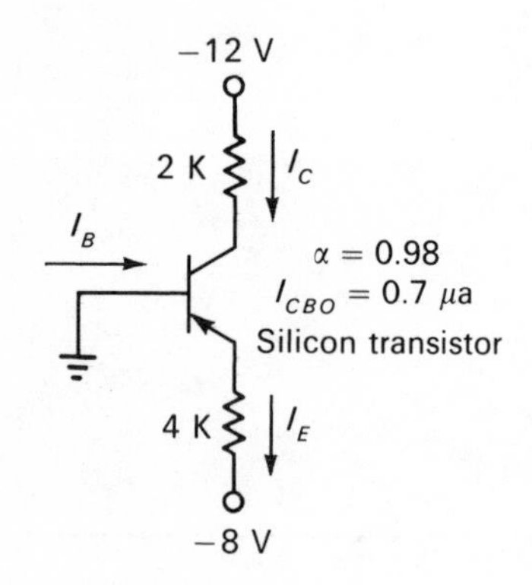

ANSWER $I_E = 0$
$V_{EC} = -12$ volts

PA 6-6 Estimate I in Fig. PA 6-6. Assume ideal diodes.

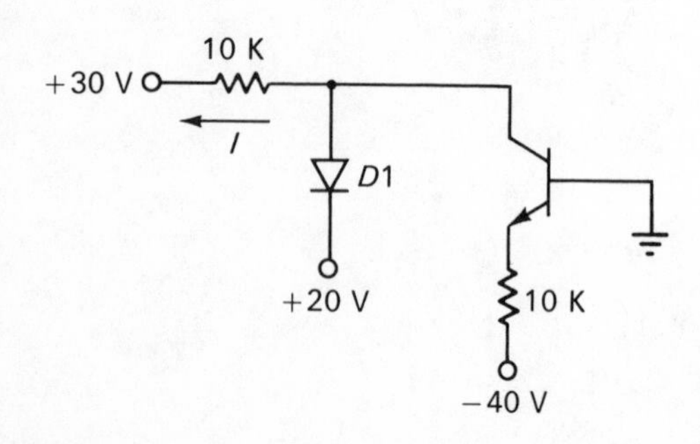

ANSWER $I = 3$ ma

PA 6-7 Roughly design the circuit of Fig. PA 6-7 so that $I_C \approx 3.5$ ma, $V_{EC} \approx 6$ volts.

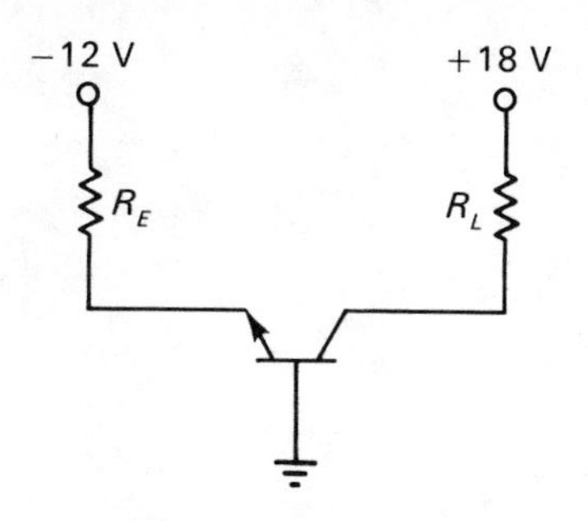

ANSWER $R_L = 5.14$ kilohms
$R_E = 3.43$ kilohms

PA 6-8 Estimate V_{EC} in Fig. PA 6-8.

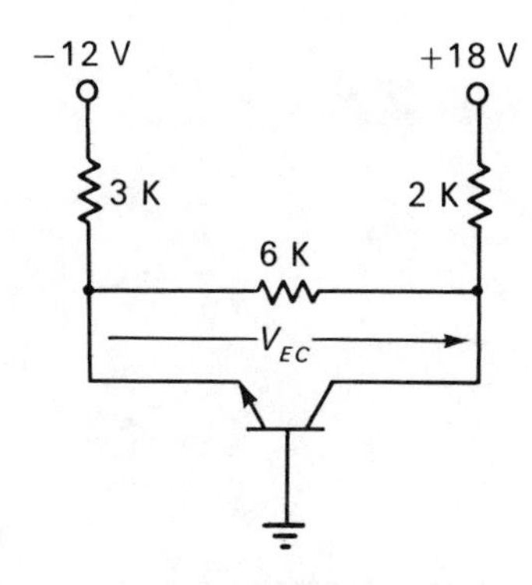

ANSWER $V_{EC} = 10$ volts

PROBLEMS WITHOUT ANSWERS

P 6-1 Sketch a simple *NPN* common-base, active-region dc model.

P 6-2 What refinements could be added to the model in P 6-1 to improve its accuracy?

P 6-3 Figure P 6-3 is an __________ transistor. To use it as an amplifier, the emitter junction should be __________ biased and the collector junction __________ biased. This means the base should be polarized __________ with respect to the emitter and __________ with respect to the collector.

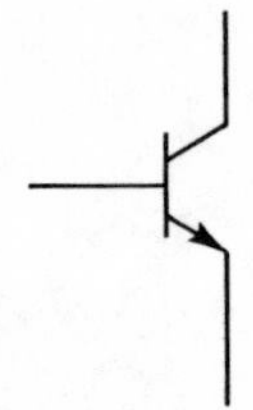

P 6-4 Determine I_C and I_B in Fig. P 6-4. Assume $\alpha = 0.98$, $I_{CBO} = 0$.

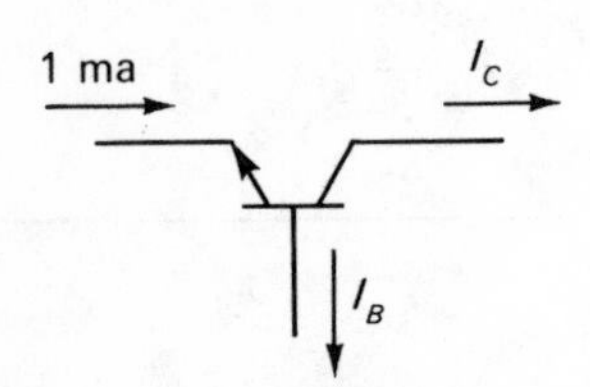

P 6-5 Estimate V_{BC} in Fig. P 6-5. No other information is given.

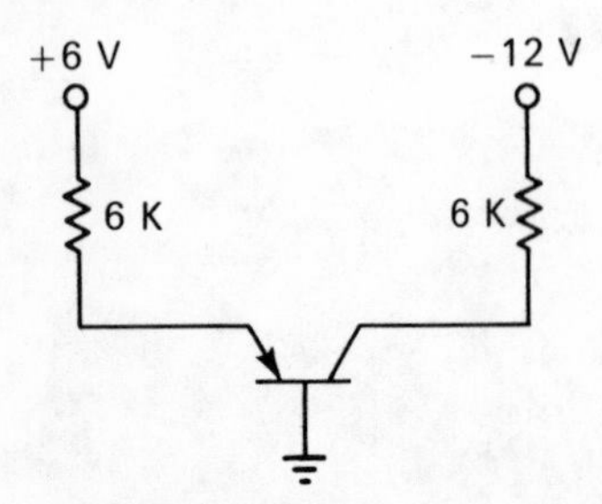

P 6-6 Estimate V_{EC} in Fig. P 6-6.

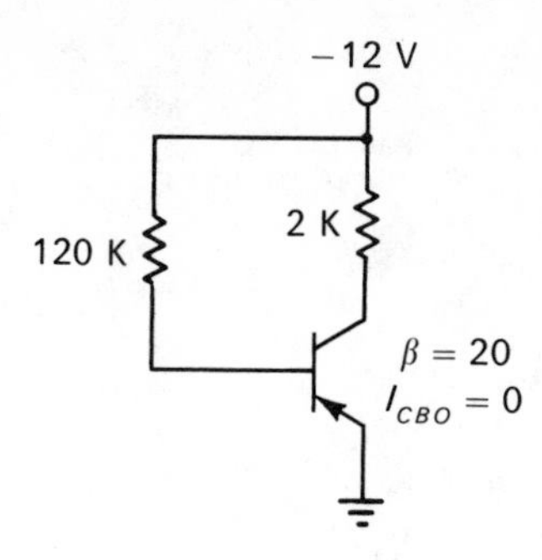

P 6-7 Estimate V_{EC} in Fig. P 6-7.

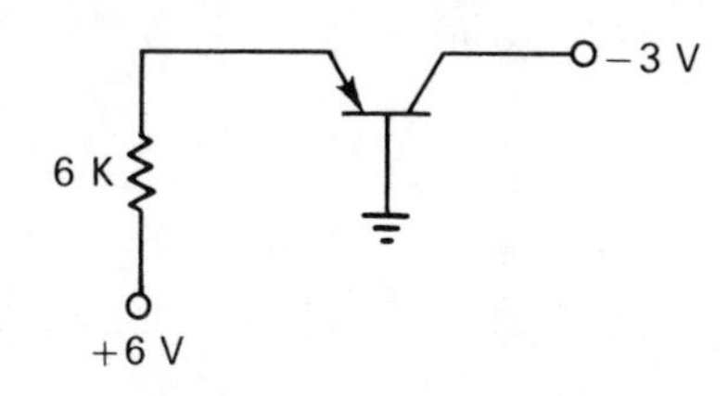

P 6-8 Estimate the minimum permissible value of R_E that will maintain active-region operation in the circuit of Fig. P 6-8.

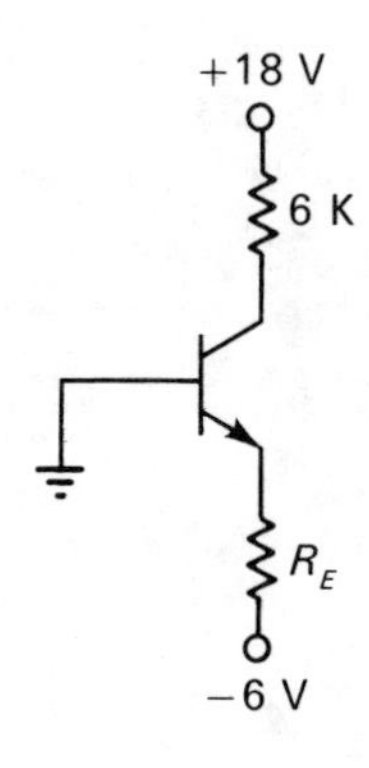

P 6-9 Using piecewise techniques, sketch the transfer functions I_C versus I_E and V_{BC} versus I_E.

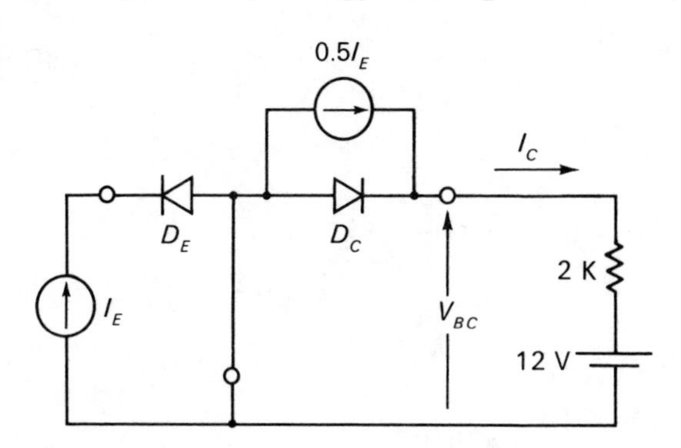

P 6-10 Determine V_{EC} in Fig. P 6-10.

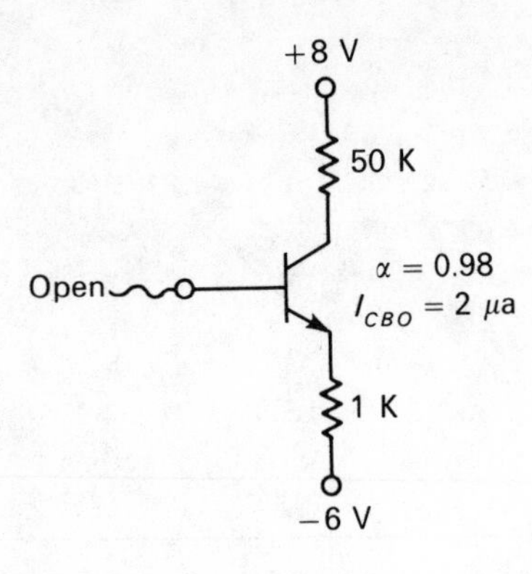

P 6-11 Determine I_C in Fig. P 6-11.

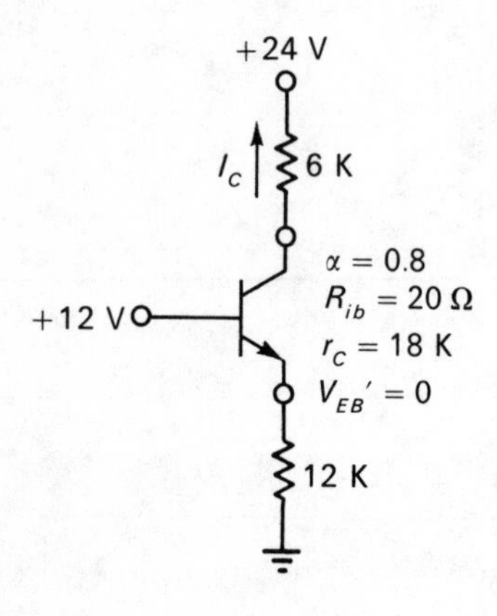

7 the common-emitter model

7-1 The common-emitter model

In the common-base model of Chapter 6, $\alpha \Delta I_E$ was the dependent collector-current source. It turns out that in most cases the base, rather than the emitter, is the desired input terminal. Thus, if the base is to be the driven input terminal, it becomes convenient to express the dependent collector-current source as a function of base current instead of emitter current. This leads to the *common-emitter model.*

One way of developing the common-emitter model is by manipulating the previously developed common-base model shown in Fig. 7-1. Fig. 7-1 may be redrawn as

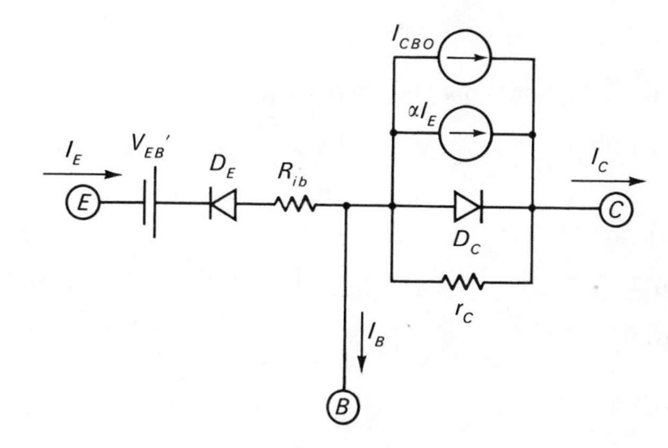

FIGURE 7-1

Fig. 7-2 if we exchange the positions of the emitter and base leads. Now assuming a current source I_B across the input terminals and a voltage source V_{EC} across the output terminals, we may write by superposition

$$1) \quad I_C = \alpha I_E + I_{CBO} - \frac{R_{ib}}{R_{ib} + r_C} I_B + \frac{V_{EC} - V_{EB}'}{R_{ib} + r_C} \qquad (7\text{-}1)$$

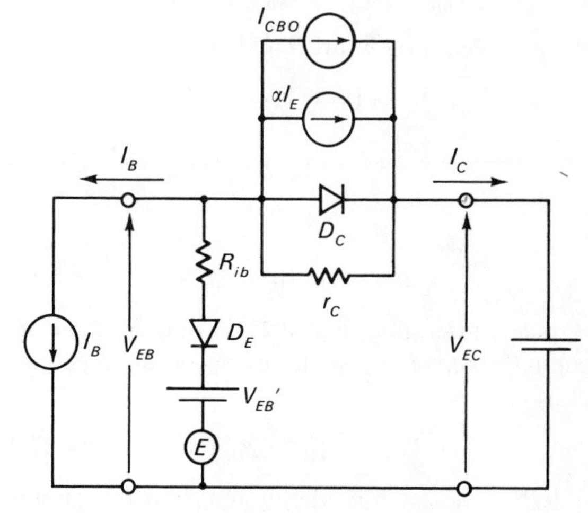

FIGURE 7-2

But

2) $$I_E = I_B + I_C$$

Substituting equation 2 into 1 and recalling that $r_C \gg R_{ib}$, we obtain, after some manipulation,

3) $$I_C \approx \frac{\alpha}{1-\alpha} I_B + \frac{I_{CBO}}{1-\alpha} - \frac{R_{ib}}{r_C(1-\alpha)} I_B + \frac{V_{EC} - V_{EB}'}{r_C(1-\alpha)} \quad (7\text{-}2)$$

But from equation 5-8

4) $$\beta = \frac{\alpha}{1-\alpha}$$

Thus equation 3 may be rewritten as

5) $$I_C \approx \beta I_B + I_{CEO} - \frac{R_{ib}}{r_D} I_B + \frac{V_{EC} - V_{EB}'}{r_D} \quad (7\text{-}3)$$

where

6) $$I_{CEO} = I_{CBO}(\beta + 1) \quad (7\text{-}4)$$

and

7) $$r_D = \frac{r_C}{\beta + 1} \quad (7\text{-}5)$$

Typically, the $(R_{ib}/r_D)I_B$ term $\ll \beta I_B$. Thus equation 5 simplifies to

8) $$I_C \approx \beta I_B + I_{CEO} + \frac{V_{EC} - V_{EB}'}{r_D} \quad (7\text{-}6)$$

which, to a somewhat poorer but still most useful approximation, is

9) $$I_C \approx \beta I_B + I_{CEO} \quad (7\text{-}7)$$

On the input side we may, for Fig. 7-2, write by superposition

10) $$V_{EB} = V_{EB}' + R_{ib}(I_B + I_C)$$

Substituting equation 8 into 9 and simplifying

11) $$V_{EB} \approx V_{EB}' + R_{ib}(\beta + 1) I_B + R_{ib} I_{CEO} + \frac{R_{ib}}{r_D}(V_{EC} - V_{EB}') \quad (7\text{-}8)$$

Let

12) $$R_{ie} = R_{ib}(\beta + 1) \quad (7\text{-}9)$$

For typical transistors and circuit conditions we may neglect in the active region the last two terms in equation 11. Thus

13) $$V_{EB} = V_{EB}' + R_{ie} I_B \quad (7\text{-}10)$$

An *NPN* transistor model designed to fit equations 8 and 13 is shown in Fig. 7-3. A *PNP* model is shown in Fig. 7-4. Note the relation between I_B and the βI_B dependent current source. They are both directed into or out of the transistor. This relationship must be maintained irrespective of any other assumed voltage and reference directions.

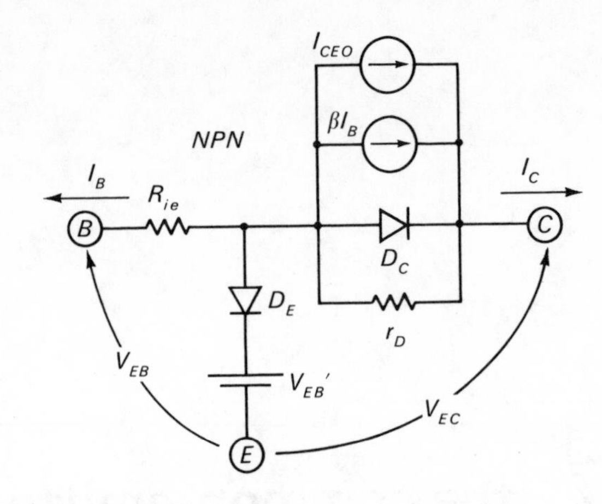

FIGURE 7-3

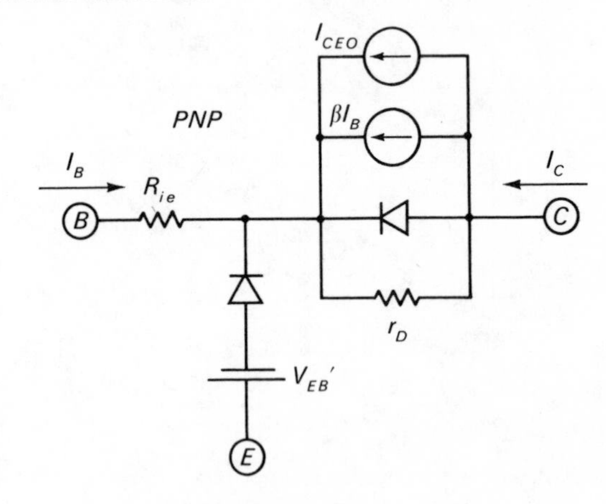

FIGURE 7-4

7-2 Piecewise development

An alternative approach to developing the piecewise common-emitter model is to piecewise linearize the common-emitter collector and base characteristics which are normally available in the forms shown in Fig. 7-5*a* and 7-5*b*. Suitable piecewise approximations are shown in Fig. 7-6*a* and 7-6*b* respectively.

It might be interesting to correlate the previously derived piecewise model of Fig. 7-3 with the piecewise characteristics of Fig. 7-6*a* and 7-6*b*. The transistor terminal condition at the boundary of the cut-off and active regions may be determined in Fig. 7-7. Here we are setting the emitter diode at its breakpoint. Thus, there are zero volts across D_E, and there is zero emitter current flowing through it. With $I_E = 0$, we have

1) $$I_C = -I_B$$

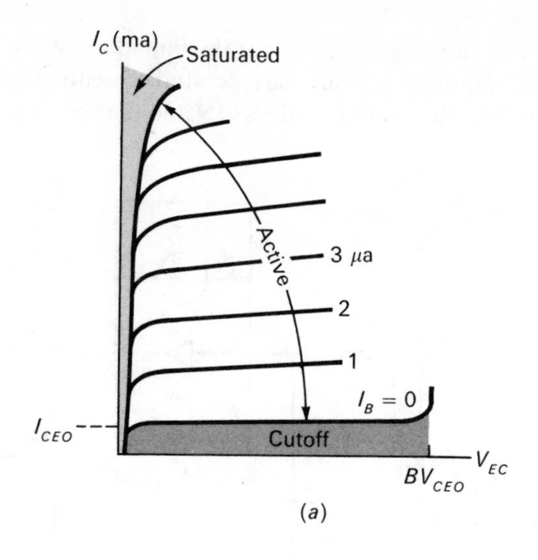

FIGURE 7-5

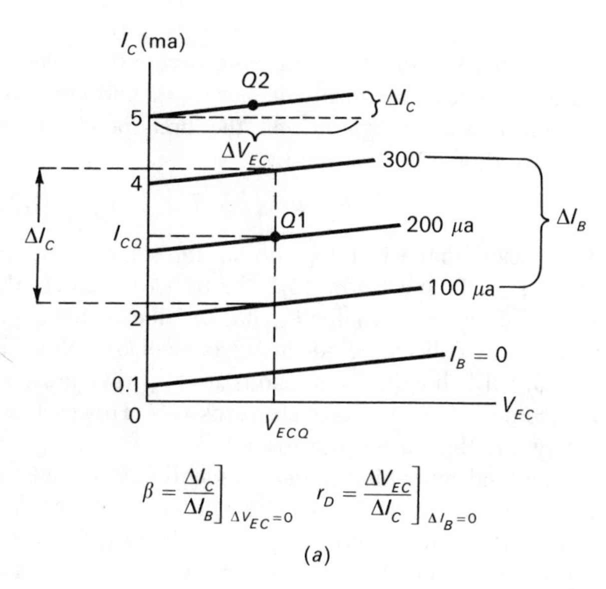

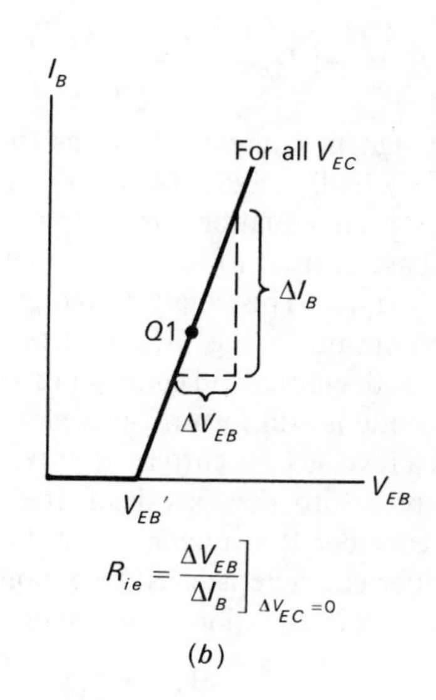

FIGURE 7-6

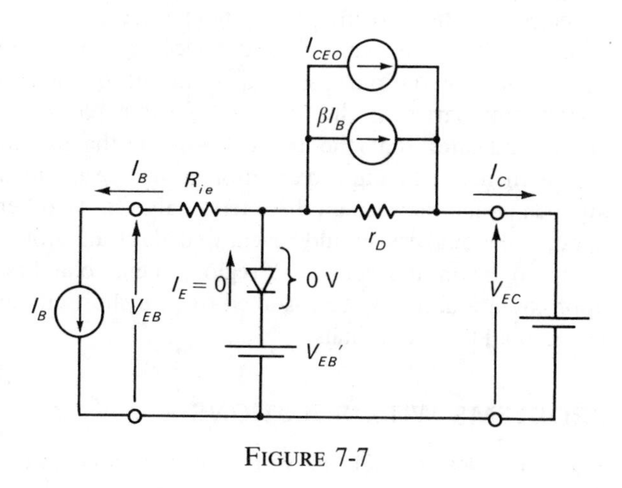

FIGURE 7-7

On the input side we may write

2) $$V_{EB} = V_{EB}' + R_{ie} I_B \qquad (7\text{-}11)$$

For $I_B = 0$, $V_{EB} = V_{EB}'$, which correlates with the break at V_{EB}' in Fig. 7-6*b*. Equation 2 is a linear equation, and, for $V_{EB} \gg V_{EB}'$, the slope is given by

3) $$\frac{\Delta V_{EB}}{\Delta I_B} = R_{ie} \qquad (7\text{-}12)$$

which is also in agreement with Fig. 7-6*b*.

On the output side we may, for Fig. 7-7, write

4) $$I_C = \beta I_B + I_{CEO} + \frac{V_{EC} - V_{EB}'}{r_D} \qquad (7\text{-}13)$$

If equation 1 is solved for I_B and then substituted into 4, there results, after some manipulation,

5) $$I_C = \frac{I_{CEO}}{\beta + 1} + \frac{V_{EC} - V_{EB}'}{r_D/(\beta + 1)} \qquad (7\text{-}14)$$

which from equations 7-4 and 7-5 may be written as

6) $$I_C = I_{CBO} + \frac{V_{EC} - V_{EB}'}{r_c} \qquad (7\text{-}15)$$

Usually r_C is so large that for a reasonable value of V_{EC} the second term in equation 6 is negligible and, hence,

7) $$I_C \approx I_{CBO} \qquad (7\text{-}16)$$

Equation 7 tells us that the collector current with the emitter lead open ($I_E = 0$) is approximately I_{CBO} and not I_{CEO}.

We could also substitute equation 1 into 4 and solve for the base current that would be required to just cut the transistor off. This leads to the result

8) $$I_B = -\left(I_{CBO} + \frac{V_{EC} - V_{EB}'}{r_C}\right) \qquad (7\text{-}17)$$

The negative sign in equation 8 means that a reverse base current is actually required to cut the transistor off. If the second term in equation 8 is again neglected, then the required base current to just achieve cut-off should be equal to $-I_{CBO}$. This analysis indicates that the region labeled cut-off in Fig. 7-5*a* isn't truly cut-off in the sense that both emitter and collector diodes are OFF. The emitter diode is still forward-biased and it will remain so until a reverse base current approximately equal to I_{CBO} is forced into the base lead. It is customary, however, to consider that region below $I_B = 0$ on the common-emitter characteristic as the cut-off region.

If we set $I_B = 0$ in equation 4, we obtain

9) $$I_C = I_{CEO} + \frac{V_{EC} - V_{EB}'}{r_D} \qquad (7\text{-}18)$$

Equation 9, to a reasonable approximation, is

10) $$I_C \approx I_{CEO}$$

In the active region, the output characteristics are respectively described by equation 4. For example, if we vary I_B and study the resultant variation in I_C while holding V_{EC} constant, we obtain

11) $$\beta = \left.\frac{\Delta I_C}{\Delta I_B}\right|_{\Delta V_{EC}=0} \qquad (7\text{-}19)$$

On the other hand, if we hold I_B constant and vary V_{EC}, the slope is

12) $$\frac{1}{r_D} = \left.\frac{\Delta I_C}{\Delta V_{EC}}\right|_{\Delta I_B=0} \qquad (7\text{-}20)$$

Equations 11 and 12 directly correlate with Fig. 7-6*a*.

The terminal conditions at the boundary of the active and saturated regions may be studied with the aid of Fig. 7-8. The emitter diode is ON and the collector diode

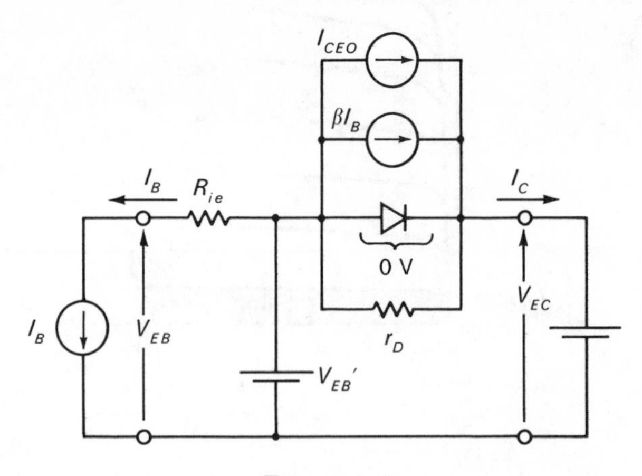

FIGURE 7-8

is at its breakpoint. With zero volts across the collector diode, there are also zero volts across r_D, and hence no current flows through r_D. At the breakpoint of the collector diode, V_{EC} is given by

13) $$V_{EC} = V_{EB}' \qquad (7\text{-}21)$$

This means that when V_{EC} should, for some reason or other, start falling below V_{EB}', the transistor enters the saturated region. Actually, V_{EC} has to fall even lower yet because the collector diode also has a *holdoff voltage* in the forward direction analogous to V_{EB}' that must be overcome before it effectively turns ON. However, we will ignore this for our purposes.

One final point that might be mentioned is that the emitter-to-collector voltage should not exceed the voltage BV_{CEO} indicated in Fig. 8-5*a* when the transistor is driven from a high-impedance (constant-current) base-current source. The voltage BV_{CEO} is a *breakdown voltage* associated with the common-emitter circuit.

By way of review, remember that the transistor models developed in this and the preceding chapter are active-region models. The emitter and collector diodes are shown only for the purpose of determining the active-region boundaries. If the analysis of some particular circuit indicates that the transistor is in the cut-off region, then, to a first approximation, it may be assumed an open circuit between all three terminals. On the other hand, if the analysis should reveal that the transistor is going to be in the saturated region, then, to a first approximation, it may be assumed to be a short circuit between all three terminals.

PROBLEMS WITH SOLUTIONS

PS 7-1 Develop an active-region common-emitter

model for the *NPN* transistor having the characteristics of Fig. PS 7-1*a* and 7-1*b*.

SOLUTION Reasonable piecewise approximations to the volt-ampere curves of Fig. PS 7-1*a* and 7-1*b* are shown in Fig. PS 7-1*c* and 7-1*d* respectively. With the parameters obtained from the curves, the model of Fig. PS 7-1*e* is constructed by analogy to Fig. 7-3.

PS 7-2 Using the model of Fig. PS 7-1*e* for the transistor in the circuit of Fig. PS 7-2*a*, determine I_C and V_{EC}.

SOLUTION We will assume the transistor is in its active region and the emitter diode is ON since the open-circuit voltage trying to turn D_E ON is 12 volts − 0.7 volt = 11.3 volts. Hence I_B = 11.3 volts/24.2 kilohms

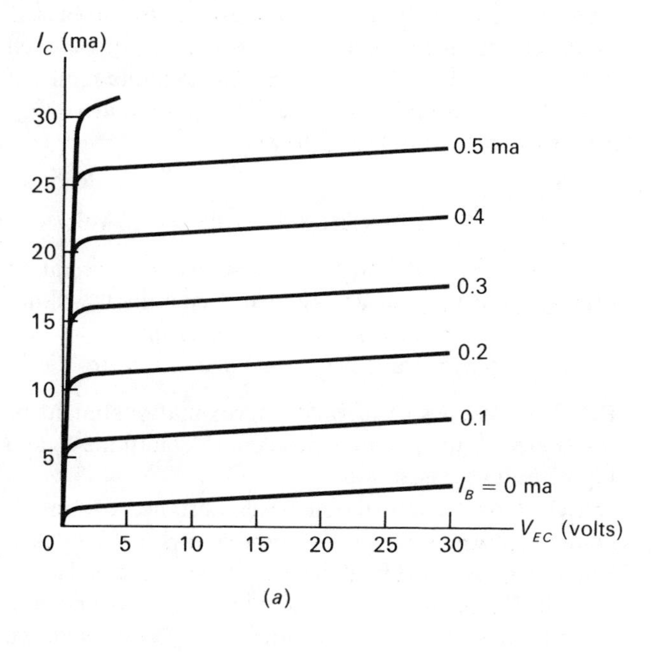

(*a*)

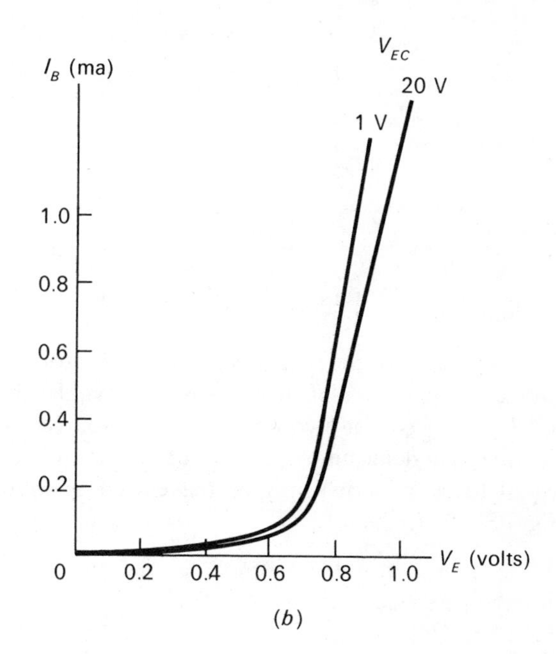

(*b*)

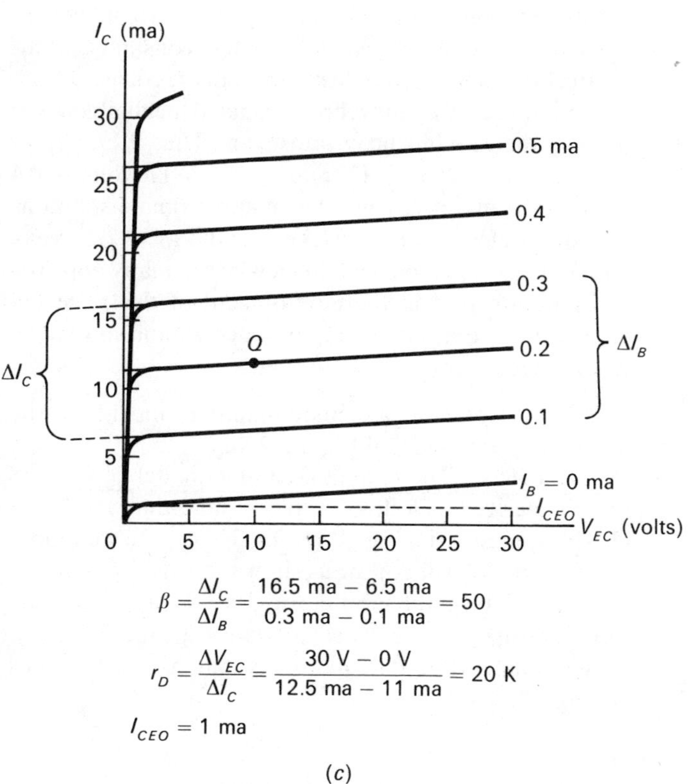

(*c*)

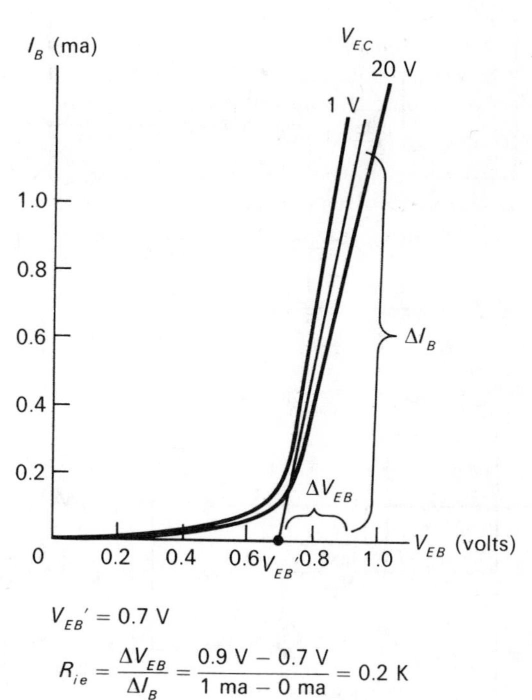

(*d*)

FIGURE PS 7-1

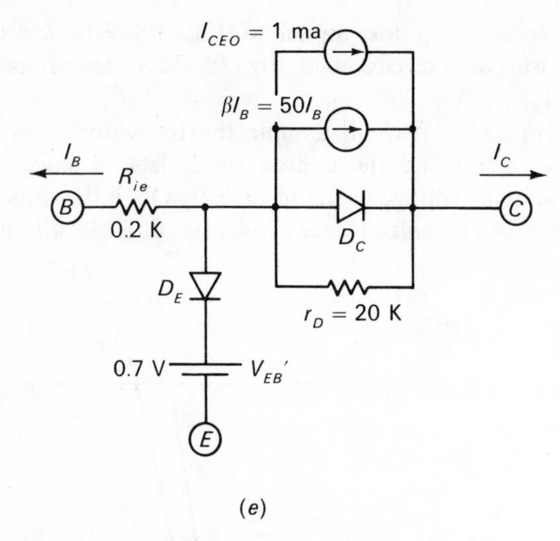

(e)

FIGURE PS 7-1 (*Continued*)

= 0.467 ma and the dependent βI_B collector-current source is equal to 50 (0.467 ma) = 23.3 ma. To this we add I_{CEO} = 1 ma for a combined current-source total of 24.3 ma. To determine I_C, this current source and r_D (equal to 20 kilohms) may be thevenized as shown in Fig. PS 7-2c to yield

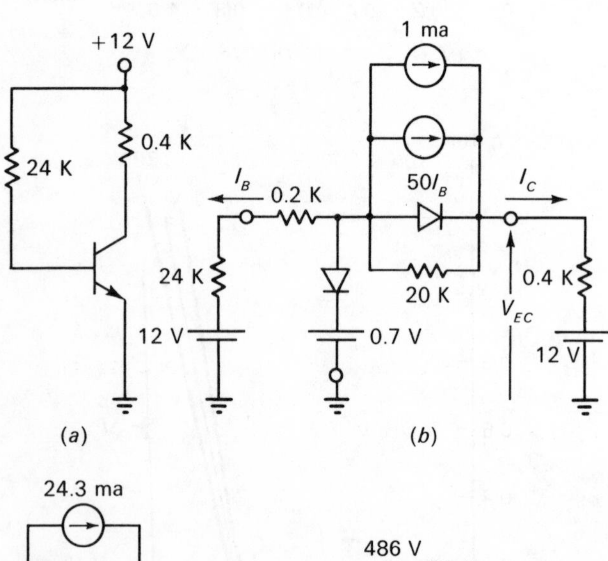

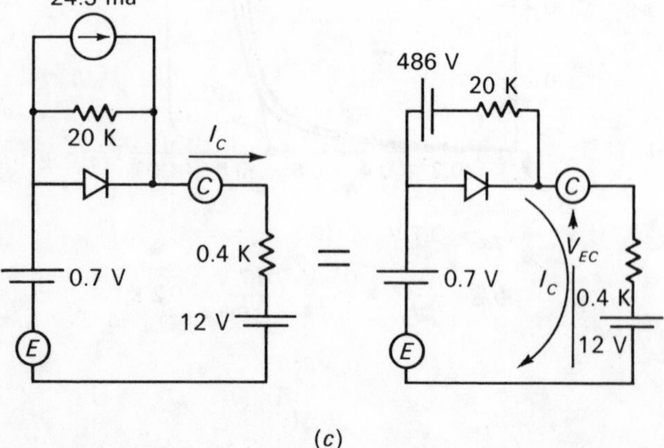

(c)

FIGURE PS 7-2

$$I_C = \frac{486 \text{ volts} + 12 \text{ volts} - 0.7 \text{ volt}}{20.4 \text{ kilohms}} = 24.4 \text{ ma}$$

Therefore,

$$\begin{aligned} V_{EC} &= 12 \text{ volts} - 0.4 \text{ kilohm } (24.4 \text{ ma}) \\ &= 12 \text{ volts} - 9.76 \text{ volts} = 2.24 \text{ volts} \end{aligned}$$

Active-region operation was assumed in this analysis. This means the collector diode must be reverse-biased. To check that the collector diode is truly reverse-biased we may, from Fig. PS 7-2c, imagine the diode removed from the circuit and determine the open-circuit voltage from cathode to anode as follows:

$$\begin{aligned} V_{DC(OC)} &= -V_{EC} + 0.7 \text{ volt} \\ &= -2.24 \text{ volts} + 0.7 \text{ volt} = 1.54 \text{ volts} \end{aligned}$$

Hence, the collector diode is reverse-biased and operation is in the active region, as assumed. Generally it will not be necessary to worry about V_{EB}'. Determining V_{EC} will usually suffice to establish the state of the collector diode.

PS 7-3 What sort of valid approximations might be considered in the previous problem to determine I_C and V_{EC} with more ease?

SOLUTION The external base resistance of the 24-kilohm resistor swamps the internal R_{ie} = 0.2 kilohm, and the 12-volt external base voltage swamps V_{EB}' = 0.7 volt. Hence, $I_B \approx$ 12 volts/24 kilohms = 0.5 ma. On the output side, r_D = 20 kilohms $\gg R_L$ = 0.4 kilohm. Therefore, considering r_D and R_L as a current divider, virtually all of the current from the constant-current sources (βI_B and I_{CEO}) will exit through the collector lead. Furthermore, I_{CEO} may be considered negligible as is usually the case in silicon transistors. Thus $I_C \approx \beta I_B$ = 50 (0.5 ma) = 25 ma. Therefore, V_{EC} = 12 volts − 0.4 kilohm (25 ma) = 2 volts. These approximate solutions are quite close to I_C = 24.4 ma and V_{EC} = 2.24 volts previously calculated, and, in view of the many approximations inherent in the development of the piecewise model itself, this simplified approach should always be considered.

PS 7-4 Construct a common-emitter model from the common-base model of Fig. PS 7-4a.

SOLUTION The common-emitter model of Fig. PS 7-4b is derived as follows. First we note that we have a *PNP* transistor in Fig. PS 7-4a. Hence the common-emitter model will appear as shown in Fig. 7-4, which is used to establish the form of Fig. PS 7-4b. The following relationships are then used to transform the common-base model of Fig. PS 7-4a to a common-emitter model of Fig. PS 7-4b.

1) $$\beta = \frac{\alpha}{1 - \alpha} = \frac{0.99}{1 - 0.99} = 99$$

2) $$I_{CEO} = I_{CBO}\,(\beta + 1) = 10\ \mu a\,(99 + 1) = 1 \text{ ma}$$

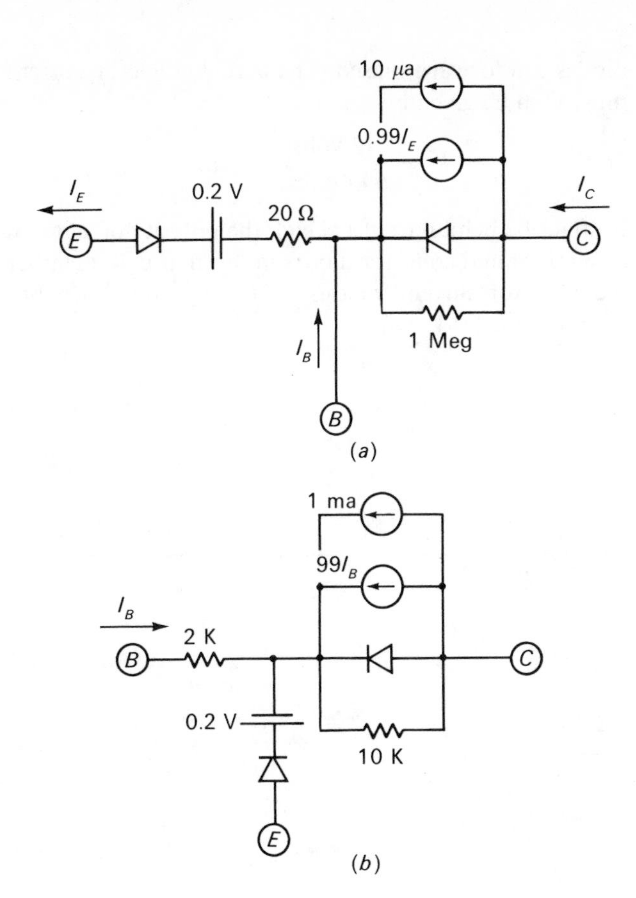

FIGURE PS 7-4

3) $R_{ie} = R_{ib}(\beta + 1) = 20 \text{ ohms } (99 + 1) = 2 \text{ kilohms}$

4) $$r_D = \frac{r_C}{\beta + 1} = \frac{1000 \text{ kilohms}}{99 + 1} = 10 \text{ kilohms}$$

PS 7-5 Estimate V_{EC} and I_C in Fig. PS 7-5*a* with no other information given.

SOLUTION. As a first step, we might simplify the circuit by thevenizing the two 6-kilohm resistors across the −12-volt supply. This procedure yields Fig. PS 7-5*b*. From Fig. PS 7-5*b* we may reason as follows. The −6-volt supply will tend to forward-bias the emitter diode. Assuming negligible voltage drop across this forward-biased emitter diode means that the emitter lead will sit at the same potential as the base lead, which is ground or zero volts. Thus the emitter end of the 3-kilohm resistor is at zero volts, and the bottom end is at minus 6 volts. This means the current through the 3-kilohm resistor is at zero volts, and the bottom end is at −6 volts. This means the current through the 3-kilohm will be approximately equal to the emitter current. Thus $I_C \approx 2$ ma also. The voltage developed across the 3-kilohm resistor in the collector lead is then 3 kilohms (2 ma) = 6 volts. Hence the voltage at the collector terminal with respect to ground is equal to $V_C = 12$ volts − 6 volts = 6 volts. Since the base is grounded this means $V_{BC} = 6$ volts also. Therefore, $V_{EC} = V_{EB} + V_{BC} = 0 \text{ volts} + 6 \text{ volts} = 6 \text{ volts}$.

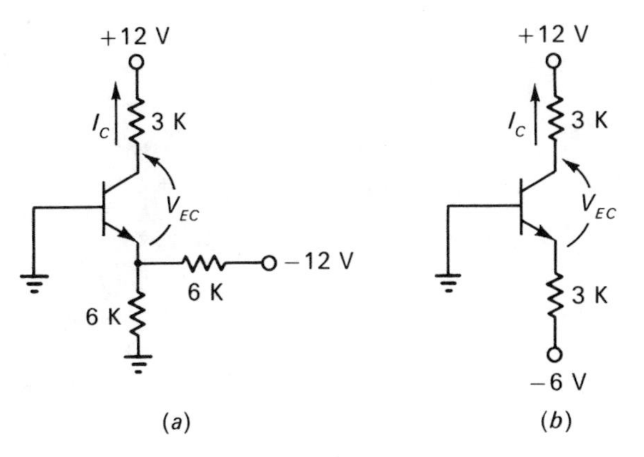

FIGURE PS 7-5

Note that this problem was solved without the use of any models. This will often be the case. The models are useful when a more detailed kind of analysis is required or when the model serves to convey a feeling for useful approximations.

PS 7-6 Estimate I_C and V_{EC} in Fig. PS 7-6*a*.

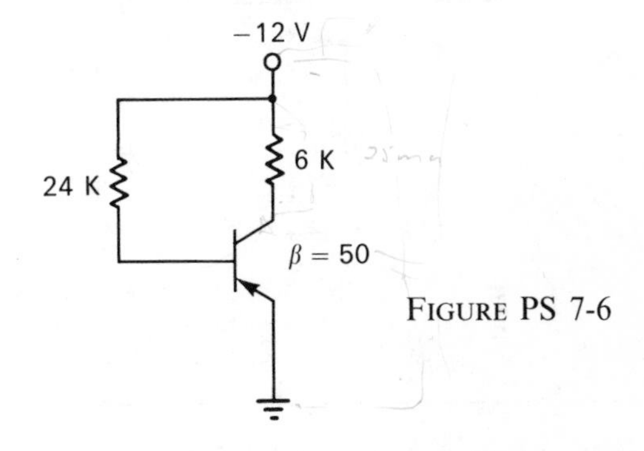

FIGURE PS 7-6

SOLUTION. Clearly the emitter diode wants to be forward-biased through the 24-kilohm resistor and −12-volt supply. Therefore, with the emitter diode ON, to a good approximation, we may write

1) $$I_B = \frac{12 \text{ volts}}{24 \text{ kilohms}} = 0.5 \text{ ma}$$

Since β is specified as being equal to 50, it follows that the βI_B current source in the collector will result in a collector current given by

2) $$I_C = \beta I_B = 50\,(0.5 \text{ ma}) = 25 \text{ ma}$$

To a good approximation, we may neglect I_{CEO} and r_D, and anyway these parameters aren't specified. Now we may calculate V_{EC} from

3) $$V_{EC} = -12 \text{ volts} + 6 \text{ kilohms } I_C$$
$$= -12 \text{ volts} + 6 \text{ kilohms } (25 \text{ ma}) = 138 \text{ volts}$$

The 138-volt value of V_{EC} might prove somewhat disconcerting since the maximum power-supply voltage applied to the system is only 12 volts. This might clue us in to suspecting that the transistor has entered some operating region other than the active region. Since V_{EC} is a positive quantity, it follows that the collector diode will want to be turned ON. Hence, the transistor must actually have gone into saturation and the solution for V_{EC} of 138 volts is in error. Instead, V_{EC} must be approximately zero volts since both the collector and emitter diodes are forward-biased. The actual collector current must then be given by

4) $$I_C = \frac{12 \text{ volts}}{6 \text{ kilohms}} = 2 \text{ ma}$$

It might be helpful to just sketch the polarity or sense of the emitter and collector diodes in this figure. Remember the collector current for this *PNP* transistor flows into the collector lead.

PROBLEMS WITH ANSWERS

PA 7-1 Estimate V_C in Fig. PA 7-1.

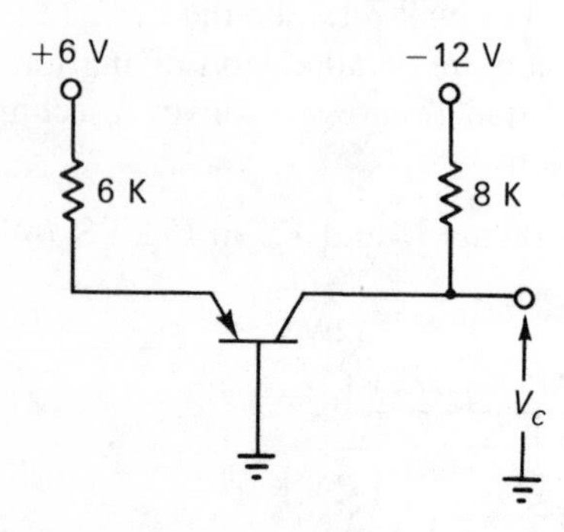

ANSWER $V_C = -4$ volts

PA 7-2 Estimate V_C in Fig. PA 7-2.

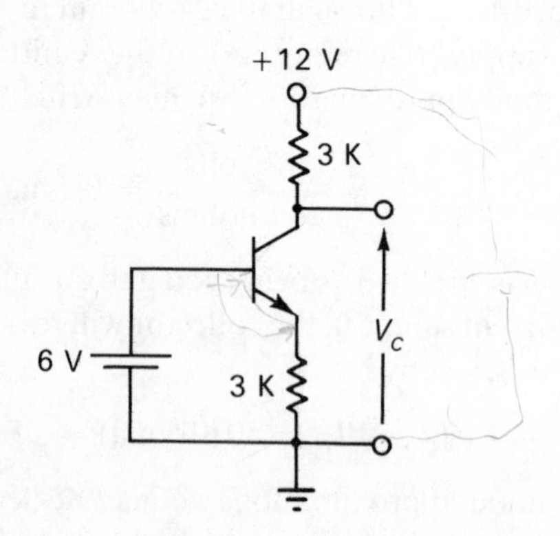

ANSWER $V_C = 6$ volts

PA 7-3 Estimate I_C in Fig. PA 7-3.

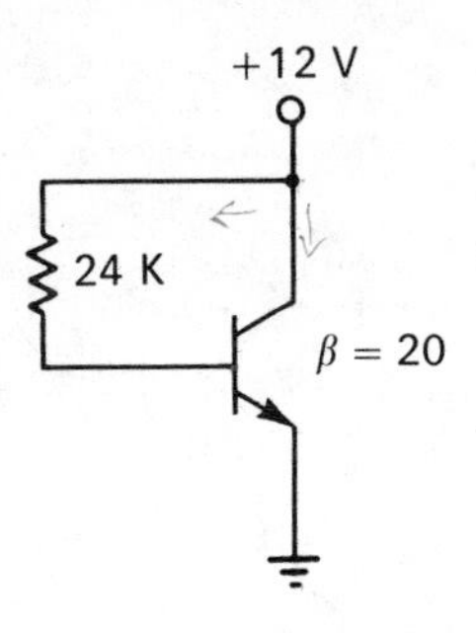

ANSWER $I_C = 10$ ma

PA 7-4 Estimate V_C in Fig. PA 7-4.

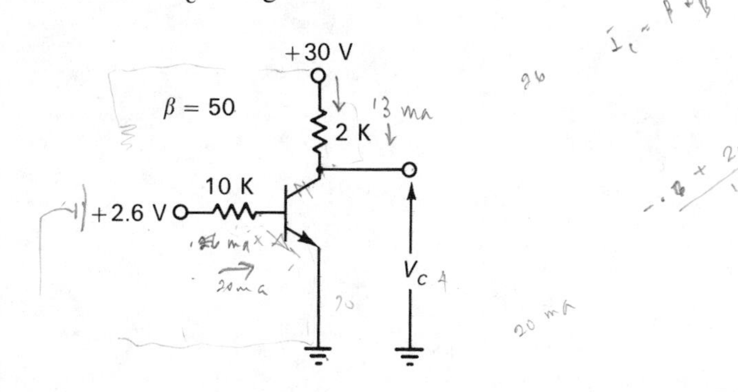

ANSWER $V_C = 10$ volts

PA 7-5 Estimate V_C in Fig. PA 7-5.

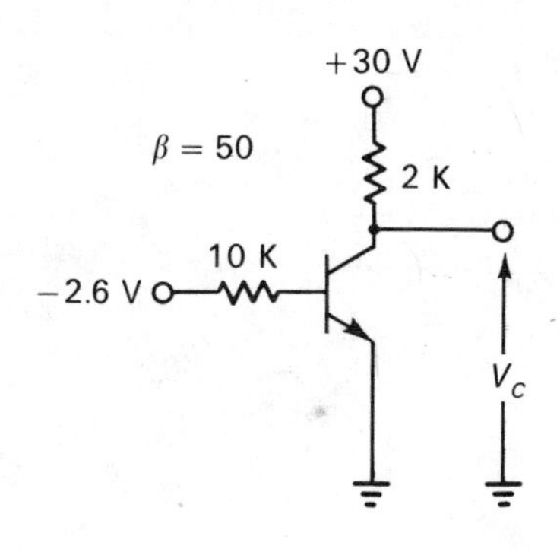

ANSWER $V_C = 30$ volts

PA 7-6 Estimate V_C in Fig. PA 7-6. Assume a silicon transistor.

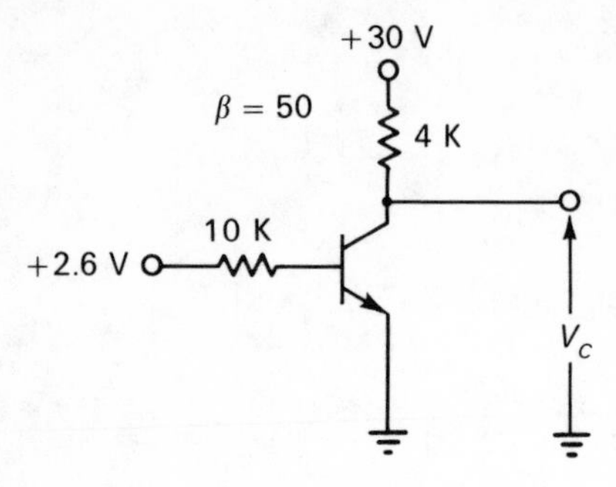

ANSWER $V_C = 0$ volts

PA 7-7 Estimate V_C in Fig. PA 7-7. Assume a silicon transistor.

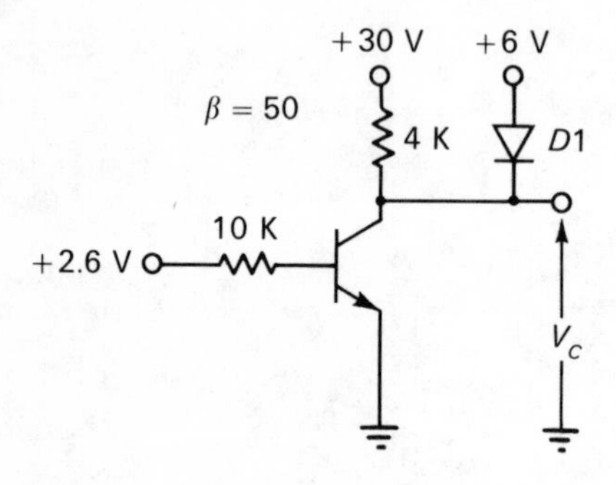

ANSWER $V_C = 6$ volts

PA 7-8 Estimate I_C in Fig. PA 7-8.

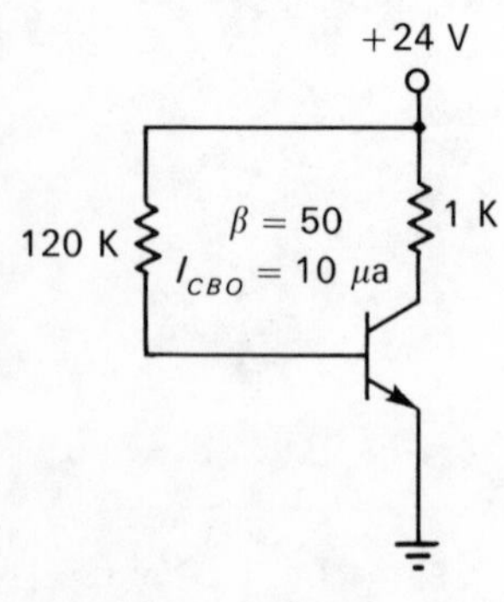

ANSWER $I_C = 10.5$ ma

PROBLEMS WITHOUT ANSWERS

P 7-1 Estimate I_C in Fig. P 7-1.

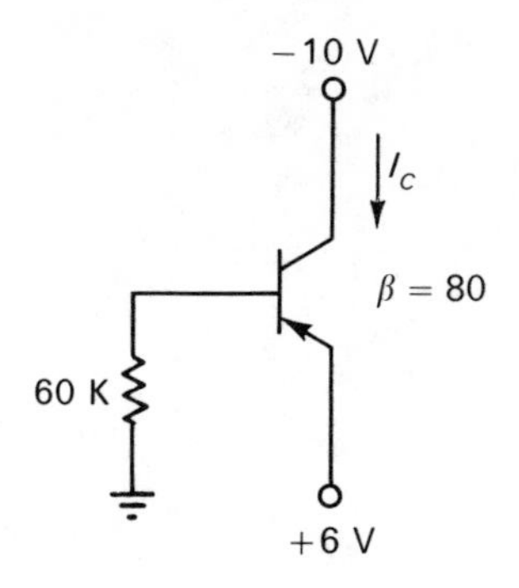

P 7-2 Determine V_C in Fig. P 7-2. Use a detailed common-emitter model.

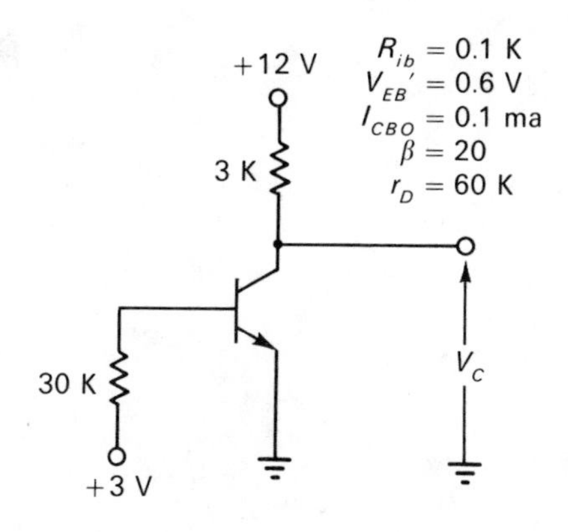

P 7-3 Estimate V_{EC} and V_C in Fig. P 7-3.

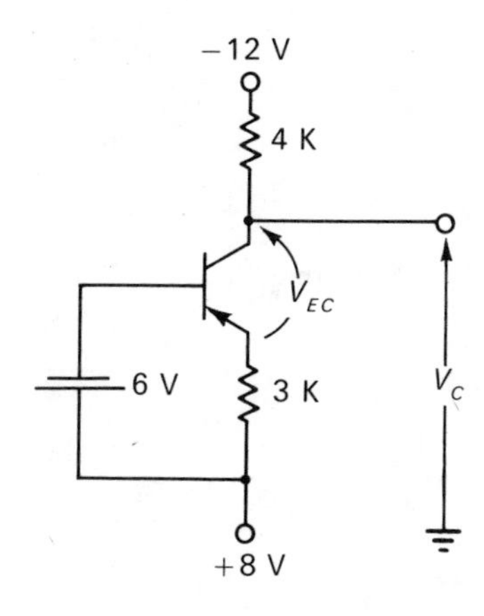

P 7-4 Estimate I_C in Fig. P 7-4.

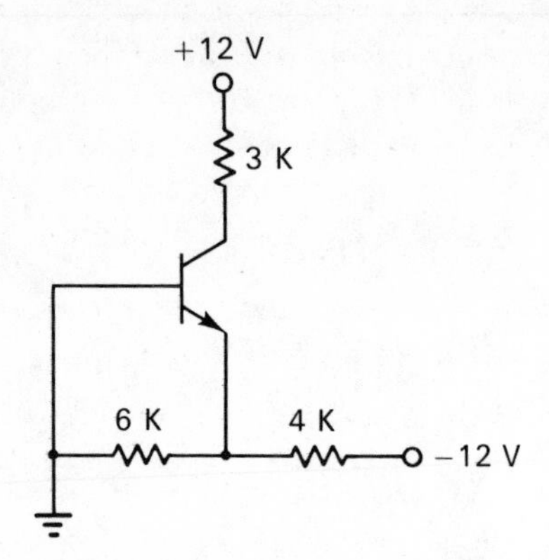

P 7-5 Describe the action of the relay contacts as the switch in the base lead of the transistor is opened and closed if the relay closes at 1.8 ma and releases at 0.8 ma.

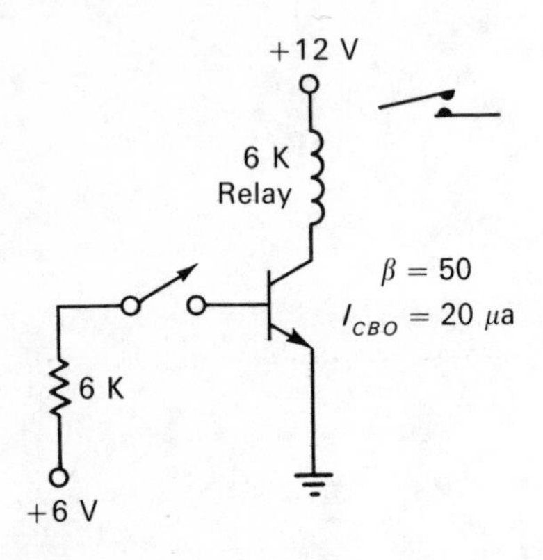

P 7-6 Estimate V_C in Fig. P 7-6.

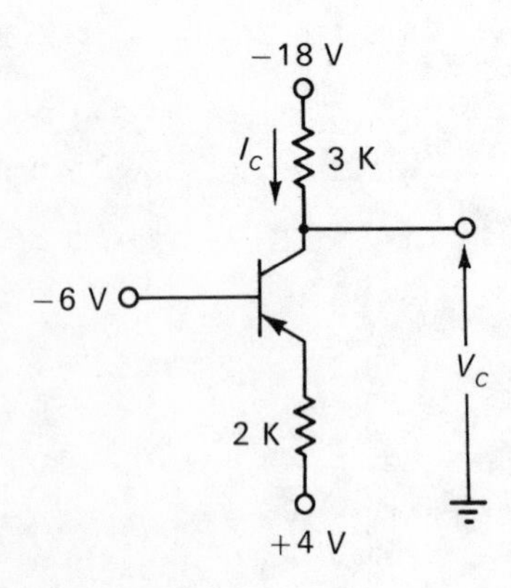

8 biasing and stabilization

The thing we want to do here is learn how to analyze transistor circuits for their quiescent operating points, and also to consider factors such as temperature effects which cause drift of the Q point. For linear amplification, the Q point is normally held at some point in the active region. Since there are many possible biasing configurations, we will consider a few basic ones here.

8-1 Constant base-current bias

Figure 8-1*a* illustrates what may be called *constant base-current bias.* Assuming active-region operation, the

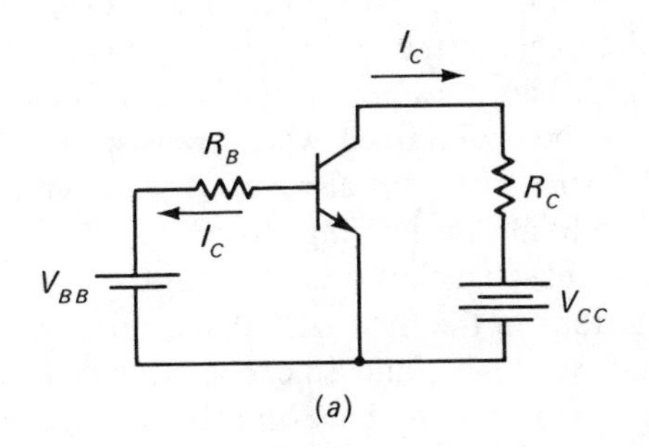

(*a*)

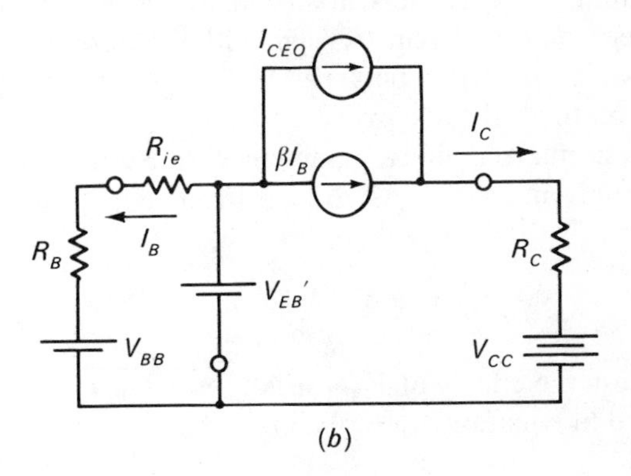

(*b*)

FIGURE 8-1

collector current may be readily determined from Fig. 8-1*b*, in which the r_D of the transistor model is neglected for simplicity. Thus

1) $$I_B = \frac{V_{BB} - V_{EB}'}{R_B + R_{ie}} \tag{8-1}$$

If $V_{BB} \gg V_{EB}'$ and $R_B \gg R_{ie}$

2) $$I_B \approx \frac{V_{BB}}{R_B} \tag{8-2}$$

Equation 2 is independent of the transistor parameters, and hence I_B is essentially constant. Thus the name constant base-current bias.

On the output side

3) $$I_C = \beta I_B + I_{CEO} \tag{8-3}$$

Substituting equation 1 into 3 yields

$$4) \qquad I_C = \frac{\beta(V_{BB} - V_{EB}')}{R_B + R_{ie}} + I_{CEO} \qquad (8\text{-}4)$$

or, if we substitute equation 2 into 3 and recall that $I_{CEO} = I_{CBO}(\beta + 1)$, we obtain

$$5) \quad I_C \approx \frac{\beta V_{BB}}{R_B} + I_{CBO}(\beta + 1)$$

$$\approx \beta\left(\frac{V_{BB}}{R_B} + I_{CBO}\right) \qquad (8\text{-}5)$$

Equation 5 reveals that I_C is exceedingly sensitive (directly proportional) to β, and since β may vary by a factor of three for the same type transistor, constant base-current bias does not seem advisable if the same Q point is to be maintained when transistors are interchanged. Furthermore β also varies with temperature, perhaps doubling in the range from $-25°C$ to $+50°C$.

Constant base-current bias also has the disadvantage that variations in the thermal leakage current, I_{CBO}, are multiplied by $\beta + 1$, and since I_{CBO} typically doubles with every 10°C rise, we have a problem, quite severe in germanium and far less in silicon, which has a much smaller leakage current to begin with. The point is that constant base-current bias exhibits the $\beta + 1$ multiplying effect on the leakage current.

It is useful to define a parameter S called the *stability factor*, given by $S = \Delta I_C/\Delta I_{CBO}$. From equation 5 we obtain

$$6) \qquad S = \frac{dI_C}{dI_{CBO}} \approx \frac{\Delta I_C}{\Delta I_{CBO}} = \beta + 1 \approx \beta \qquad (8\text{-}6)$$

If just the effects of V_{EB}' variations upon I_C are considered in equation 4, we obtain

$$7) \qquad S_1 = \frac{dI_C}{dV_{EB}'} \approx \frac{\Delta I_C}{\Delta V_{EB}'} = \frac{-\beta}{R_B + R_{ie}} \qquad (8\text{-}7)$$

and if $R_B \gg R_{ie}$ as is often true

$$8) \qquad S_1 = \frac{\Delta I_C}{\Delta V_{EB}'} \approx \frac{-\beta}{R_B} \qquad (8\text{-}8)$$

The negative sign in equation 8 indicates that I_C decreases with an increase in V_{EB}', but since V_{EB}' typically decreases between 2 to 3 mv/C°, the collector current will actually increase from this cause as temperature increases. Note also that a large R_B minimizes S_1, the change in I_C due to V_{EB}'.

8-2 Constant emitter-current bias

An alternative biasing scheme called *constant emitter-current bias* is shown in Fig. 8-2*a*. The active-region equivalent circuit is shown in Fig. 8-2*b* with r_C omitted for simplicity. From Fig. 8-2*b*

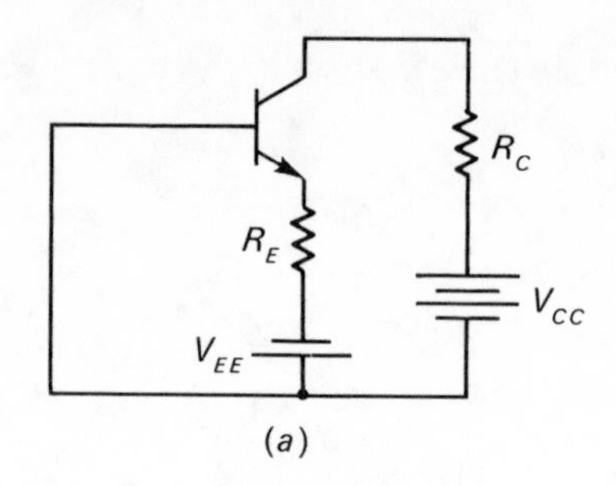

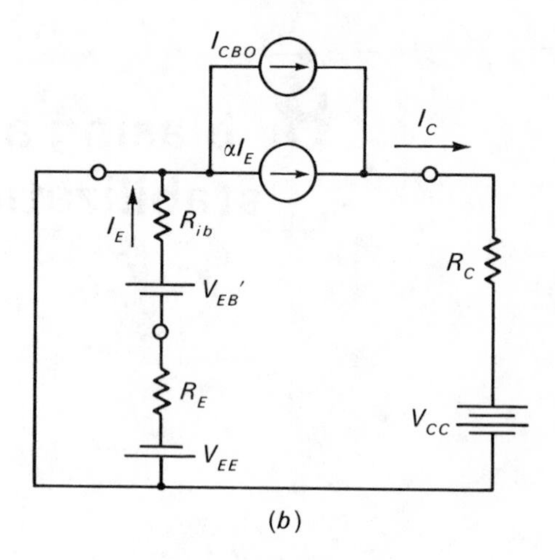

FIGURE 8-2

$$1) \qquad I_E = \frac{V_{EE} - V_{EB}'}{R_E + R_{ib}} \qquad (8\text{-}9)$$

For $V_{EE} \gg V_{EB}'$ and $R_E \gg R_{ib}$

$$2) \qquad I_E \approx \frac{V_{EE}}{R_E} \qquad (8\text{-}10)$$

Equation 2 indicates that if the approximations involved are valid (as they usually are) the emitter current is constant and essentially determined by V_{EE} and R_E.

On the output side,

$$3) \qquad I_C = \alpha I_E + I_{CBO}$$

Substituting equation 1 into 3 and recalling $\alpha = \beta/(\beta + 1) \approx 1$

$$4) \qquad I_C = \frac{\beta}{\beta + 1}\,\frac{V_{EE} - V_{EB}'}{R_E + R_{ib}} + I_{CBO} \qquad (8\text{-}11)$$

and, from equation 2,

$$5) \qquad I_C \approx \frac{V_{EE}}{R_E} + I_{CBO} \qquad (8\text{-}12)$$

Note from equation 5 that with constant emitter-current bias, I_C is essentially independent of β variations and also there is no β multiplication of I_{CBO}. Thus, to this extent, constant emitter-current bias is superior to constant base-current bias. From equation 4 we obtain

6) $$S = \frac{\Delta I_C}{\Delta I_{CBO}} = 1 \qquad (8\text{-}13)$$

To study V_{EB}' effects upon I_C with this bias method, we have from equation 4

7) $$S_1 = \frac{\Delta I_C}{\Delta V_{EB}'} = \frac{-\beta}{\beta + 1} \frac{V_{EB}'}{R_E + R_{ib}} \qquad (8\text{-}14)$$

Since $\beta/(\beta + 1) \approx 1$, and for $R_E \gg R_{ib}$

8) $$S_1 = \frac{\Delta I_C}{\Delta V_{EB}'} \approx -\frac{V_{EB}'}{R_E} \qquad (8\text{-}15)$$

Equation 8 indicates that a large R_E minimizes variations in I_C due to V_{EB}'.

We may then conclude that constant emitter-current bias is, in general, preferable to constant base-current bias. Since a large R_E promotes a constant emitter current whereas a large R_B promotes a constant base current, we might intuitively reason that a large R_E and a small R_B tend to promote stabilization of the operating point.

8-3 Combination bias

It turns out that the biasing circuitry for most practical linear amplifier stages may be reduced to the form shown in Fig. 8-3*a* which is a little of both biasing methods. Assuming active-region operation, we obtain from Fig. 8-3*b* the following equations.

1) $$I_C = \beta I_B + I_{CEO}$$

2) $$V_{EE} + V_{BB} - V_{EB}' = R_E I_E + (R_B + R_{ie}) I_B$$

3) $$I_E = I_B + I_C$$

Substituting equations 1 and 3 into 2 and simplifying, we obtain after some manipulation the following equations for all the terminal currents.

4) $$I_E = \frac{(\beta + 1)(V_{EE} - V_{BB} - V_{EB}') + (R_B + R_{ie}) I_{CEO}}{R_B + R_{ie} + R_E(\beta + 1)} \qquad (8\text{-}16)$$

Recalling that $I_{CEO} = I_{CBO}(\beta + 1)$ and dividing both numerator and denominator of equation 4 by $\beta + 1$ yields

5) $$I_E = \frac{V_{EE} - V_{BB} - V_{EB}'}{[(R_B + R_{ie})/(\beta + 1)] + R_E} + \frac{(R_B + R_{ie}) I_{CBO}}{(R_B + R_{ie} + R_E)/(\beta + 1)} \qquad (8\text{-}17)$$

Equation 5 reveals a most powerful method of estimating I_E in a circuit of the form shown in Fig. 8-3*a* if we assume the I_{CBO} term is negligible. Note that the numerator of

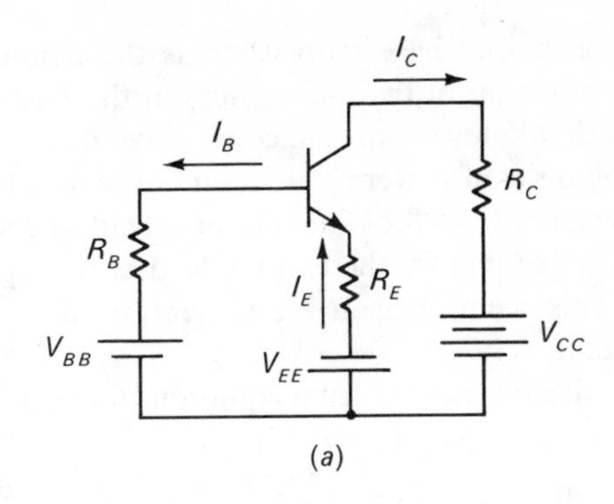

(*a*)

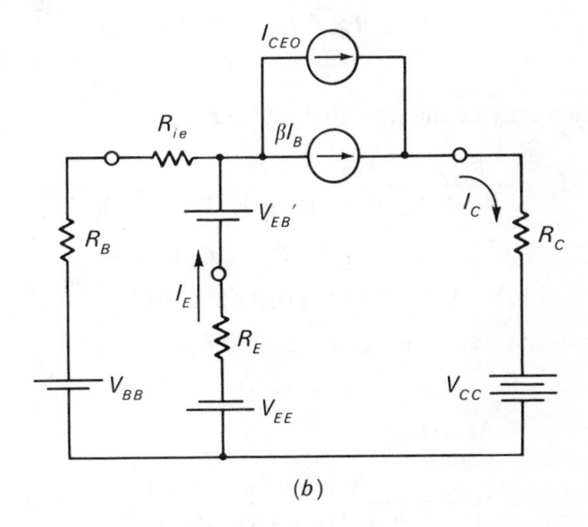

(*b*)

FIGURE 8-3

the first term is the algebraic sum of all the driving voltages around the emitter-base loop. The denominator must then be the impedance presented to the flow of emitter current, I_E. This impedance consists of R_E (as might be expected since I_E directly flows through R_E) in series with whatever is in the base lead (both internal and external) looking as if it is divided by $\beta + 1$. In other words, when we look into the emitter of the transistor, any impedance in the base lead is reflected as if it is $\beta + 1$ times smaller. This result isn't too startling if we consider that the base current is $\beta + 1$ times smaller than the emitter current, and hence, any voltage dropped across an R_B will be $\beta + 1$ times smaller than that dropped across an R_E.

If equations 1 and 4 are substituted into 3, we may obtain, after manipulation,

6) $$I_B = \frac{V_{EE} + V_{BB} - V_{EB}'}{R_B + R_{ie} + R_E(\beta + 1)} - \frac{R_E I_{CEO}}{R_B + R_{ie} + R_E(\beta + 1)} \qquad (8\text{-}18)$$

Neglecting the I_{CEO} term, we see from the first term that the base current may be quickly estimated by noting that the numerator is again the algebraic sum of the voltages about the emitter-base loop whereas the denominator contains the sum of the impedances in the base lead in series with whatever impedance we have in the emitter lead looking as if it were $\beta + 1$ times greater. In other words, whenever we look into the base lead to determine I_B, any impedance in the emitter lead is multiplied by $\beta + 1$. This seems reasonable since I_E is $\beta + 1$ times greater than I_B.

The collector current from equations 6 and 1 is given by

$$7) \quad I_C = \frac{\beta(V_{EE} + V_{BB} - V_{EB}') + (R_B + R_{ie} + R_E)\, I_{CEO}}{R_B + R_{ie} + R_E(\beta + 1)} \qquad (8\text{-}19)$$

which may be manipulated into the form

$$8) \quad I_C = \frac{\beta(V_{EE} + V_{BB} - V_{EB}')}{R_B + R_{ie} + R_E(\beta + 1)} + \frac{1 + [R_E/(R_B + R_{ie})]}{(1/\beta + 1) + [R_E/(R_B + R_{ie})]}\, I_{CBO} \qquad (8\text{-}20)$$

From the second term in equation 8 we obtain

$$9) \quad S = \frac{\Delta I_C}{\Delta I_{CBO}} = \frac{1 + [R_E/(R_B + R_{ie})]}{[1/(\beta + 1)] + [R_E/(R_B + R_{ie})]} \qquad (8\text{-}21)$$

as the collector-current stability factor for the generalized circuit of Fig. 8-3*a*. Note that if $R_B \gg R_E$, we favor constant base-current bias and in the limit equation 9 approaches $\beta + 1$ as might be expected.

From the first term of equation 8

$$10) \quad S_1 = \frac{\Delta I_C}{\Delta V_{EB}'} = \frac{-\beta}{R_B + R_{ie} + R_E(\beta + 1)} \qquad (8\text{-}22)$$

Note from equation 10 that if $R_E(\beta + 1) \gg R_B + R_{ie}$

$$11) \quad S_1 \approx -\frac{1}{R_E} \qquad (8\text{-}23)$$

which again emphasizes the desirability of a large R_E if V_{EB}' variations are to be minimized.

From another viewpoint we may reason that the presence of the emitter-resistor R_E introduces negative feedback. If the collector current, say, increases, it must be due to an increase in the emitter current. Any increase in emitter current causes an increase in the voltage drop across R_E. This voltage drop across R_E subtracts from the driving voltages around the emitter-base loop, which tends to leave less voltage available to force current through the forward-biased emitter junction. Thus R_E senses changes in I_E which are related to changes in I_C and introduces a bucking voltage into the emitter-base loop so as to compensate in part for the mechanism which tends to cause the change in I_C.

8-4 Constant collector-voltage biasing

Another biasing scheme that is occasionally encountered is shown in Fig. 8-4*a*. Stabilization of the collector

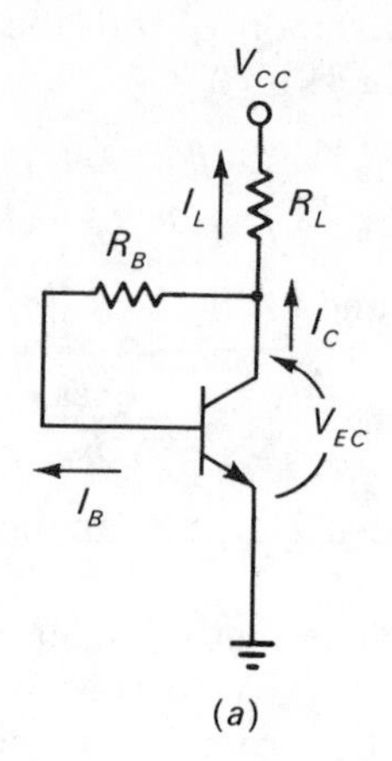

(*a*)

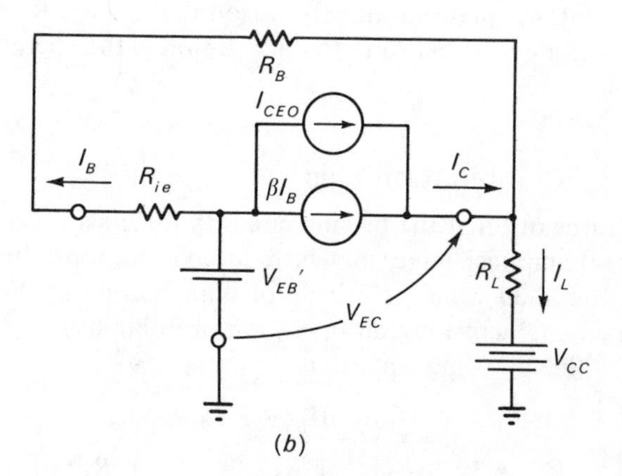

(*b*)

FIGURE 8-4

voltage at the operating point is obtained by *negative voltage feedback*. If the term negative voltage feedback is unfamiliar at this time, just ignore it and consider the following circuit description. Inspection of Fig. 8-4*a* indicates that the base current is dependent upon V_{EC}. If the temperature should, say, rise, then I_C would increase due to the increase in I_{CBO} and decrease in V_{EB}'. However, if I_C increases, the voltage drop across R_L increases, which tends to decrease V_{EC}. If V_{EC} tends to decrease, there is less voltage available to force I_B through R_B. If I_B tends to decrease, the collector current, which depends upon the βI_B current source, will also tend to decrease. Thus we see that if I_C increases, an opposing (negative) feedback mechanism comes into play which tends to minimize the increase in I_C.

The operating point and stability factors in the circuit of Fig. 8-4*a* may be determined from the equivalent circuit of Fig. 8-4*b*. Active-region operation is assumed and r_D is assumed large enough to be neglected. We may write

1) $$I_C = \beta I_B + I_{CEO}$$

2) $$I_B = \frac{V_{EC} - V_{EB}'}{R_B + R_{ie}}$$

3) $$V_{EC} = V_{CC} - R_L I_L$$

4) $$I_L = I_B + I_C$$

Substituting equations 4 and 3 into 2 yields

5) $$I_B = \frac{V_{CC} - V_{EB}' - R_L I_C}{R_B + R_{ie} + R_L}$$

Substituting equation 5 into 1 and solving for I_C, we obtain

6) $$I_C = \frac{\beta V_{CC}}{D} - \frac{\beta V_{EB}'}{D} + \frac{(R_B + R_{ie} + R_L)}{D} I_{CEO} \quad (8\text{-}24)$$

where

7) $$D = R_B + R_{ie} + R_L(\beta + 1) \quad (8\text{-}25)$$

The stability factor S may then be easily derived from the last term in equation 6 if we let $I_{CEO} = I_{CBO}(\beta + 1)$:

8) $$S = \frac{\Delta I_C}{\Delta I_{CBO}} = \frac{(R_B + R_{ie} + R_L)(\beta + 1)}{R_B + R_{ie} + R_L(\beta + 1)} \quad (8\text{-}26)$$

which may be manipulated into the form

9) $$S = \frac{\Delta I_C}{\Delta I_{CBO}} = \frac{1 + [R_L/(R_B + R_{ie})]}{1/(\beta + 1) + [R_L/(R_B + R_{ie})]} \quad (8\text{-}27)$$

Inspection of equation 9 indicates that making R_L large and R_B small tends to minimize the change in I_C due to I_{CBO} variations.

From the second term in equation 6 we obtain

10) $$S_1 = \frac{\Delta I_C}{\Delta V_{EB}'} = -\frac{\beta}{R_B + R_{ie} + R_L(\beta + 1)} \quad (8\text{-}28)$$

Inspection of equation 10 indicates that making either R_L and/or R_B large minimizes changes in I_C due to variation in V_{EB}'. Considering the results of equations 9 and 10, we may conclude the all around stability of the operating point is achieved by making R_L large relative to R_B.

PROBLEMS WITH SOLUTIONS

PS 8-1 Estimate I_C and V_{EC} in the silicon transistor circuit of Fig. PS 8-1*a*.

SOLUTION. The circuit is simple enough to analyze without drawing an equivalent circuit. For example, if we just remember that the polarity of the emitter and collector diodes is as shown in Fig. PS 8-1*b*, we see that the emitter diode is forward-biased. Therefore, assuming $V_{EB}' = 0.65$ volts and $R_{ie} \ll 24$ kilohms,

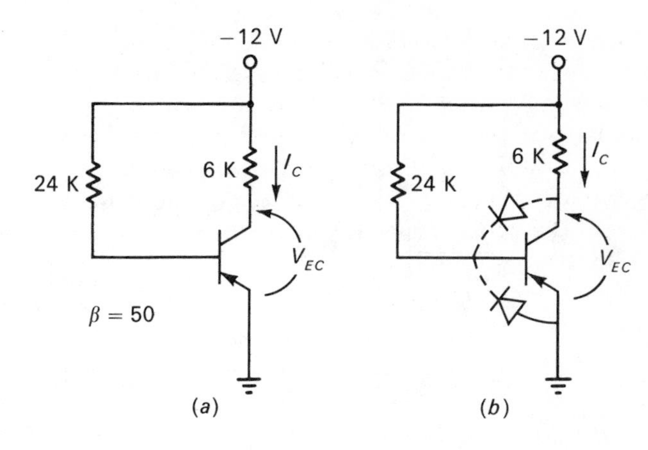

FIGURE PS 8-1

1) $$I_B = \frac{12 \text{ volts} - 0.65 \text{ volt}}{24 \text{ kilohm}} \approx 0.47 \text{ ma}$$

Now neither I_{CBO} or I_{CEO} is given, but since the transistor is silicon these quantities are usually negligible. Thus

2) $$I_C = \beta I_B + I_{CEO} = 50\,(0.47 \text{ ma}) + 0 = 23.6 \text{ ma}$$

Hence,

3) $$V_{EC} = -12 \text{ volts} + 6 \text{ kilohms}\,(23.6 \text{ ma}) = -12 \text{ volts} + 142 \text{ volts} = 130 \text{ volts}$$

At this point we might be inclined to suspect something is weird since the supply voltage is only 12 volts in magnitude and yet we developed 142 volts across the 6-kilohm resistor. Nevertheless, let us continue.

Equation 2 is based on the assumption that the transistor is in its active region, which means D_E is ON and D_C is OFF. For D_C to be OFF it must be reverse-biased. However, equation 3 indicates that D_C is forward-biased by 130 volts − 0.65 volt ≈ 130 volts. Now D_C cannot be forward-biased and still maintain 130 volts across it. Evidently, then, our initial assumption that the transistor is in its active region was wrong. Instead, both diodes are ON and the transistor is actually in a saturated region. Therefore, the 6-kilohm resistor has −12 volts at one end and 0 volts at its other end. Thus with 12 volts across it, the current through this resistor is 12 volts/6 kilohms = 2 ma, and this must also be the collector current in the saturated region.

PS 8-2 Develop some design equations for the circuit of Fig. PS 8-2 to determine R_L and R_B to bias the transistor at a specific Q point (V_{EC}, I_C) in the circuit shown. Assume V_{CC} and the range of β or h_{FE} (same thing) values are also given.

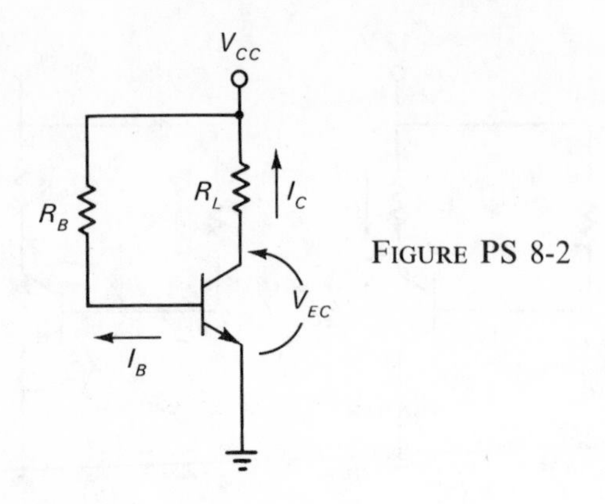

FIGURE PS 8-2

SOLUTION

1) $$R_L = \frac{V_{CC} - V_{EC}}{I_C}$$

Neglecting r_D, the approximate equation for I_C is

2) $$I_C = h_{FE} I_B + I_{CBO}(h_{FE} + 1)$$

Solving equation 2 for I_B yields

3) $$I_B = \frac{I_C - I_{CBO}(h_{FE} + 1)}{h_{FE}} \approx \frac{I_C}{h_{FE}} - I_{CBO}$$

The approximate form of equation 3 is valid for $h_{FE} \geqq 10$ which is usually true. Assuming this condition is satisfied, then we see I_B depends on h_{FE} and I_{CBO}. Remember also that h_{FE} depends upon I_C to some extent. In silicon transistors the I_{CBO} term in equation 3 may usually be neglected unless I_C is very small, say 0.1 ma or less, and h_{FE} is very large, say > 100. Therefore, equation 3 for silicon transistors reduces to

4) $$I_B = \frac{I_C}{h_{FE}}$$

Now the production tolerance in h_{FE} may be 3:1. Therefore,

5) $$I_{B(\max)} = \frac{I_C}{h_{FE(\min)}}$$

6) $$I_{B(\min)} = \frac{I_C}{h_{FE(\max)}}$$

If I_B has a range of values due to the spread in h_{FE}, so will the calculated R_B since

7) $$R_B = \frac{V_{CC} - V_{EB}}{I_B}$$

where $V_{EB} \approx V_{EB}' \approx 0.65$ volt in silicon. Therefore,

8) $$R_{B(\min)} = \frac{V_{CC} - V_{EB}}{I_{B(\max)}} = \frac{(V_{CC} - V_{EB})\, h_{FE(\min)}}{I_C}$$

9) $$R_{B(\max)} = \frac{V_{CC} - V_{EB}}{I_{B(\min)}} = \frac{(V_{CC} - V_{EB})\, h_{FE(\max)}}{I_C}$$

Assuming $h_{FE(\min)} = 30$ and $h_{FE(\max)} = 90$, $V_{CC} = 12$ V, $V_{EC} = 6$ V, $I_C = 1$ ma, and $V_{EB} = 0.6$ volt,

10) $$R_{B(\min)} = \frac{(12 \text{ volts} - 0.6 \text{ volt})30}{1 \text{ ma}} = 342 \text{ kilohms}$$

11) $$R_{B(\max)} = \frac{(12 \text{ volts} - 0.6 \text{ volt})90}{1 \text{ ma}} = 1{,}026 \text{ kilohms}$$

The spread in the calculated values of R_B indicates why constant base-current bias doesn't lend itself to predictable design. The correct value of R_B would have to be hand-tailored to match the transistor in question.

The load resistor, R_L, may be determined from equation 1. Thus

12) $$R_L = \frac{12 \text{ volts} - 6 \text{ volts}}{1 \text{ ma}} = 6 \text{ kilohms}$$

PS 8-3 Assume in the circuit of Fig. PS 8-3 that $R_{ie} = 0.375$ kilohm, $\beta = 54$, $r_D = 5.9$ kilohms, $I_{CEO} = 1$ ma, and $V_{EB}' = 0.3$ volt. Estimate I_C.

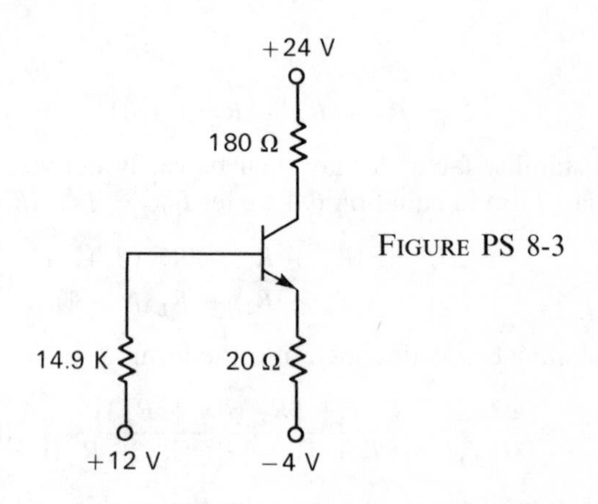

FIGURE PS 8-3

SOLUTION Since r_D is relatively large compared to the external load resistance, we may, to a first approximation, neglect it. Next we may estimate the component of base current due to the external voltages and V_{EB}' in the emitter-base loop. Recalling that any impedance in the emitter lead is reflected $\beta + 1$ times greater when looking into the base, we may write

1) $$I_B = \frac{12 \text{ volts} + 4 \text{ volts} - 0.3 \text{ volt}}{0.375 \text{ kilohms} + 14.9 \text{ kilohms} + 0.02 \text{ kilohm } (54 + 1)}$$

$$= \frac{15.7 \text{ volts}}{16.38 \text{ kilohms}} = 0.959 \text{ ma}$$

Therefore, I_C just due to these voltages is

2) $$I_C = \beta I_B = 54\,(0.959 \text{ ma}) = 51.8 \text{ ma}$$

To this might be added the effect of I_{CBO}. If the base lead were left open, then I_{CEO} would be the additional collector current so that I_C couldn't exceed 51.8 ma + 1 ma = 52.8 ma. However, the effect of I_{CBO} would be less than I_{CEO} since the base isn't open. Specifically we may apply equation 8-19 to obtain

3) $I_C =$

$$\frac{54(12\text{ V} + 4\text{ V} - 0.3\text{ V}) + (14.9\text{ K} + 0.375\text{ K} + 0.02\text{ K})\,1\text{ ma}}{14.9\text{ K} + 0.375\text{ K} + 0.02\text{ K}\,(54 + 1)}$$

$$= \frac{847\text{ volts} + 15.3\text{ volts}}{16.38\text{ kilohms}} = \frac{863\text{ volts}}{16.4\text{ kilohms}} = 52.7\text{ ma}$$

Note that the approximate solution in step 2 is in close agreement with the more exact solution of step 3.

PS 8-4 Estimate I_C at 25°C and 45°C in the circuit of Fig. PS 8-4. The parameter values indicated are for 25°C.

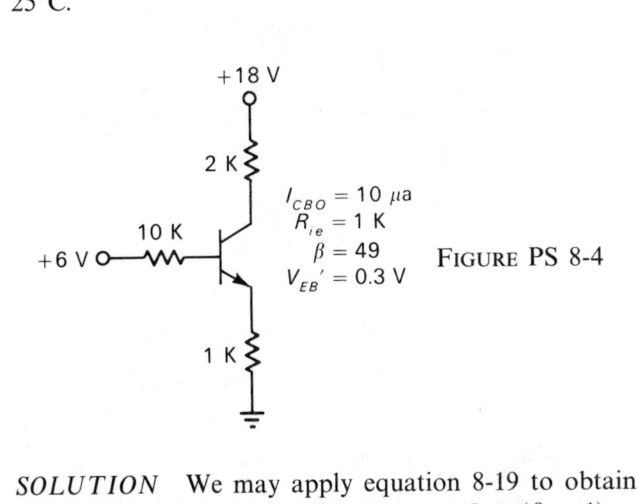

FIGURE PS 8-4

SOLUTION We may apply equation 8-19 to obtain I_C at 25°C if we first determine $I_{CEO} = I_{CBO}(\beta + 1) =$ 10 μa (50) = 0.5 ma. Thus

1) $I_C =$

$$\frac{49(6\text{ V} - 0.3\text{ V}) + (10\text{ K} + 1\text{ K} + 1\text{ K})0.5\text{ ma}}{10\text{ K} + 1\text{ K} + 1\text{ K}\,(49 + 1)}$$

$$= \frac{279\text{ volts} + 6\text{ volts}}{61\text{ kilohms}} = 4.68\text{ ma} \quad (\text{at } 25°\text{C})$$

At 45°C, I_{CBO} will have increased to 40 μa. This may be determined from the rule of thumb that I_{CBO} roughly doubles for every 10°C rise. Thus I_{CBO} = 20 μa at 35°C and 40 μa at 45°C. In equation form this may be expressed as

2) $$I_{CBO} = I_{CBO(\text{ref})}\,2^{(T - T_{\text{ref}})/10}$$

where the ref subscript pertains to the reference value of 25°C. This is a useful equation to memorize and you should make every effort to do so. Substituting values we obtain

$$I_{CBO} = 10\mu\text{a}\; 2^{(45°\text{C} - 25°\text{C})/10} = 10\;\mu\text{a}\; 2^2 = 40\;\mu\text{a}$$

Good design practice would allow for about a 50 percent increase in I_{CBO} to allow for the resistive component of collector leakage current through r_D which is usually ignored. Thus for a conservative estimate, we will use an I_{CBO} at 45°C of 40 μa + $(\Delta I_{CBO}/2)$ = 40 μa + (40 μa − 10 μa)/2 = 55 μa and therefore, I_{CEO} = 55 μa (50) = 2.75 ma. Note how drastically I_{CEO} has increased in the temperature rise from 25°C to 45°C.

We must also allow for the decrease in V_{EB}' with increasing temperature. Thus at 45°C

3) $$V'_{EB(45)} = V'_{EB(25)} + \frac{dV_{EB}'}{dT}\Delta T$$

where $dV_{EB}'/dT \approx -2.5$ mv/°C and ΔT is the temperature change. Substituting values

4) V_{EB}' = 0.3 volt + (−2.5 mv/°C) (45°C − 25°C)

= 0.3 volt − 0.05 volt = 0.25 volt

Therefore, again applying equation 8-19 but with the 45°C values,

5) $I_C =$

$$\frac{49(6\text{ V} - 0.25\text{ V}) + (10\text{ K} + 1\text{ K} + 1\text{ K})(2.75\text{ ma})}{10\text{ K} + 1\text{ K} + 1\text{ K}\,(49 - 1)}$$

$$= \frac{281.8\text{ volts} + 33\text{ volts}}{61\text{ kilohms}} = 5.16\text{ ma}$$

This problem could also have been solved using the stability factors S and S_1. First, I_C could have been determined at 25°C as in equation 1. Then the change in I_C due to I_{CBO} could have been determined from equation 8-21, where ΔI_{CBO} = 55 μa − 10 μa = 45 μa from equation 2. Thus

6) $$S = \frac{\Delta I_C}{\Delta I_{CBO}}$$

$$= \frac{1 + [1\text{ kilohm}/(10\text{ kilohms} + 1\text{ kilohm})]}{[1/(49 + 1)] + [1\text{ kilohm}/(10\text{ kilohms} + 1\text{ kilohm})]}$$

and therefore

7) $$\Delta I_C = S\Delta I_{CBO} = 9.84\,(45\;\mu\text{a}) = 443\;\mu\text{a}$$

From equation 8-22 we obtain

8) $$S_1 = \frac{\Delta I_C}{\Delta V_{EB}'}$$

$$= -\frac{49}{10\text{ kilohms} + 1\text{ kilohm} + 1\text{ kilohm}\,(49 + 1)}$$

$$= -\frac{0.803}{\text{kilohm}}$$

But $\Delta V_{EB}' \approx -2.5$ mv/°C (45°C − 25°C) = −0.05 volt. Therefore, ΔI_C due to $\Delta V_{EB}'$ is

9) $\Delta I_C = S_1 \Delta V_{EB}' = -\dfrac{0.803}{\text{kilohm}}(-0.05 \text{ volt}) = 0.04 \text{ ma}$

The total change in I_C is then

10) $$\Delta I_C = S_1 \Delta V_{EB}' + S\Delta I_{CBO}$$
$$= 0.04 \text{ ma} + 0.443 \text{ ma} = 0.483 \text{ ma}$$

The I_C at 45°C should then be given by

$$I_{C(\text{at } 45°\text{C})} = I_{C(\text{at } 25°\text{C})} + \Delta I_C$$
$$= 4.68 \text{ ma} + 0.483 \text{ ma} = 5.16 \text{ ma}$$

which is in agreement with the result obtained in equation 5.

PS 8-5 Determine R_1, R_2, and R_E in the circuit of Fig. PS 8-5*a* so that the transistor *Q* point lies at V_{EC} = 6 volts, I_C = 1 ma, and the stability factor S = 5. Assume a silicon transistor with β = 49 and I_{CBO} = 0.01 ma.

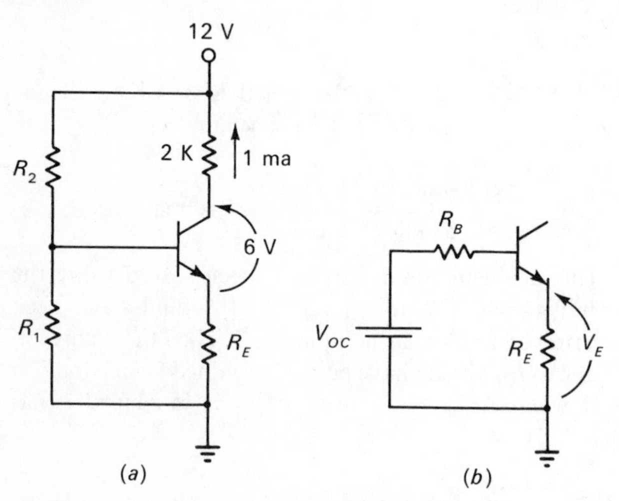

FIGURE PS 8-5

SOLUTION

1) $$I_C = \alpha I_E + I_{CBO}$$

from which

2) $I_E = \dfrac{I_C - I_{CBO}}{[\beta/(\beta+1)]} = \dfrac{1 \text{ ma} - 0.01 \text{ ma}}{49/50} = 1.01 \text{ ma}$

Now

3) $$V_E = V_{CC} - R_L I_C - V_{EC}$$
$$= 12 \text{ volts} - 2 \text{ kilohms (1 ma)} - 6 \text{ volts}$$
$$= 4 \text{ volts}$$

Therefore

4) $$R_E = \frac{V_E}{I_E} = \frac{4 \text{ volts}}{1.01 \text{ ma}} = 3.96 \text{ kilohms}$$

Also

5) $I_B = I_E - I_C = 1.01 \text{ ma} - 1.00 \text{ ma} = 0.01 \text{ ma}$

Next we may thevenize the R_1, R_2 voltage divider (sometimes called a base bleeder) to obtain Fig. PS 8-5*b*. Therefore,

6) $$V_{OC} = V_E + 0.65 \text{ volt} + R_B I_B$$

We do not know V_{OC} or R_B in equation 6. We do, however, know that

7) $$V_{OC} = \frac{R_2}{R_1 + R_2} V_{CC}$$

and

8) $$R_B = \frac{R_1 R_2}{R_1 + R_2}$$

Also we are given

9) $$S = 5 = \frac{1 + (R_E/R_B)}{[1/(\beta+1)] + (R_E/R_B)}$$

If equation 9 is solved for the ratio R_E/R_B, we obtain

10) $$\frac{R_E}{R_B} = \frac{\beta + 1 - S}{(\beta+1)(S-1)}$$
$$= \frac{49 + 1 - 5}{(49+1)(5-1)} = 0.225$$

Substituting 4 into 10 and solving for R_B yields

11) $$R_B = R_E/0.225 = 3.96 \text{ kilohms}/0.225$$
$$= 17.6 \text{ kilohms}$$

Substituting 3, 5, and 11 into 6 yields

12) V_{OC} = 4 volts + 0.65 volt + 17.6 kilohms (0.01 ma) = 4.83 volts

Substituting 12 into 7 yields

13) $$4.83 \text{ volts} = \frac{R_2}{R_1 + R_2} 12 \text{ volts}$$

Solving 13 for $R_2/(R_1 + R_2)$ yields

14) $$\frac{R_2}{R_1 + R_2} = \frac{4.83 \text{ volts}}{12 \text{ volts}} = 0.402$$

Substituting 11 and 14 into 8, we obtain

15) $$17.6 \text{ kilohms} = R_1 (0.402)$$

from which we obtain

16) $$R_1 = \frac{17.6 \text{ kilohms}}{0.402} = 43.8 \text{ kilohms}$$

Solving 8 for R_2 yields

17) $$R_2 = \frac{R_1 R_B}{R_1 - R_B}$$

$$= \frac{43.8 \text{ kilohms } (17.6 \text{ kilohms})}{43.8 \text{ kilohms} - 17.6 \text{ kilohms}} = 29.4 \text{ kilohms}$$

The circuit has now been designed to meet the required specifications.

PS 8-6 Estimate the Q point (V_{EC} and I_C) of the transistor in the circuit of Fig. PS 8-6*a*.

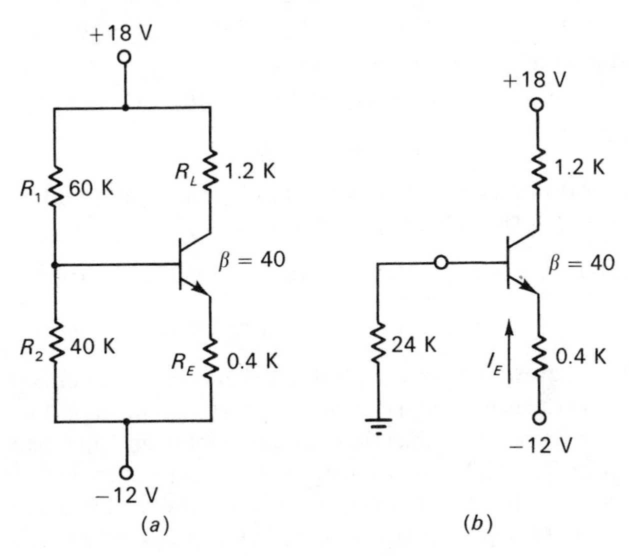

FIGURE PS 8-6

SOLUTION First thevenize the 60-kilohm and 40-kilohm base bleeder resistors to simplify the circuit as shown in Fig. PS 8-6*b*. If you are rusty on Thévenin's theorem, review it in "Outline for DC Circuit Analysis" in this series of texts. From Fig. PS 8-6*b* we obtain

1) $$I_E = \frac{12 \text{ volts} - 0.65 \text{ volt}}{0.4 \text{ kilohm} + [24 \text{ kilohms}/(40 + 1)]}$$
$$= 11.52 \text{ ma}$$

2) $$I_C = {}^{40}/_{41}\,(11.52 \text{ ma}) = 11.24 \text{ ma}$$

3) $$V_{EC} = -R_E I_E - V_{EE} + V_{CC} - R_L I_C = -4.61 \text{ volts} + 12 \text{ volts} + 18 \text{ volts} - 13.48 \text{ volts}$$
$$= 11.9 \text{ volts}$$

PS 8-7 Determine the transistor Q point in the circuit of Fig. PS 8-7*a* if the nonlinear load resistance R_{NL} has the *V-I* curve of Fig. PS 8-7*b*. Assume a silicon transistor.
SOLUTION Thévenize the base circuit to obtain Fig. PS 8-7*c*. Assuming active-region operation,

1) $$I_E = \frac{6 \text{ volts} - 0.65 \text{ volt}}{2 \text{ kilohms} + [30 \text{ kilohms}/(59 + 1)]}$$
$$= \frac{5.35 \text{ volts}}{2.5 \text{ kilohms}} = 2.14 \text{ ma}$$

2) $$I_C = \alpha I_E = \frac{59}{59 + 1}(2.14 \text{ ma}) = 2.10 \text{ ma}$$

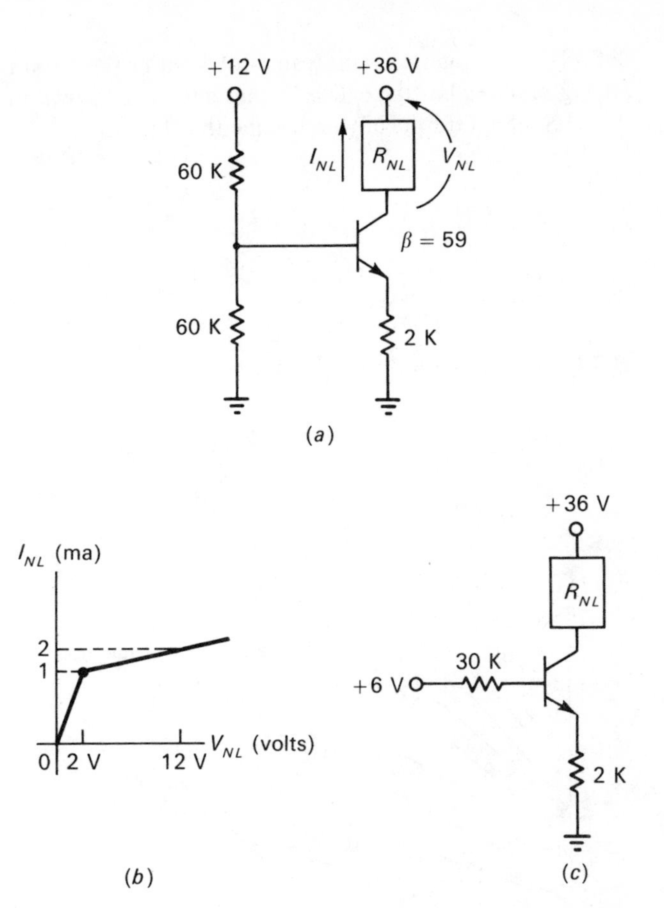

FIGURE PS 8-7

This collector current will approximate a constant-current source to the load. This means we are entering the *V-I* characteristic of Fig. PS 8-7*b* horizontally at a current value of $I_{NL} = 2.1$ ma. To determine the corresponding voltage V_{NL}, we note that for $I_{NL} > 1$ ma the incremental resistance is

3) $$r_{NL} = \frac{12 \text{ volts} - 2 \text{ volts}}{2 \text{ ma} - 1 \text{ ma}} = 10 \text{ kilohms}$$

Therefore for $I_{NL} > 1$ ma,

4) $$V_{NL} = 2 \text{ volts} + r_{NL}(\Delta I_{NL})$$
$$= 2 \text{ volts} + 10 \text{ kilohms } (2.1 \text{ ma} - 1 \text{ ma})$$
$$= 13 \text{ volts}$$

We may therefore determine

5) $$V_{EC} = -R_E I_E + V_{CC} - V_{NL}$$
$$= -2 \text{ kilohms } (2.14 \text{ ma}) + 36 \text{ volts} - 13 \text{ volts}$$
$$= 18.7 \text{ volts}$$

Since V_{EC} is a positive quantity, the collector diode in this *NPN* transistor is reverse-biased and active-region operation is assured.

If the concept of incremental resistance is unclear, review Chapter 4 in "Outline for DC Circuit Analysis" in this series of texts.

PS 8-8 The germanium transistor used in the circuit of Fig. PS 8-8*a* has the collector characteristics shown in Fig. PS 8-8*b*. Graphically determine the Q point.

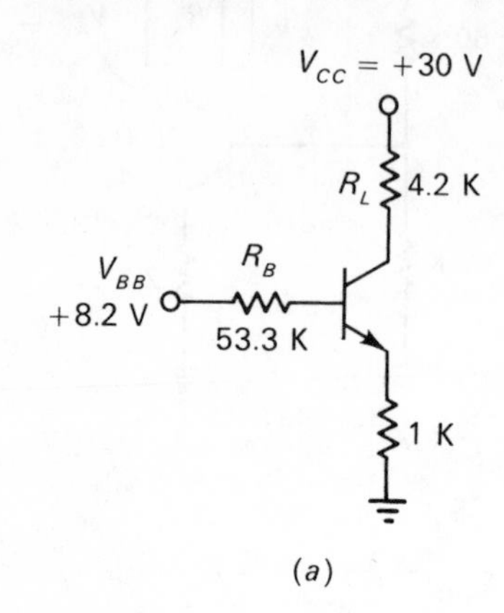

(*a*)

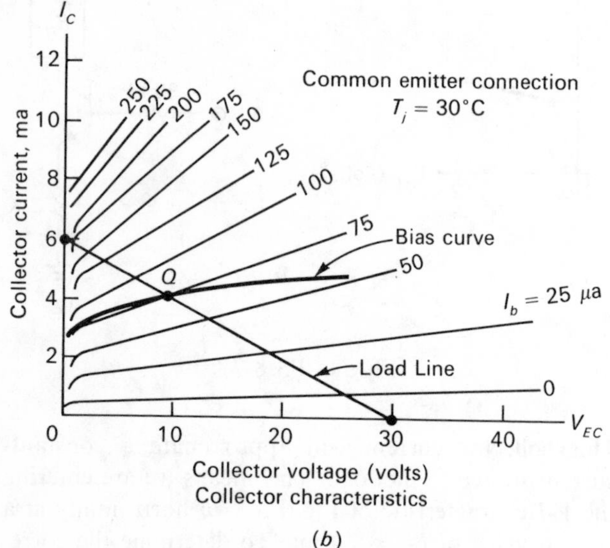

(*b*)

FIGURE PS 8-8

SOLUTION We may write

1) $$V_{EC} = V_{CC} - R_L I_C - R_E I_E$$

but $I_E \approx I_C$, therefore

2) $$V_{EC} \approx V_{CC} - (R_L + R_E) I_C$$

Equation 2 is a linear equation called the load line equation. It may be conveniently plotted on the collector characteristics by first assuming $I_C = 0$, which means $V_{EC} = V_{CC}$. This is the V_{EC} axis intercept of the load line in Fig. PS 8-8*b*. Next let $V_{EC} = 0$ and solve for I_C to obtain

3) $$I_C = \frac{V_{CC}}{R_L + R_E} = \frac{30 \text{ volts}}{4.2 \text{ kilohms} + 1 \text{ kilohm}}$$
$$= 5.77 \text{ ma} \approx 6 \text{ ma}$$

We use the approximation because in graphic work there are so many inherent errors anyway. The operating point may lie anywhere on this load line. To determine just where on the load line the Q point sits, we may construct a bias curve based on the equation

4) $$V_{BB} = R_B I_B + R_E I_E + V_{EB} \approx R_B I_B + R_E I_C + V_{EB}$$

Equation 4 may be solved for I_C to obtain

5) $$I_C = \frac{V_{BB} - V_{EB} - R_B I_B}{R_E} = \frac{V_{BB} - V_{EB}}{R_E} - \frac{R_B}{R_E} I_B$$

Assuming a germanium transistor and $V_{EB} = 0.2$ volt, and substituting values, equation 5 reduces to

6) $$I_C = 8 \text{ ma} - 53.3 I_B$$

Substituting I_B values and calculating the corresponding I_C values yields the following table

(ma) I_B	0	0.025	0.05	0.075	0.10
(ma) I_C	8	6.67	5.34	4.00	2.67

If the above values are plotted (whenever possible) and then connected to form a smooth curve as shown in Fig. PS 8-8*b*, we note that the bias curve intercepts the load line at $V_{EC} \approx 9.2$ volts and $I_C \approx 4$ ma. Since these values represent the simultaneous solution to the load line and bias curve equations (equations 2 and 5 respectively) this locates the Q point.

Actually a graphic solution of this nature doesn't mean very much due to the large spread in actual transistor characteristics from the nominal curve shown.

PS 8-9 Estimate V_{EC} in Fig. PS 8-9. Assume a silicon transistor.

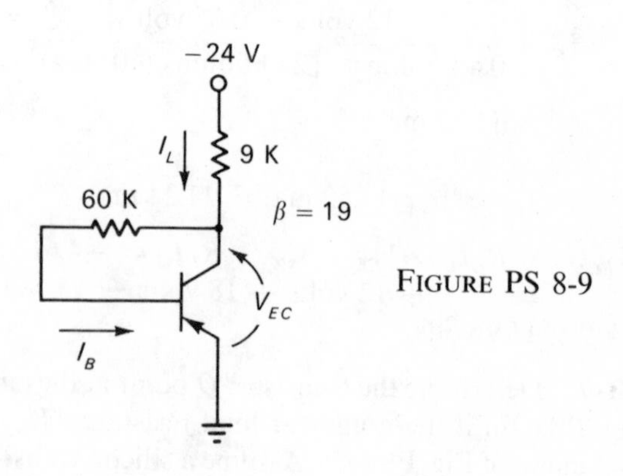

FIGURE PS 8-9

SOLUTION

1) $$V_{EC} = -24 \text{ volts} + 9 \text{ kilohms } I_L$$

2) $$I_L = I_B + I_C$$

3) $$I_B = \frac{-V_{EC} + 0.65 \text{ volt}}{60 \text{ kilohms}}$$

4) $$I_C = 19\, I_B$$

Substituting 4 into 2 yields

5) $$I_L = 20\,I_B$$

Substituting 3 into 5 yields

6) $$I_L = \frac{20}{60\text{ kilohms}}(-V_{EC} + 0.65\text{ volt})$$

Substituting 6 into 1 and solving for V_{EC} yields

7) $$V_{EC} = -24\text{ volts} + 9\text{ kilohms}\left[\frac{20}{60\text{ kilohms}}(-V_{EC} + 0.65\text{ volt})\right]$$

which simplifies to

8) $$V_{EC} = -5.5\text{ volts}$$

If the V_{EB} of 0.65 volt is neglected, the result is $V_{EC} \approx -6$ volts, which is a good enough estimate for most applications.

PROBLEMS WITH ANSWERS

PA 8-1 Estimate I_C in Fig. PA 8-1 at 25°C and 65°C.

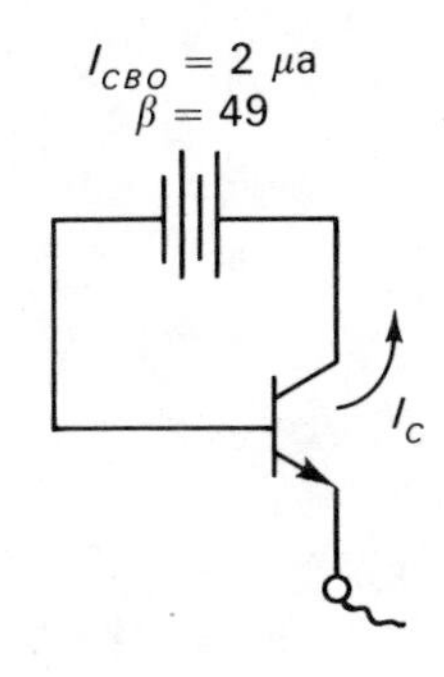

ANSWER 2μa at 25°C; 32 μa at 65°C

PA 8-2 Estimate I_C in Fig. PA 8-2 at 25°C and 65°C.

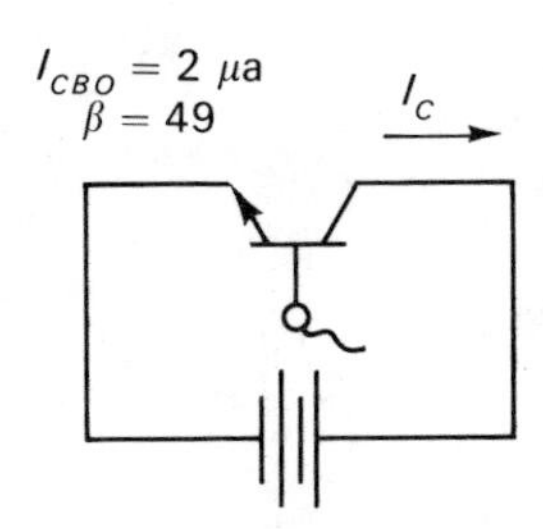

ANSWER 0.1 ma at 25°C; 1.6 ma at 65°C

PA 8-3 Estimate I_C and V_{EC} in Fig. PA 8-3.

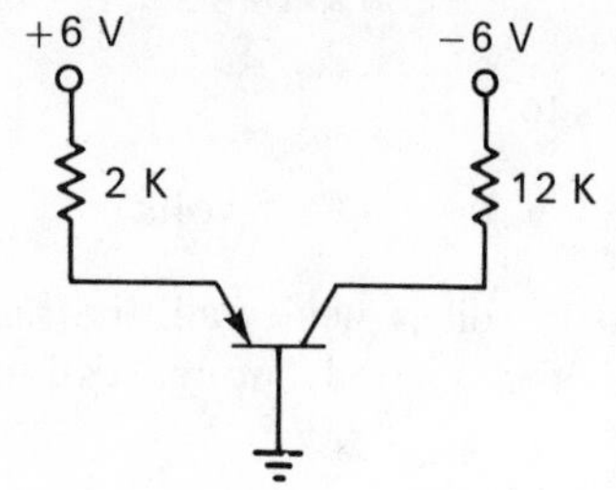

ANSWER

$I_C \approx 0.5$ ma
$V_{EC} \approx 0$

PA 8-4 Estimate I_C and V_{EC} in Fig. PA 8-4.

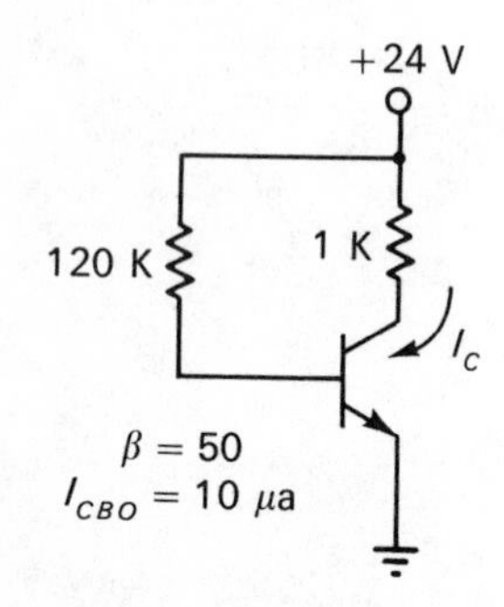

ANSWER

$I_C = -10.5$ ma
$V_{EC} = 13.5$ volts

PA 8-5 Estimate I_C and V_{EC} in Fig. PA 8-5.

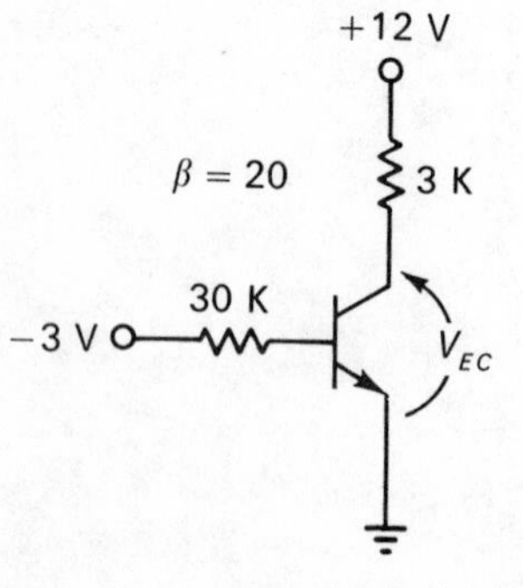

ANSWER

$I_C \approx 0$
$V_{EC} = 12$ volts

PA 8-6 Estimate I_C and V_{EC} in Fig. PA 8-6.

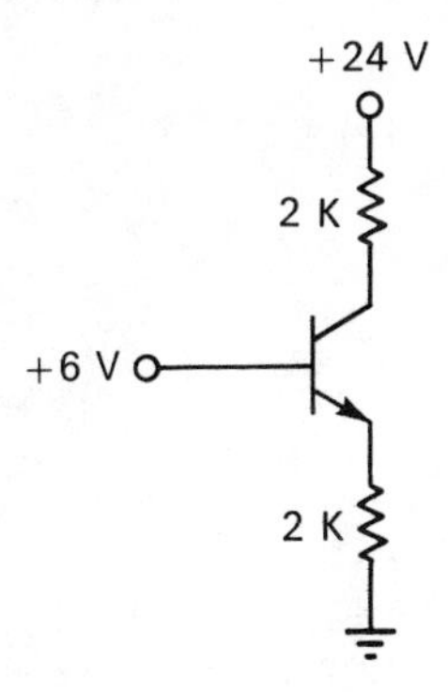

ANSWER $I_C \approx 3$ ma
$V_{EC} \approx 12$ volts

PA 8-7 Estimate V_{EC} in Fig. PA 8-7.

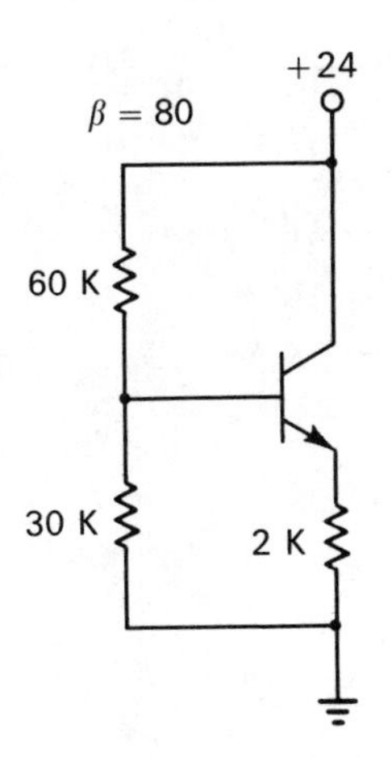

ANSWER $I_C \approx 3.6$ ma
$V_{EC} \approx 16.9$ volts

PA 8-8 Estimate V_B and V_{EC} in Fig. PA 8-8. Assume $V_{EB} = 0.6$ volt.

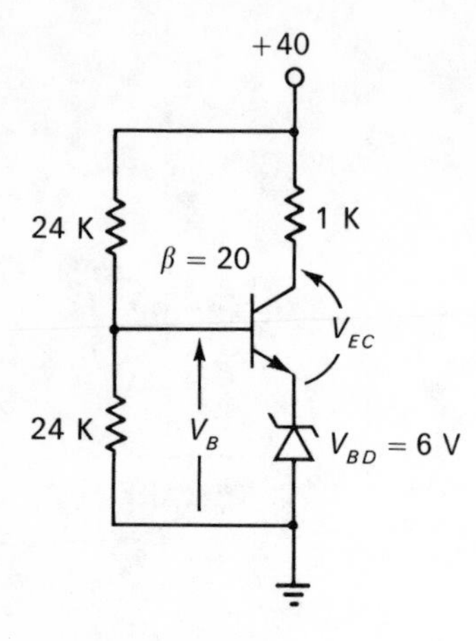

ANSWER $V_B = 6.6$ volts
$V_{EC} = 11.7$ volts

PA 8-9 Estimate E_O in Fig. PA 8-9.

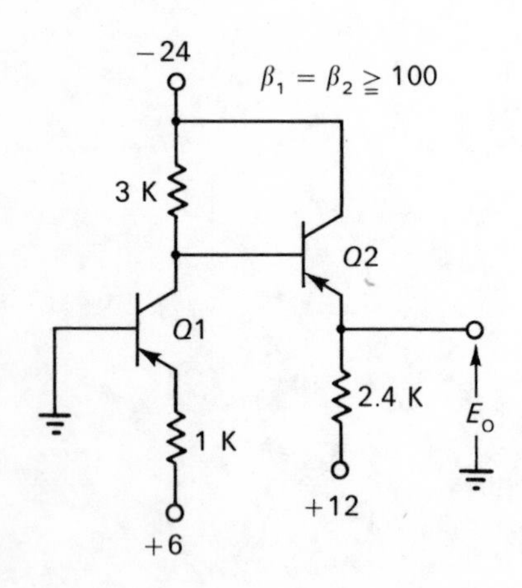

ANSWER $E_O = -6$ volts

PA 8-10 Estimate V_{EC} in Fig. PA 8-10.

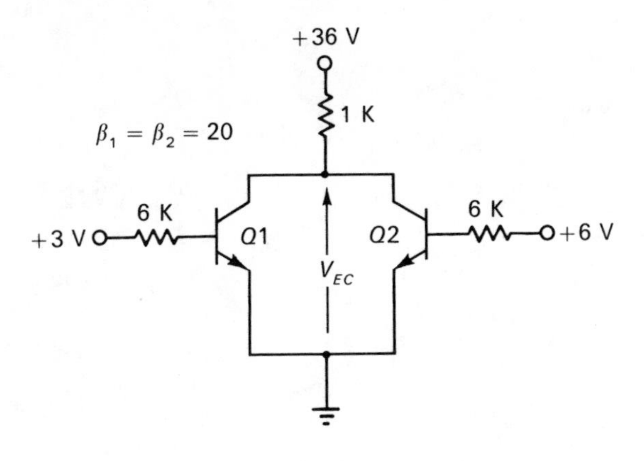

ANSWER V_{EC} = 6 volts

PA 8-11 What is the stability factor in Fig. PA 8-11? What is the change in S if R_L is doubled?

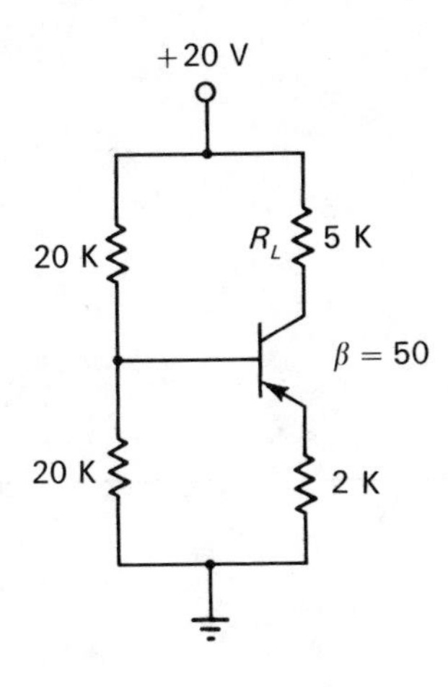

ANSWER S = 5.46; no change in S

PA 8-12 If $I_{CBO} = 2\ \mu a$, $\beta = 50$, and $V_{EB}' = 0.24$ volt at 25°C, estimate ΔV_{EC} if the temperature rises to 45°C.

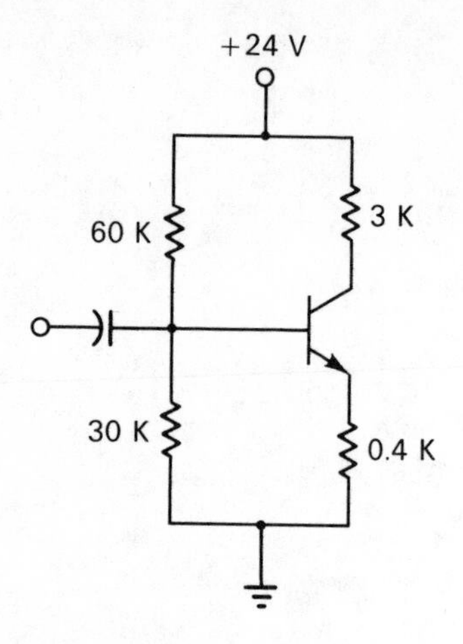

ANSWER $\Delta V_{EC} \approx -0.74$ volt

PROBLEMS WITHOUT ANSWERS

P 8-1 Determine V_{EC} in Fig. P 8-1.

P 8-2 Determine I in Fig. P 8-2.

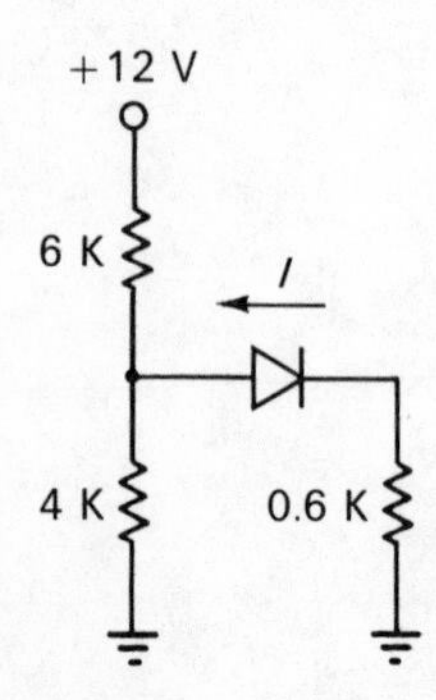

P 8-3 Suppose the stability factor of an amplifier stage is 10 and $I_{CBO} = 5\ \mu a$ at room temperature. At some elevated temperature $I_{CBO} = 10\ \mu a$. What is the resultant change in collector current equal to?

P 8-4 What is the impedance seen looking into the A-B terminals in Fig. P 8-4?

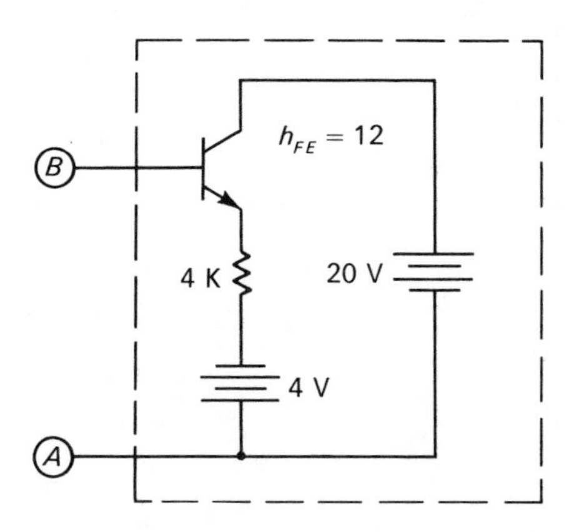

P 8-5 Determine V_{EC} and I_C in Fig. P 8-5.

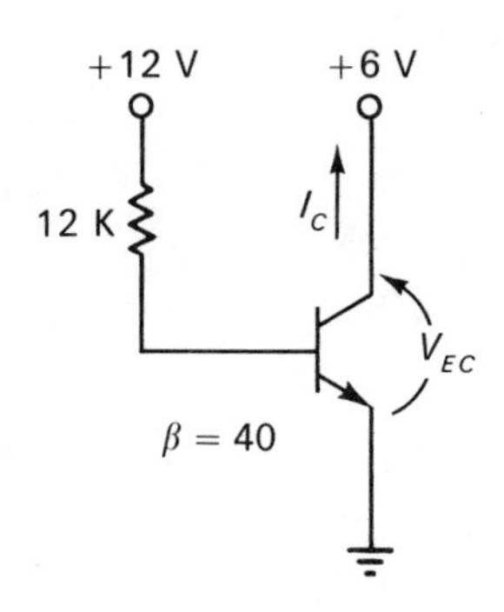

P 8-6 Estimate V_{EC} in Fig. P 8-6.

+6 V

6 K

V_{EC}

4 K

−12 V

P 8-7 Estimate I_C in Fig. P 8-7.

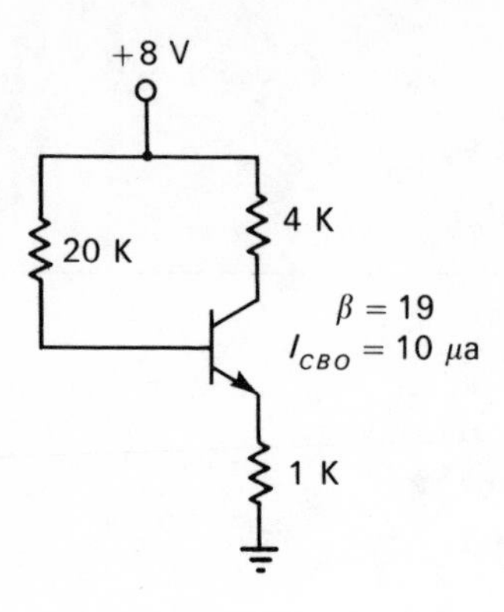

P 8-8 Estimate I_C in Fig. P 8-8.

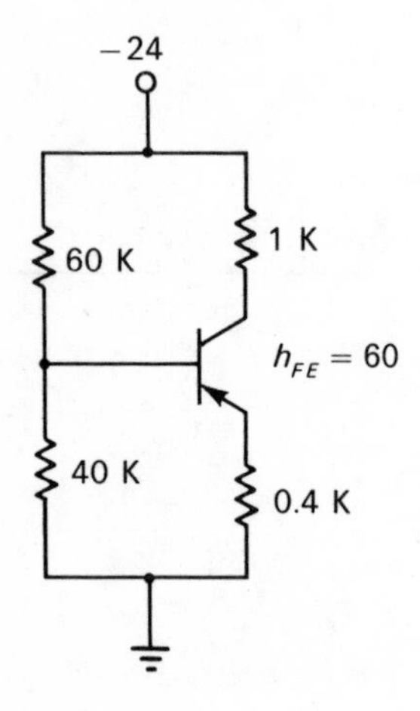

P 8-9 Estimate I_C in Fig. P 8-9.

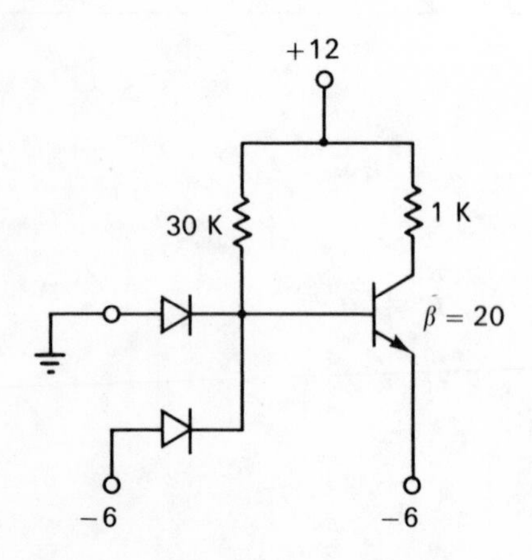

P 8-10 Estimate the maximum value of R_L that may be used in this circuit without the transistor going into saturation.

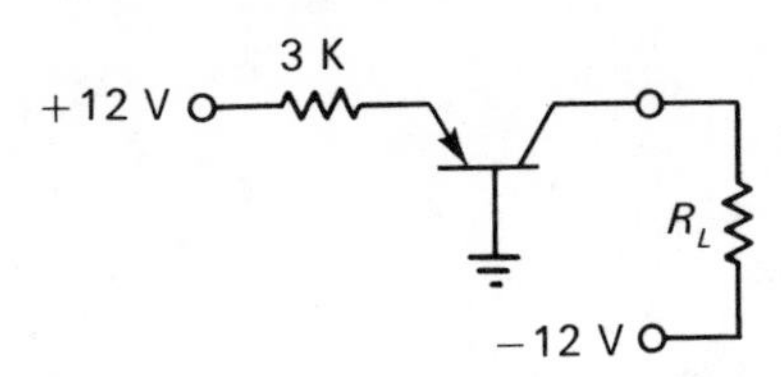

P 8-11 Estimate E_O.

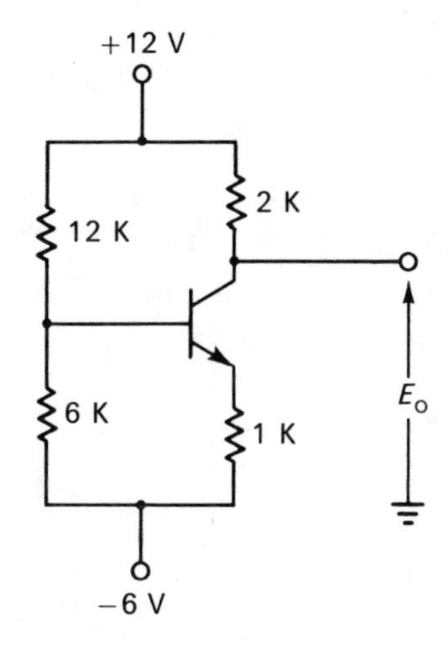

P 8-12 Determine R_B and R_L for $V_{EC} = 5$ volts and $I_C = 1$ ma, assuming $\beta = 50$ and $I_{CBO} = 5\ \mu$a. Assume a silicon transistor.

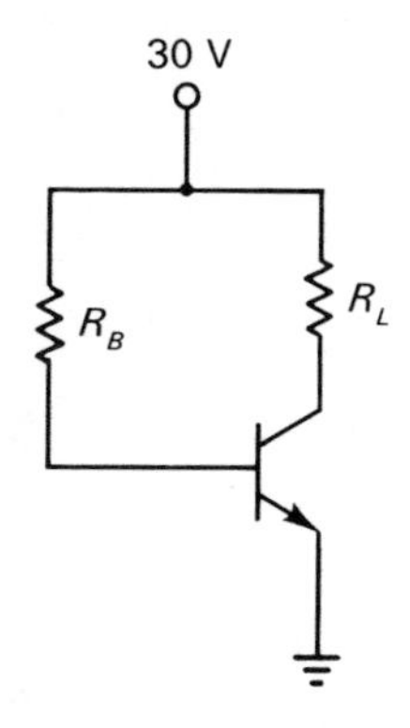

9 transistor incremental models

In Chapters 6 and 7 we developed the common-base and common-emitter models of the transistor. These models were primarily based upon piecewise linear approximations to the transistor volt-ampere characteristics. The parameters of the resultant model therefore inherently represent average values that approximately fit the nonlinear device characteristics over a large operating region. For this reason the piecewise derived models may also be considered *large-signal models.* By large-signal models we mean models that may be used to study the device behavior when it is driven by some excitation that causes the operating point to swing over a large region of the volt-ampere characteristic.

In this chapter we will consider some small-signal incremental models of the transistor. Although some of these small-signal models may, in some respects, look similar to the large-signal models, there are important differences. First of all these are incremental models or ac equivalent circuits of the transistor. This means only parameters which are involved in determining the device response to *changes* in voltage or current are relevant. Thus the piecewise model parameters such as V_{EB}' and I_{CBO} will be omitted since these are static (dc) quantities and have no significance insofar as any signal variations are concerned. Likewise the V_{CC}, V_{EE}, and other power supplies may be replaced by their ac impedance which is usually a short-circuit for most practical purposes. In essence, the superposition theorem is being applied and only the ac or incremental parameters are being considered.

A second difference between the large-signal piecewise models and the small signal incremental models is that the parameter values may grossly differ at some particular operating point. For example, the slope of a piecewise linear approximation to the V-I curve of a forward-biased diode might correspond to a forward resistance $r_F = 10$ ohms. However, we know that on a small-signal basis the ac forward resistance is $r_f \approx 26$ mv/I_F, so that if the dc forward current $I_F = 1$ ma, $r_f = 26$ ohms, but at $I_F = 10$ ma, $r_f = 2.6$ ohms. Clearly r_F and r_f may vary considerably.

Note that lowercase subscripts are generally reserved for small-signal incremental models. In some cases the same symbol, for instance α and β, is used in both small- and large-signal models. However, the nature of the problem or other symbols used will clearly indicate whether small- or large-signal values are implied. It should be understood that there is no sharp distinction between small and large signals. Common sense and the parameter values available usually determine the choice of model used. Large-signal parameters may usually be determined from piecewise approximations to V-I curves whereas small-signal parameters are usually measured.

For detailed information on transistor small-signal models the reader is referred to Chapter 5 of "Semiconductor Circuit Analysis" by Phillip Cutler, published by McGraw-Hill. Otherwise, the following condensation of transistor small signal models will generally prove adequate.

9-1 The common-base *T* parameter model

Figure 9-1*a* shows an intuitive incremental (small-signal) transistor model known as the common-base

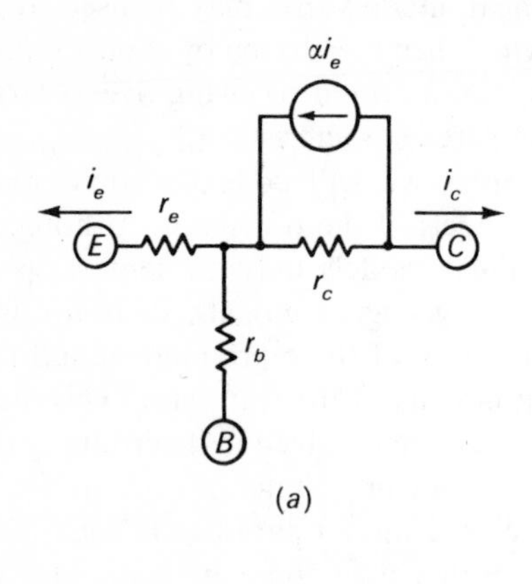

(*a*)

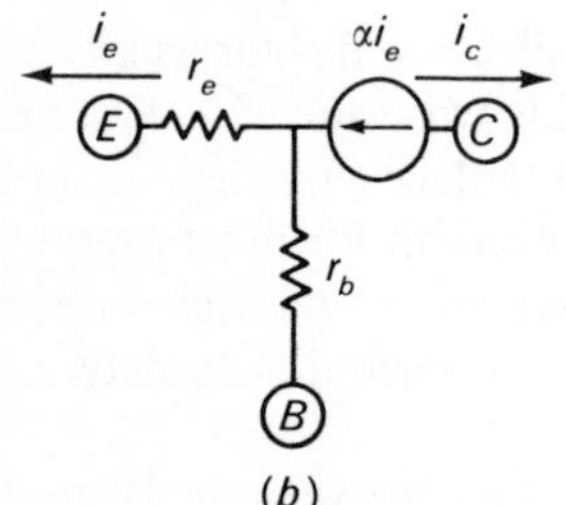

(*b*)

FIGURE 9-1

model. This and all subsequent incremental models are valid for both *PNP* and *NPN* transistors. Since these are ac equivalent circuits we don't care which way the dc currents are flowing so *PNP* or *NPN* doesn't matter. All we are concerned with are current and voltage *variations* about some quiescent operating point.

The common-base model may be recognized by the αi_e dependent current source and its shunt resistance r_c. The current directions for i_e, i_b, and i_c are purely arbitrary, but it is fairly common practice to choose them all flowing out of the transistor when electron flow is used or all into the transistor with conventional current flow. Since we are using electron flow, we have the currents i_e and i_c directed out of the device. It is not necessary to indicate i_b in the common-base model.

A most important point is that, although the external current reference directions for the device may be chosen in a purely arbitrary manner, the αi_e *dependent* current source direction is *not* arbitrary since it depends on i_e. Thus if i_e is directed counterclockwise as shown in Fig. 9-1*a*, so must the αi_e current source be similarly directed. This must be so because that component of injected emitter current that crosses the collector junction follows the emitter current in phase if we neglect transit time delays in the device.

In fact, the whole common-base incremental model bears a close resemblance to the construction and physical behavior of the device. For example, starting at the emitter we should expect some incremental resistance r_e associated with the bulk resistance of the emitter region and the dynamic resistance of the emitter junction. We might even somewhat correctly suspect that since the emitter junction approximates a forward-biased diode in the active region, that $r_e \approx (kT/q)/I_E$ where $kT/q = 26$ mv at 27°C and where I_E is the dc emitter forward current.

Certainly we would expect to see some incremental base resistance r_b, if only because the base region is lightly doped to begin with. We would also expect some apparent incremental leakage resistance r_c to manifest itself across the reverse-biased collector diode. Actually r_b and r_c are somewhat more complicated than in the simple explanation given here. However, this will do for our purposes as all we need to accept at this time is the presence of some r_b and r_c. The αi_e dependent current source simply allows for the fact that if you force more or less i_e into the transistor, more or less of the injected emitter current will correspondingly cross the collector junction.

It turns out that in most cases r_c is so very large compared to any external circuit parameters that for all practical purposes it may be neglected as shown in Fig. 9-1*b*. This means that looking into the collector lead we see an essentially constant-current source.

The incremental *T* parameters (sometimes called the *r* parameters) are rather difficult to measure and modern data sheets seldom list them. They may, however, be determined from other readily measured parameters. For low-power transistors, say 500 mw or less, typical values of the common-base *T* parameters at a *Q* point at $I_C = 1$ ma and $V_{BC} = 5$ volts might be: $r_e = 15$ ohms, $r_b = 400$ ohms, $r_c = 2$ megohms, $\alpha = 0.98$. Note r_e may be about half the value predicted by 26 mv/I_E.

9-2 The common-emitter *T* parameter model

Quite frequently it is the base current rather than the emitter current which is readily identified. In that case it is more convenient to have a transistor incremental model in which the dependent current source is a function of the base rather than the emitter current. Such a model is shown in Fig. 9-2*a* and it is called the common-emitter model. The external device current reference directions in this model are again purely arbitrary although we will stick to choosing all the currents flowing out of the device. However, i_b and the βi_b current sources are always

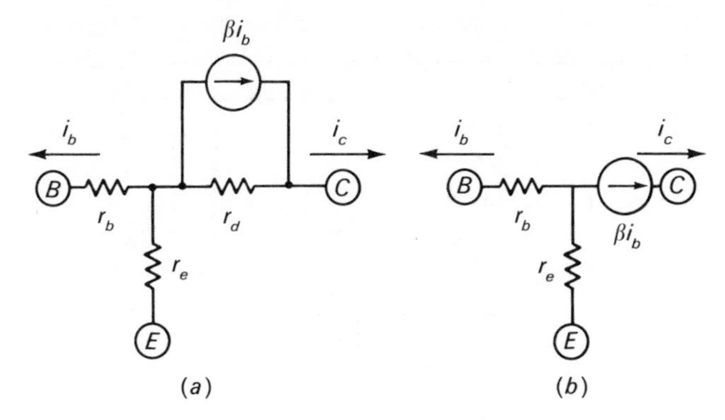

FIGURE 9-2

oppositely directed. This seems reasonable since the injected emitter current divides into two internal components, one which exits out the base lead and another which exits out the collector due to the βi_b current source.

Resistors r_e and r_b in this T model are the same as in the common-base T model. However, the collector-current source shunt resistance r_d in this common-emitter T model is not the same as r_c in the common-base version. It turns out that when the βi_b current source is used, the shunt resistance changes from r_c to r_d. For the moment, this is offered without proof. We might, however, feel more secure since

$$\beta = \frac{\alpha}{1 - \alpha} \tag{9-1}$$

$$\beta + 1 = \frac{1}{1 - \alpha} \tag{9-2}$$

$$r_d = r_c(1 - \alpha) = \frac{r_c}{\beta + 1} \tag{9-3}$$

and these relationships for the small-signal T model are quite similar in form to their large-signal counterparts that were previously developed.

Regardless of the fact that r_d is considerably less than r_c, r_d will normally be sufficiently large with respect to any collector external impedances so as to cause the collector to appear as a constant-current source to the load of the common-emitter model shown in Fig. 9-2*b*. Considerable labor may be saved with this approximation, and in view of the uncertainties involved in specifying these parameters, the approximation seems justified.

PROBLEM Given the relation $\beta = \alpha/(1 - \alpha)$, prove that $r_d = r_c/(\beta + 1)$ in the common-emitter T parameter model.

SOLUTION As a starting point, we might reason that irrespective of which model we use to portray the transistor, the emitter-to-collector voltage v_{ec} must be same in both cases. In other words v_{ec} in the common-base model of Fig. 9-1*a* should be identical to v_{ec} in the common-emitter model of Fig. 9-2*a* since the terminal voltages must be the same if each model is truly an equivalent circuit of the device. Thus by converting the αi_e current source and r_c to an equivalent voltage source, we may write for Fig. 9-1*a*

1) $$v_{ec} = -r_e i_e + r_c \alpha i_e + r_c i_c$$

but

2) $$i_e = -(i_b + i_c)$$

Substituting 2 into 1 we obtain, after some manipulation,

3) $$v_{ec} = r_e(i_b + i_c) + r_c(1 - \alpha)\, i_c - r_c \alpha i_b$$

Substituting equations 9-1 and 9-2 into 3 yields

4) $$v_{ec} = r_e(i_b + i_c) + \frac{r_c}{\beta + 1} i_c - \frac{r_c \beta}{\beta + 1} i_b$$

Substituting 9-3 into 4 we obtain

5) $$v_{ec} = r_e(i_b + i_c) + r_d i_c - r_d \beta i_b$$

6) $$v_{ec} = r_e(i_b + i_c) + r_d(i_c - \beta i_b)$$

Now if equation 6 is given a physical interpretation, it would mean that starting at the emitter and algebraically summing voltages as we follow a path to the collector, we first go up in potential by $r_e(i_b + i_c)$ volts. Looking at Fig. 9-2*a* we see that if i_b and i_c are considered as loop currents, they would flow through r_e in the same direction and give rise to a voltage across r_e given by the first term of equation 5.

Now the second term of equation 6 implies a voltage rise across r_d due to a net current, $i_c - \beta i_b$, through it. If we look at Fig. 9-2*a* and apply Kirchhoff's current law at the collector node, it follows that the net current through r_d is $i_c - \beta i_b$. Hence, Fig. 9-2*a* mechanizes equation 6 which was derived from Fig. 9-1*a* and thereby proves equation 9-3.

9-3 The *h* parameters

As previously indicated the T parameters of the transistor happen to be quite awkward to measure. It turns out instead that, because the transistor input terminals present a low impedance (due to the forward-biased emitter diode) and the output terminals a high impedance (due to the reverse-biased collector diode), it becomes convenient to characterize it in terms of the *h* or hybrid parameters.[1]

To briefly acquaint ourselves with the hybrid parameters consider Fig. 9-3*a*. Here we have a circuit or device abstractly represented as a box with a pair of input (1-1) and output (2-2) terminals. Since the transistor

[1] See Phillip Cutler, "Outline for DC Circuit Analysis," chap. 15, and/or "Semiconductor Circuit Analysis," chap. 5, McGraw-Hill Book Company, New York, 1968 and 1964.

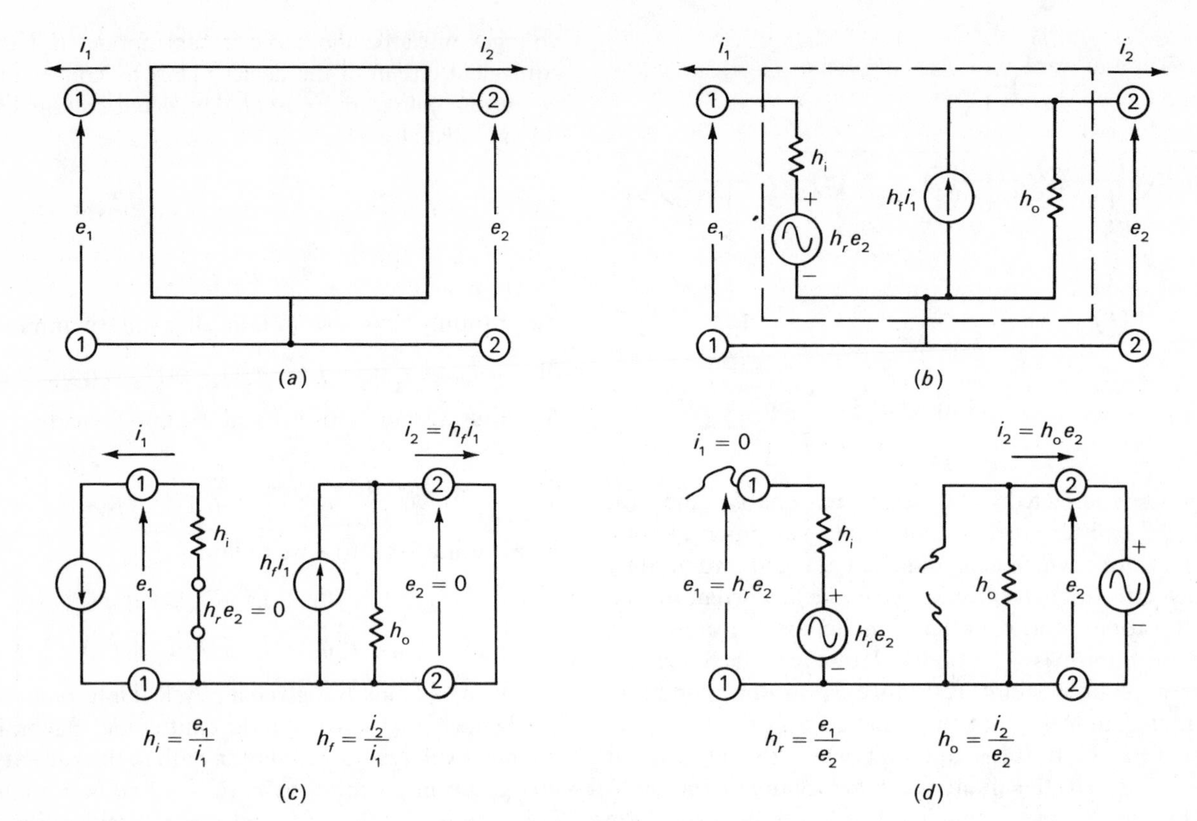

FIGURE 9-3

has only three terminals we will consider one of its terminals (the shaded one) common to both input and output.

Now, no matter what is in the box, it can be represented by the h parameter equivalent circuit of Fig. 9-3*b*. The terminal conditions for this model may be described by the following equations at the input and output terminals respectively:

1) $$e_1 = h_i i_1 + h_r e_2 \tag{9-4}$$

2) $$i_2 = h_f i_1 + h_o e_2 \tag{9-5}$$

If we just knew what the h parameter values were for a transistor at a desired quiescent operating point, we could analyze the circuit involved by well-known methods of circuit analysis.

Although this model may look strange at first, a little insight and practice in utilizing it will soon make it familiar. For a start, consider equation 1 in relation to Fig. 9-3*b*. Equation 1 states that if some current i_1 is made to flow out of the upper input terminal while some voltage e_2 appears across the output terminals, the voltage e_1 which appears across the input terminals will depend upon some impedance h_i through which i_1 flows and also upon some voltage $h_r e_2$.

The quantity h_i must have the dimensions of an impedance since multiplying h_i by i_1 in equation 1 must yield a voltage. From a physical viewpoint, we should certainly expect to see some impedance looking into the input terminals of the box, and h_i in part allows for this. However, h_i is not necessarily the same as the input impedance seen looking into the 1-1 terminals, since the $h_r e_2$ voltage source will also be a factor in determining conditions at the input terminals. The $h_r e_2$ voltage source allows for internal feedback in the device. For example, in a transistor, a change in V_{BC} causes the width of the collector depletion region to vary and this in turn means the base width is changing. If the base width varies, the apparent base resistance and α of the transistor vary and this should surely tend to be reflected at the input terminals in some manner. Since h_r is the coefficient of a voltage in equation 1, it must be a numeric or voltage gain as the product of h_r and e_2 must be a voltage. Whereas h_i is an impedance, h_r is a reverse (feedback) voltage gain.

The $h_f i_1$ dependent current source of Fig. 9-3*b* allows for the fact that a current forced into the input terminals will contribute to the current i_2 at the output terminals. From equation 2 we see that h_f must also be a numeric and it is called the forward current gain.

The parameter h_o (output admittance) in equation 2 must have the dimensions of an admittance since multiplying e_2 by h_o yields a current. Intuitively we might sense that whereas the $h_f i_1$ current source is somehow related to an αi_e or βi_b current source, h_o would be analogous to r_c or r_d, the shunt resistance across the current source.

Now how do we measure these h parameters? If in Fig. 9-3*a* a current source is connected across the input terminals to force a current i_1, and a short-circuit is connected across the output terminals to force $e_2 = 0$, we have the circuit of Fig. 9-3*c*. With $e_2 = 0$, $h_r e_2 = 0$ and $e_1 = h_i i_1$. Thus, if we force a known i_1 and measure the resultant e_1, we may evaluate $h_i = e_1/i_1$. On the output side, $i_2 = h_f i_1$, since all of the $h_f i_1$ current source will be dumped into the short across the output terminals. Thus we need only measure i_2 and i_1 to determine $h_f = i_2/i_1$. Note that h_i is the input impedance, and h_f is the current gain, *only* when the output is shorted.

If the input terminals are opened so that $i_1 = 0$, and an e_2 voltage source is placed across the output terminals, we obtain Fig. 9-3*d*. With $i_1 = 0$, it follows that $e_1 = h_r e_2$. Since we can measure e_1 and e_2, we can compute the reverse voltage gain $h_r = e_1/e_2$. Also in Fig. 9-3*d* we see that with $i_1 = 0$ the $h_f i_1$ current source is also zero. Replacing this current source with its internal impedance, we have an open circuit. Therefore, $i_2 = h_o e_2$. and since i_2 and e_2 are readily measured, we may determine the output admittance h_o from $h_o = i_2/e_2$. Note that h_r is the reverse voltage gain, and h_o is the output admittance, only with the input terminals open so that $i_1 = 0$. Be sure you remember that the *e*'s and *i*'s in this discussion only represent ac or incremental components. Thus, for example, when i_1 is set equal to zero it means there is no change in the static or dc value of I_1 which may be flowing through the input terminals.

9-4 The common-base *h* parameters

The h parameter equivalent circuit of Fig. 9-3*b* is perfectly general. If, however, we specifically wish to discuss the

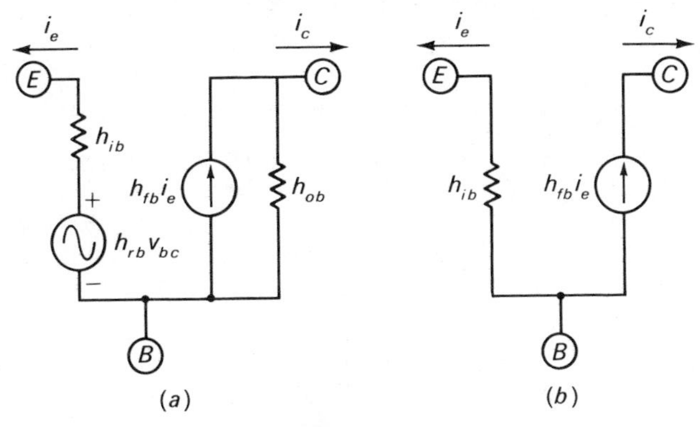

FIGURE 9-4

h parameters of a transistor connected in the common-base configuration, we append b subscripts to the h parameters, and thus $e_1 = v_{be}$, $i_1 = i_e$, $e_2 = v_{bc}$, and $i_2 = i_c$ as shown in Fig. 9-4*a*. Now it turns out that h_{rb} is typically on the order of 0.001 which means $h_{rb}v_{bc}$ may usually be assumed negligible. Also, h_{ob} is typically one thousand times smaller a conductance than any load the collector current might flow into. Hence h_{ob} may usually be neglected and the $h_{fb}i_e$ current source determines i_c as shown in Fig. 9-4*b*.

9-5 The common-emitter *h* parameters

If we wish to specifically discuss the h parameters of a transistor connected in the common-emitter configuration we append e subscripts to the h parameters, and $e_1 = v_{eb}$, $i_1 = i_b$, $e_2 = v_{ec}$, and $i_2 = i_c$ as shown in Fig. 9-5*a*. Typically, h_{re} and h_{oe} are so small that these

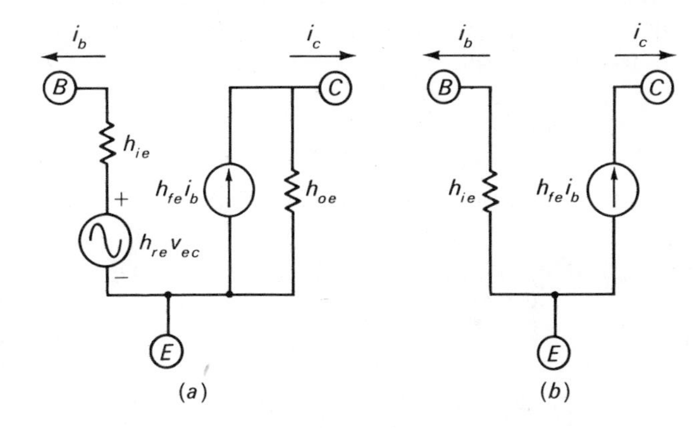

FIGURE 9-5

quantities may usually be neglected for the same reason they were in the common-base approximation of Fig. 9-4*b*. The simplified common-emitter h parameter model will then appear as shown in Fig. 9-5*b*.

For a more detailed discussion of these small-signal equivalent circuits see Chapter 5 of "Semiconductor Circuit Analysis" by Phillip Cutler, published by McGraw-Hill.

Be sure to do all the problems with solutions as they illustrate how the h and T parameters may be interrelated.

PROBLEMS WITH SOLUTIONS

PS 9-1 Assume you have just measured the common-base h parameters but actually want to construct the common-base T parameter equivalent circuit. How could you express these T parameters in terms of the h parameters?

SOLUTION One approach would be to somehow write some equations representing conditions at the terminals of the T model of Fig. 9-1*a* and then to manipulate the T equations so that they have the same independent and dependent variables as the h model equations 9-4 and 9-5. Once this is done we can compare the coefficients of the variables to relate the T and h parameters. For example, for Fig. 9-1*a* we may write the following loop equations if we mentally thevenize the αi_e current source and r_c, and do some simplifying.

1) $$v_{be} = (r_b + r_e)\, i_e + r_b i_c$$

2) $$v_{bc} = (r_b + \alpha r_c)\, i_e + (r_b + r_c)\, i_c$$

Now equations 9-4 and 9-5 are the general h parameter equations. For the common-base circuit they become (see Fig. 9-4*a*)

3) $$v_{be} = h_{ib} i_e + h_{rb} v_{bc}$$

4) $$i_c = h_{fb} i_e + h_{ob} v_{bc}$$

Comparing 1 and 3 we see that the second term in 1 requires v_{bc} instead of i_c as a variable. We can express i_c in terms of v_{bc} from equation 2 to obtain

5) $$i_c = \frac{v_{bc} - (r_b + \alpha r_c)\, i_e}{r_b + r_c}$$

Substituting 5 into 1 and simplifying yields

6) $$v_{be} = (r_b + r_e)\, i_e + \frac{r_b}{r_b + r_c}\left[v_{bc} - (r_b + \alpha r_c)\, i_e\right]$$

or

7) $$v_{be} = \left[r_e + r_b - \frac{r_b}{r_b + r_c}(r_b + \alpha r_c)\right] i_e + \frac{r_b}{r_b + r_c} v_{bc}$$

Next we may rewrite 5 into the form of 4 to obtain

8) $$i_c = -\left(\frac{r_b + \alpha r_c}{r_b + r_c}\right) i_e + \frac{1}{r_b + r_c} v_{bc}$$

We have now wrestled the T parameter equations into h parameter form. Comparing coefficients of the variables in the T equations, 7 and 8, with the h equations, 3 and 4, respectively, we may conclude that

9) $$h_{ib} = r_e + r_b - \frac{r_b}{r_b + r_c}(r_b + \alpha r_c) \approx r_e + r_b(1 - \alpha) \qquad (9\text{-}6)$$

10) $$h_{rb} = \frac{r_b}{r_b + r_c} \qquad (9\text{-}7)$$

11) $$h_{fb} = -\left(\frac{r_b + \alpha r_c}{r_b + r_c}\right) \approx -\alpha \qquad (9\text{-}8)$$

12) $$h_{ob} = \frac{1}{r_b + r_c} \approx \frac{1}{r_c} \qquad (9\text{-}9)$$

Equations 9-6 through 9-9 may be used to determine the h parameters from the T parameters. The approximate forms of 9-6 and 9-8 are based on the fact that in a practical transistor $\alpha r_c \gg r_b$, and $r_c \gg r_b$.

Our goal, however, was to express the T parameters in terms of the h parameters. Thus from 10 and 12 we obtain

13) $$r_c = \frac{1 - h_{rb}}{h_{ob}} \approx \frac{1}{h_{ob}} \qquad (9\text{-}10)$$

14) $$r_b = \frac{h_{rb}}{h_{ob}} \qquad (9\text{-}11)$$

Substituting 13 and 14 into the exact form of 11 and solving for α we obtain, after some manipulation,

15) $$\alpha = -\left(\frac{h_{fb} + h_{rb}}{1 - h_{rb}}\right) \approx -h_{fb} \qquad (9\text{-}12)$$

since $h_{fb} \gg h_{rb}$. Substituting 13, 14, and 15 into 9 and solving for r_e eventually yields

16) $$r_e = h_{ib} - \frac{h_{rb}}{h_{ob}}(1 + h_{fb}) \qquad (9\text{-}13)$$

PS 9-2 Write the loop equations for the h parameter model of Fig. 9-4*a* and compare coefficients with the loop equations for Fig. 9-1*a* to verify equations 9-10 through 9-13.

SOLUTION From Fig. 9-4*a* we may write, upon thevenizing the current source,

1) $$v_{be} = h_{ib} i_e + h_{rb} v_{bc}$$

2) $$v_{bc} = \frac{-h_{fb}}{h_{ob}} i_e + \frac{i_c}{h_{ob}}$$

Remember, h_{ob} is an admittance. Substituting 2 into 1 yields, after some manipulation,

3) $$v_{be} = \left(h_{ib} - \frac{h_{rb} h_{fb}}{h_{ob}}\right) i_e + \frac{h_{rb}}{h_{ob}} i_c$$

The loop equations for Fig. 9-1*a* are

4) $$v_{be} = (r_b + r_e)\, i_e + r_b i_c$$

5) $$v_{bc} = (r_b + \alpha r_c)\, i_e + (r_b + r_c)\, i_c$$

Comparing coefficients of 4 and 5 with 3 and 2, respectively, we obtain

6) $$r_b + r_e = h_{ib} - \frac{h_{rb} h_{fb}}{h_{ob}}$$

7) $$r_b = \frac{h_{rb}}{h_{ob}}$$

8) $$r_b + \alpha r_c = \frac{-h_{fb}}{h_{ob}}$$

9) $$r_b + r_c = \frac{1}{h_{ob}}$$

Equations 6 through 9 may be solved for r_e, r_c, and α to obtain solutions similar to those in PS 9-1.

PS 9-3 Still another approach for obtaining the interrelation between the T and h parameters would be to perform the h parameter measurements upon the T model of Fig. 9-1a. Do this and see if you can derive equations 9-6 through 9-9. Once these equations are obtained, they could be solved as in PS 9-1 to express the T parameters in terms of the h parameters.

SOLUTION The h parameters, you will recall, are, in general, measured by driving the input with a current source and the output with a voltage source. Thus the h parameter equations are

1) $$e_1 = h_i i_1 + h_r e_2$$

2) $$i_2 = h_f i_1 + h_o e_2$$

Inspection of these equations indicates that if we set $e_2 = 0$, meaning a short circuit across the output, we may obtain from 1 and 2, respectively,

3) $$h_i = \left.\frac{e_1}{i_1}\right|_{e_2=0}$$

4) $$h_f = \left.\frac{i_2}{i_1}\right|_{e_2=0}$$

And if we set $i_1 = 0$, we may obtain

5) $$h_r = \left.\frac{e_1}{e_2}\right|_{i_1=0}$$

6) $$h_o = \left.\frac{i_2}{e_2}\right|_{i_1=0}$$

Applying equations 3 and 4 to the common-base T model circuit, we obtain Fig. PS 9-3a which yields, by superposition,

$$v_{be} = [r_e + (r_b \parallel r_c)]\, i_e - (r_b \parallel r_c)\, \alpha i_e$$

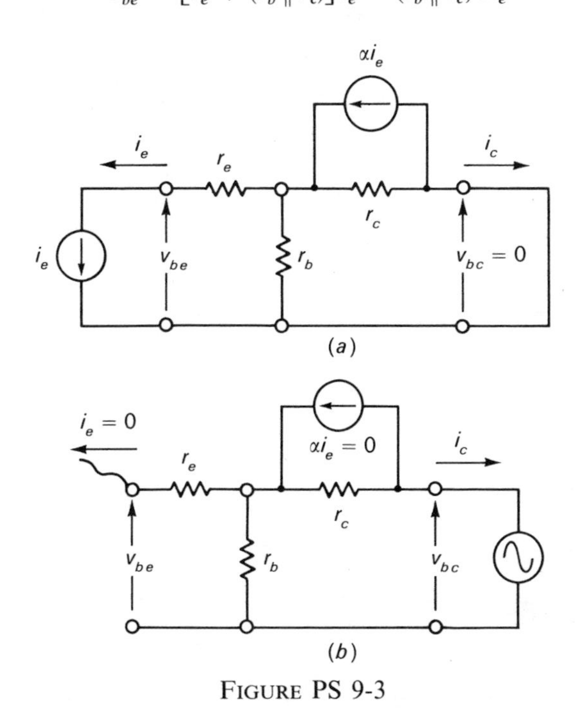

FIGURE PS 9-3

which in turn yields the ratio v_{be}/i_e or h_{ib}. Thus

7) $$h_{ib} = \frac{v_{be}}{i_e} = r_e + [r_b \parallel r_c(1 - \alpha)]$$
$$\approx r_e + r_b(1 - \alpha) \qquad \text{since } r_c \gg r_b$$

If 7 is expanded, it will prove to be identical with equation 9-6. Note, however, that this approach gives more physical insight into the nature of the h parameters as related to the T parameters.

Since $h_{fb} = i_c/i_e$ in Fig. PS 9-3a, we may write by superposition

8) $$i_c = \frac{-r_b}{r_b + r_c}\, i_e - \frac{r_c \alpha i_e}{r_b + r_c} = -\left(\frac{r_b + \alpha r_c}{r_b + r_c}\right) i_e$$

which yields

9) $$h_{fb} = \frac{i_c}{i_e} = -\left(\frac{r_b + \alpha r_c}{r_b + r_c}\right) \approx -\alpha$$

verifying equation 9-8.

Now if we open the input terminals so that $i_e = 0$ and place a v_{bc} voltage source across the output, we set up the circuit of Fig. PS 9-3b for evaluating h_{rb} and h_{ob}. Thus, with i_e equal to zero, and therefore $\alpha i_e = 0$, we have from the voltage-divider relationship

10) $$h_{rb} = \frac{v_{be}}{v_{bc}} = \frac{r_b}{r_b + r_c}$$

and by Ohm's law

11) $$h_{ob} = \frac{i_c}{v_{bc}} = \frac{1}{r_b + r_c}$$

which verifies equations 9-7 and 9-9.

PS 9-4 The common-base h parameters of a 2N43A *PNP* transistor at a Q point of $I_E = 1$ ma and $V_{BC} = -5$ volts are $h_{ib} = 28$ ohms, $h_{rb} = 4 \times 10^{-4}$, $h_{fb} = -0.9775$, and $h_{ob} - 0.5$ μmhos. Determine the exact and approximate T parameters.

SOLUTION From equation 9-10

$$r_c = \frac{1 - 0.0004}{0.5 \times 10^{-6}\text{ mho}} = 1.999 \text{ megohms} \qquad \text{(exactly)}$$

or

$$r_c = \frac{1}{0.5 \times 10^{-6}\text{ mho}} = 2 \text{ megohms} \qquad \text{(approximately)}$$

From equation 9-11

$$r_b = \frac{4 \times 10^{-4}}{0.5 \times 10^{-6}\text{ mho}} = 800 \text{ ohms}$$

From equation 9-12

$$\alpha = -\left(\frac{-0.9775 + 0.0004}{1 - 0.0004}\right) = 0.97749 \qquad \text{(exactly)}$$

or

$$\alpha = 0.9775 \qquad \text{(approximately)}$$

The fact that α is the negative of h_{fb} is due solely to the choice of reference directions for i_c and i_e. From equation 9-13

$$r_e = 28 - \frac{4 \times 10^{-4}}{0.5 \times 10^{-6}}(1 - 0.9775) = 10 \text{ ohms}$$

Note in particular how good the approximations for r_c and α are.

PS 9-5 Given the T parameter values of the previous problem, determine h_{ib} in its exact and approximate forms.
SOLUTION From equation 9-6

$$h_{ib} = 10 + 800 - \frac{800[800 + 0.97749(1{,}999{,}000)]}{800 + 1{,}999{,}000}$$

$$= 28 \text{ ohms}$$

for an exact solution, and

$$h_{ib} \approx 10 + 800\,(1 - 0.97749) = 28 \text{ ohms}$$

as an approximation. Since the approximation to h_{ib} is, in general, very close and considerably simpler, you should *memorize*

$$h_{ib} \approx r_e + r_b\,(1 - \alpha)$$

PS 9-6 Determine the h parameters for the network of Fig. PS 9-6*a*.
SOLUTION From Fig. PS 9-6*b*

$$h_i = \left.\frac{e_1}{i_1}\right|_{e_2=0} = \frac{3 \text{ kilohms } i_1}{i_1} = 3 \text{ kilohms}$$

$$h_f = \left.\frac{i_2}{i_1}\right|_{e_2=0} = \frac{20i_1 - i_1}{i_1} = 19$$

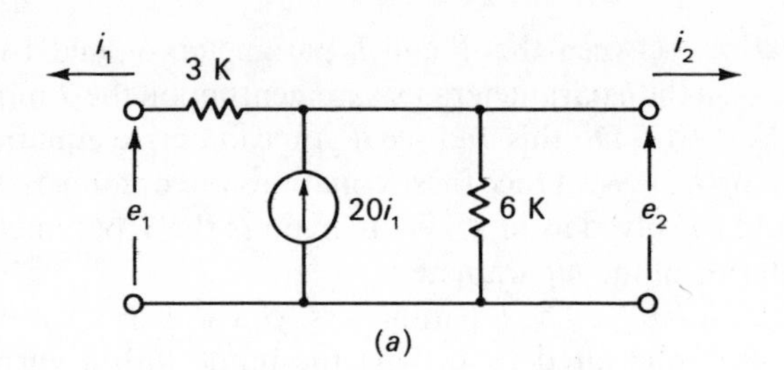

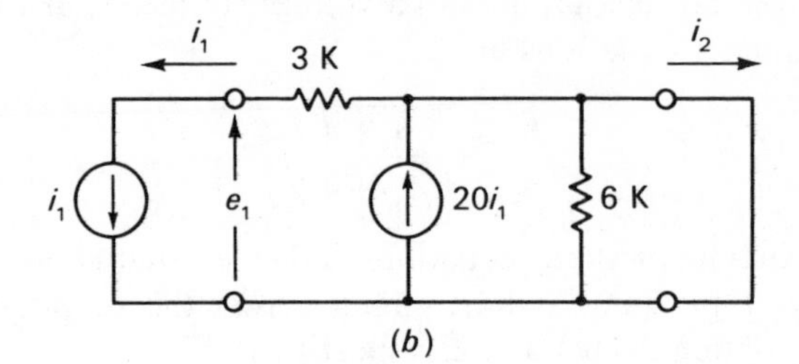

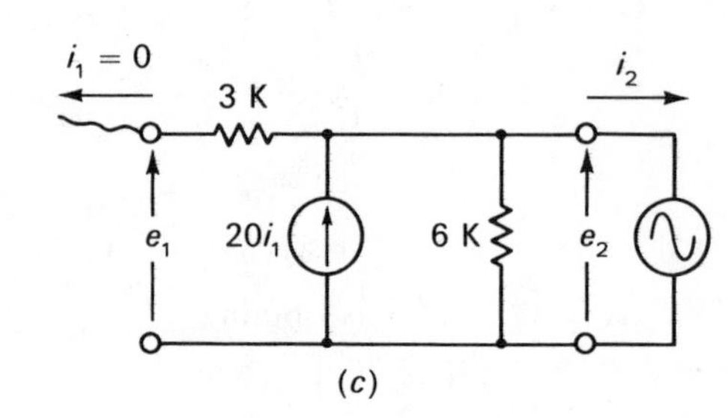

FIGURE PS 9-6

From Fig. PS 9-6*c*

$$h_r = \left.\frac{e_1}{e_2}\right|_{i_1=0} = \frac{e_2}{e_2} = 1$$

$$h_o = \left.\frac{i_2}{e_2}\right|_{i_1=0} = \frac{e_2/6 \text{ kilohms}}{e_2} = \frac{1}{6 \text{ kilohms}}$$

PROBLEMS WITH ANSWERS

PA 9-1 Using the approximate common-base Tmodel of Fig. 9-1*b* derive its h parameters.
ANSWER

$$h_{ib} = r_e + r_b\,(1 - \alpha)$$
$$h_{rb} = 0, \; h_{fb} = -\alpha$$
$$h_{ob} = 0$$

PA 9-2 Relate the common-emitter T parameters in the approximate model of Fig. 9-2b to the h parameters of Fig. 9-5b.

ANSWER

$$h_{ie} = r_b + r_e(\beta + 1)$$
$$h_{re} = 0, \; h_{fe} = \beta$$
$$h_{oe} = 0$$

PA 9-3 The common-base h parameters of a transistor are $h_{ib} = 31$ ohms, $h_{rb} = 5 \times 10^{-4}$, $h_{fb} = -0.978$, $h_{ob} = 0.6$ μmho. Estimate the common-base T parameters.

ANSWER

$$r_c = 1.67 \text{ megohms}$$
$$r_b = 833 \text{ ohms}$$
$$\alpha = 0.978$$
$$r_e = 12.7 \text{ ohms}$$

PA 9-4 Given the common-base h parameters of the previous problem, estimate the common-emitter h parameters.

ANSWER

$$h_{ie} = 1.4 \text{ kilohms}$$
$$h_{re} = 3.47 \times 10^{-4}$$
$$h_{fe} = 44$$
$$h_{oe} = 27.3 \; \mu\text{mho}$$

PA 9-5 Determine the h parameters for the network of Fig. PA 9-5.

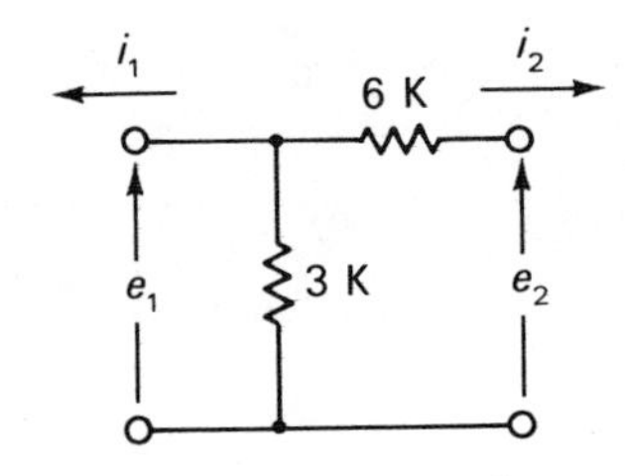

ANSWER

$$h_i = 2 \text{ kilohms}, \; h_r = 0.33$$
$$h_f = -0.33$$
$$h_o = \tfrac{1}{9} \text{ kilohm}$$

PA 9-6 Determine the h parameters for the network of Fig. PA 9-6.

ANSWER

$$h_i = R_1 (1 + k)$$
$$h_r = 0$$
$$h_f = k$$
$$h_o = 0$$

PA 9-7 Determine the output voltage, e_2, if the network within the box is described by the following h parameters: $h_i = 1$ kilohm, $h_r = 0$, $h_f = 0$, $h_o = {}^1\!/_{100}$ kilohm.

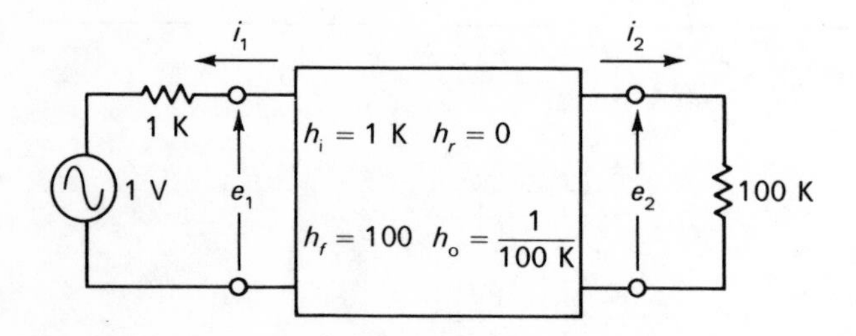

ANSWER -2500 volts

PROBLEMS WITHOUT ANSWERS

P 9-1 What are the h parameters of the network shown in Fig. P 9-1?

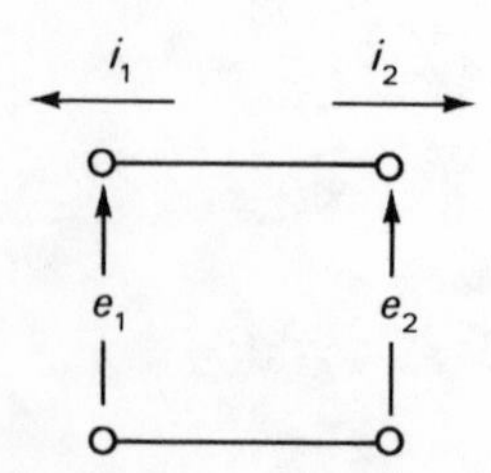

P 9-2 What is the relation between h_{fe} and h_{fb}?

P 9-3 Why is h_{fb} equal to the *negative* of α?

P 9-4 What effect would increasing the quiescent (dc) emitter current have upon h_{ie}?

P 9-5 What h parameter is being measured in Fig. P 9-5? The biasing details are omitted.

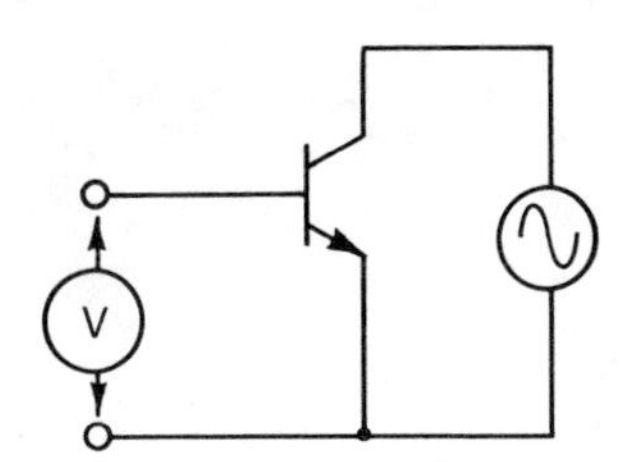

P 9-6 What are the h parameters of the network in Fig. P 9-6?

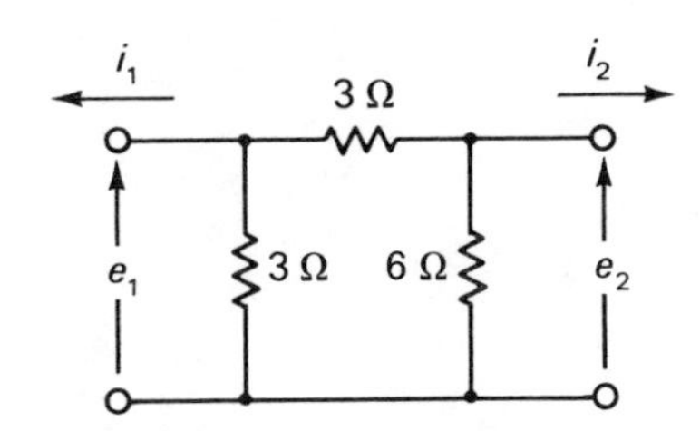

P 9-7 What are the h parameters for the network of Fig. P 9-7?

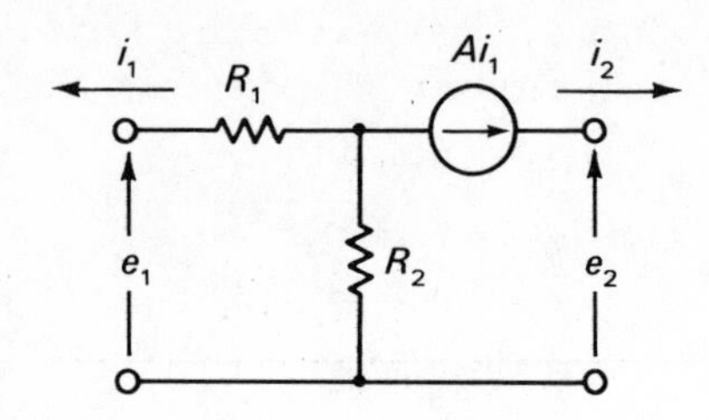

P 9-8 Determine i_1 and e_2 in Fig. P 9-8 if $h_i = 2$ ohms, $h_r = 0$, $h_f = 5$, $h_o = 0.5$ mho, $R_L = 2$ ohms, $R_g = 2$ ohms, and $E_g = 5$ volts.

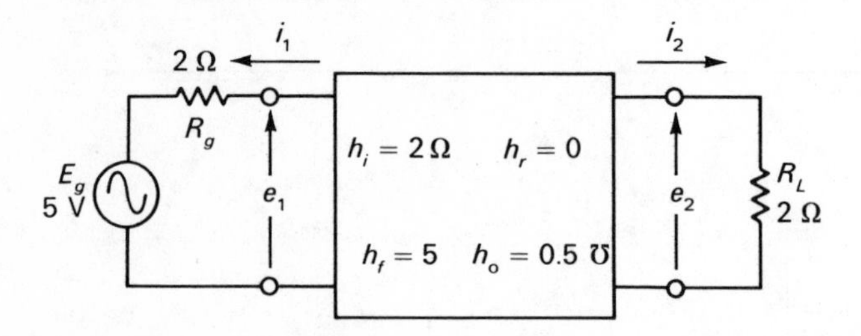

10 transistor high-frequency models

Frequently it is necessary to characterize the transistor in a manner which enables us to determine its circuit behavior at relatively high frequencies. This may be due to an application in which the device characteristics limit the high-frequency response of the circuit.

10-1 The hybrid-π model

Figure 10-1 illustrates what is known as the *hybrid-π equivalent circuit* of a transistor. (For a detailed discussion

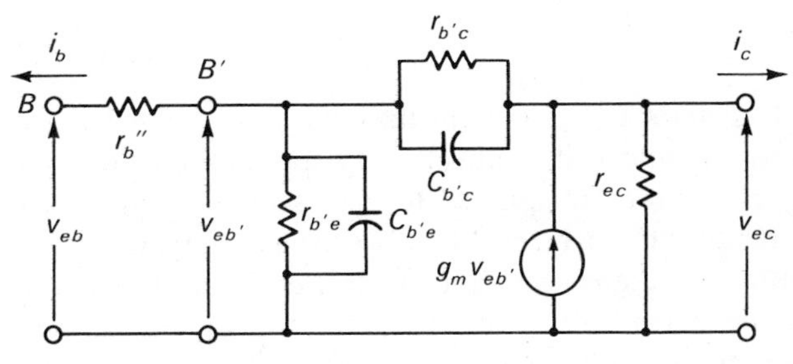

FIGURE 10-1

of the hybrid-π equivalent circuit, the reader is referred to Chapter 6 of "Semiconductor Circuit Analysis.") This circuit is essentially a common-emitter model which includes those transistor reactive elements that limit the high-frequency response. The hybrid-π model is quite applicable over the useful frequency range of the transistor. The physical significance of the hybrid-π parameters is as follows: r_b'' represents the resistance of the relatively lightly doped base region between the base lead and the base-emitter junction (represented by B'). Resistance $r_{b'e}$ is the dynamic impedance presented by the emitter junction to the flow of base current. A good approximation to $r_{b'e}$ may be obtained by recalling that the dynamic impedance of a forward-biased diode is given by $r_f = V_T/I$. Since the emitter junction essentially behaves as a forward-biased diode, we may, by analogy, write the approximate relation

1) $$r'_e = \frac{V_T}{I_E} \tag{10-1}$$

where r_e' is the dynamic impedance presented to the flow of incremental emitter current and I_E is the dc emitter current. The numerator V_T, you will recall, is about 26 mv at 27°C, which is generally considered to be room temperature. Since r_e' is the impedance presented to the flow of i_e and i_b is $h_{fe} + 1$ times smaller than i_e, it follows that the dynamic impedance presented to the flow of base current is $h_{fe} + 1$ times greater than it is to the emitter current.[1] Therefore we may write

2) $$r_{b'e} = r_e'(h_{fe} + 1) \tag{10-2}$$

[1] You will recall from Chapter 9 that $h_{fe} = \beta$.

The $r_{b'c}$ parameter is somewhat analogous to h_{re} of the h parameter model in that it allows for internal feedback, which enables the output voltage to modify the input characteristics. Parameter r_{ec} is somewhat analogous to $1/h_{oe}$ in that it enables a change in v_{ec} to effect the required change in i_c as called for by the slope of the collector characteristics. The dependent current source in the hybrid-π model is labeled $g_m v_{eb'}$ where $v_{eb'}$ is the internal voltage indicated in Fig. 10-1. The g_m parameter is essentially given by

$$g_m = \frac{1}{r_e'} \tag{10-3}$$

Capacitor $C_{b'c}$ represents the *collector-junction barrier capacity*; and if it is not directly specified on the data sheet, it may be assumed to be essentially equal to a parameter C_{ob}, which is usually specified. Capacitor $C_{b'e}$ represents the *emitter-junction barrier capacity* plus the diffusion capacity associated with the forward-biased emitter junction. In most transistors, except for some very high-frequency types, the emitter-diffusion capacity is generally much larger than the barrier capacity. The emitter-junction diffusion capacity may be assumed to increase in direct proportion to the dc emitter current. It is interesting to note that since r_e' essentially decreases in direct proportion to the emitter current, the time constant $r_{b'e}C_{b'e}$ tends to remain substantially constant. The value of $C_{b'e}$ is usually not specified on the transistor data sheet. It may, however, be approximated in terms of another parameter, f_T, which is frequently specified. This parameter, f_T, is defined as the frequency at which the common-emitter, short-circuit current gain is equal to unity. We may consider f_T as a figure of merit since it turns out to be indicative of the *gain-bandwidth product* given by

$$f_T \approx \frac{g_m}{2\pi C_{b'e}} \tag{10-4}$$

Since g_m is defined by Eq. (10-3), we obtain from Eq. 4

$$C_{b'e} = \frac{1}{2\pi f_T r_e'} \tag{10-5}$$

Some of the older data sheets specify a parameter called the *alpha-cutoff frequency*, instead of f_T. For all practical purposes these two quantities may be considered equal.

We have not yet indicated how r_b'' is determined. In some cases it is indicated on the data sheet. If r_b'' is not a frequency-sensitive parameter, as is usually true except in the case of some grown junction transistors, it may conveniently be measured as follows. A signal generator is connected across the emitter-to-base terminals with the transistor properly biased, and the frequency is adjusted to some high value such that $C_{b'e}$ looks like a short circuit. The input impedance is then essentially r_b''. An alternative means of estimating r_b'' if it is not frequency-sensitive is to use the relationship

$$r_b'' = h_{ie} - r_{b'e} \tag{10-6}$$

where h_{ie} is obtained from the common-emitter h parameters.

Figure 10-2a is an approximate low-frequency hybrid-π equivalent circuit. This approximation is justified by

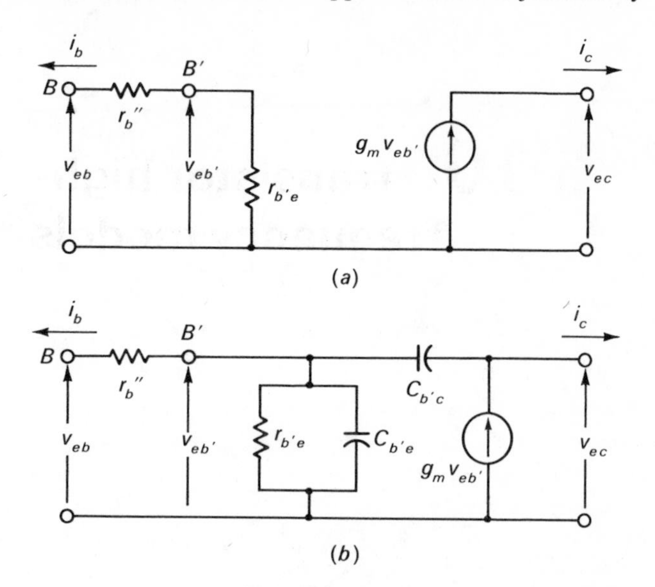

FIGURE 10-2

the fact that in a good-quality transistor, $r_{b'c}$ will typically be several megohms while r_{ec} might be several hundred kilohms. Of course, these values are dependent upon the type of transistor and operating point. However, they are representative of small-signal transistors in typical applications. Neglecting r_{ec} assumes that the external load is much smaller than r_{ec}. This is generally true, but caution should be exercised if the load is a high-Q tuned circuit with a high antiresonant impedance. The reactance of $C_{b'c}$ and $C_{b'e}$ are assumed so high as to be negligible in this low- and mid-frequency model. Figure 10-2b illustrates a high-frequency approximation to Fig. 10-1.

10-2 The Miller effect

As with the common-cathode, vacuum-tube amplifier, we encounter a *Miller-effect* problem in the common-emitter transistor amplifier. The Miller effect manifests itself as a variation of input impedance with the amplifier gain. It may be demonstrated that any impedance connected between the hot (ungrounded) input and output terminals is reflected across the input terminals as if it were essentially divided by the amplifier voltage gain. Since our primary concern with the Miller effect is in the high-frequency range where $C_{b'c}$ effectively shunts $r_{b'c}$, we shall work with the model of Fig. 11-2.

Our goal is to study the input impedance seen looking into this model when the output is terminated in some load Z_L. To simplify the problem let us break the circuit at $C_{b'c}$ and forget everything to the left of the break while the impedance seen looking to the right of the break is investigated. The equivalent circuit to the right of the break will then appear as shown in Fig. 10-3. A

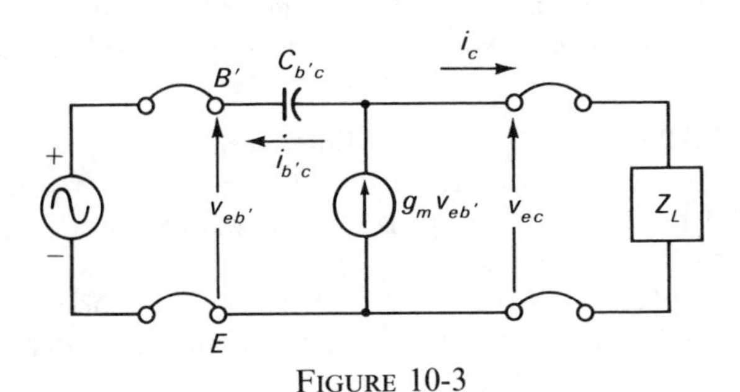

FIGURE 10-3

fictitious $v_{eb'}$ source is assumed, and the current it forces into $C_{b'c}$ is to be evaluated. To simplify the analysis we will assume that over the useful operating range of the transistor, the collector current is primarily furnished by the $g_m v_{eb'}$ current source rather than through the leakage path furnished by $C_{b'c}$. Hence, we may write

1) $$v_{ec} \approx -Z_L g_m v_{eb}$$

2) $$i_{b'c} = (v_{eb'} - v_{ec}) j\omega C_{b'c}$$

Substituting equation 1 into 2 and evaluating z_{in} yields

3) $$z_{\text{in}} = \frac{v_{eb'}}{i_{b'c}} = \frac{1}{j\omega C_{b'c}(1 + g_m Z_L)}$$

Inspection of equation 3 indicates that, insofar as the input impedance is concerned, $C_{b'c}$ looks as if it were larger by a factor $1 + g_m Z_L$. With these considerations we may now construct Fig. 10-4 where the reflected Miller-effect capacitance is given by

4) $$C_M = C_{b'c}(1 + g_m Z_L) \qquad (10\text{-}7)$$

Figure 10-4 is based on the assumption that i_c is essentially furnished by the $g_m v_{eb'}$ current source.

Inspection of equation 4 indicates that if Z_L is purely

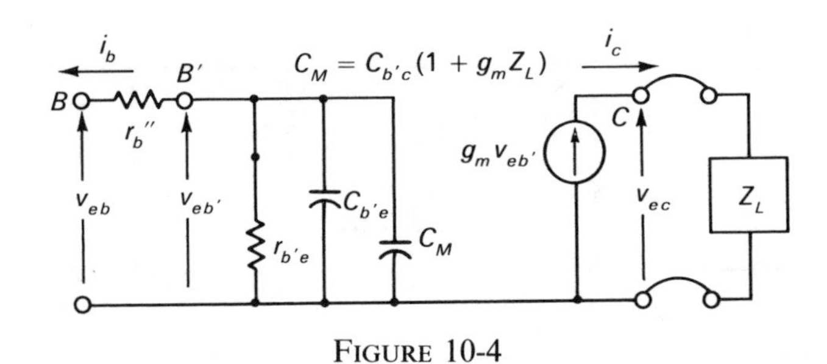

FIGURE 10-4

resistive, C_M will be equivalent to a pure capacitor. However, if Z_L is reactive, $X_{C(M)}$ will have both real and reactive components. A detailed discussion of this problem is beyond the scope of this text. (For a detailed discussion of the Miller effect and some pertinent problems, the reader is referred to Chapter 6 of "Semiconductor Circuit Analysis.")

PROBLEMS WITH SOLUTIONS

PS 10-1 Develop an expression for the short-circuit current gain in terms of the hybrid-π parameters shown in Fig. PS 10-1.

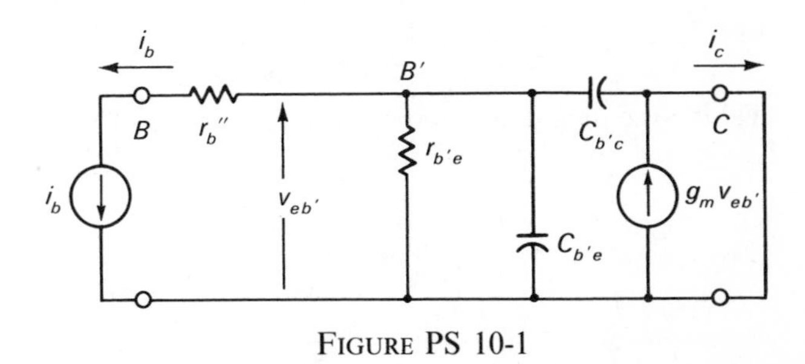

FIGURE PS 10-1

SOLUTION Our initial goal will be to manipulate Fig. PS 10-1 into the form of Fig. 10-4. Since $Z_L = 0$, the reflected Miller capacity, by equation 10-7, is just $C_{b'c}$. Inspection of Fig. 10-4 indicates

1) $$i_c = g_m v_{eb'}$$

Let

2) $$C_{eq} = C_{b'e} + C_M = C_{b'e} + C_{b'c}$$

Also let

3) $$z_{eq} = r_{b'e} \parallel C_{eq} = \frac{r_{b'e}}{1 + j\omega C_{eq} r_{b'e}}$$

4) $$v_{eb'} = z_{eq} i_b$$

Substituting equation 4 into 1 and simplifying, we obtain

5) $$A_i = \frac{i_c}{i_b} = g_m z_{eq}$$

But

6) $$r_{b'e} \approx h_{fe} r_e'$$

and

7) $$g_m = \frac{1}{r_e'}$$

Substituting equations 3, 6, and 7 into 5 yields

8) $$A_i = \frac{i_c}{i_b} = \frac{h_{fe}}{1 + j\omega/\omega_{\text{high}}} \qquad (10\text{-}8)$$

where h_{fe} is the low-frequency, short-circuit current gain and

9) $$\omega_{\text{high}} = \frac{1}{C_{eq} r_{b'e}} \quad (10\text{-}9)$$

A Bode plot of equation 8 would show a downward break at a 3-db point indicated by ω_{high}.

PS 10-2 Prove that $\omega_T = 2\pi f_T \approx h_{fe}\omega_{\text{high}}$ where ω_{high} is defined in equation 10-9.

SOLUTION You will recall that f_T is defined as the frequency at which the common-emitter, short-circuit current gain is equal to unity. Hence, with the aid of equation 10-8, we may write

1) $$A_i = \frac{i_c}{i_b} = \frac{h_{fe}}{1 + j\omega/\omega_{\text{high}}} = 1$$

Taking the magnitude of the denominator, we obtain

2) $$\sqrt{1 + \frac{\omega}{\omega_{\text{high}}}^{2}} = h_{fe}$$

or

3) $$\left(\frac{\omega}{\omega_{\text{high}}}\right)^2 = h_{fe}^2 - 1 \approx h_{fe}^2$$

Solving for ω yields

4) $$\omega = h_{fe}\omega_{\text{high}}$$

which concludes the proof. The frequency ω_{high}, as defined in equation 10-9, is sometimes called the *β-cut-off frequency*, while f_T is sometimes called the *α-cut-off frequency*.

PS 10-3 Express the hybrid-π parameters in terms of the common-emitter h and T parameters. Assume that the frequency range is low enough to consider all capacitors in the models as open circuits.

SOLUTION We can kill two birds with one stone by finding the h parameters of the common-emitter T and hybrid-π models. This will give us equations relating the T and h parameters and the hybrid-π and h parameters. By equating the corresponding h parameters for the

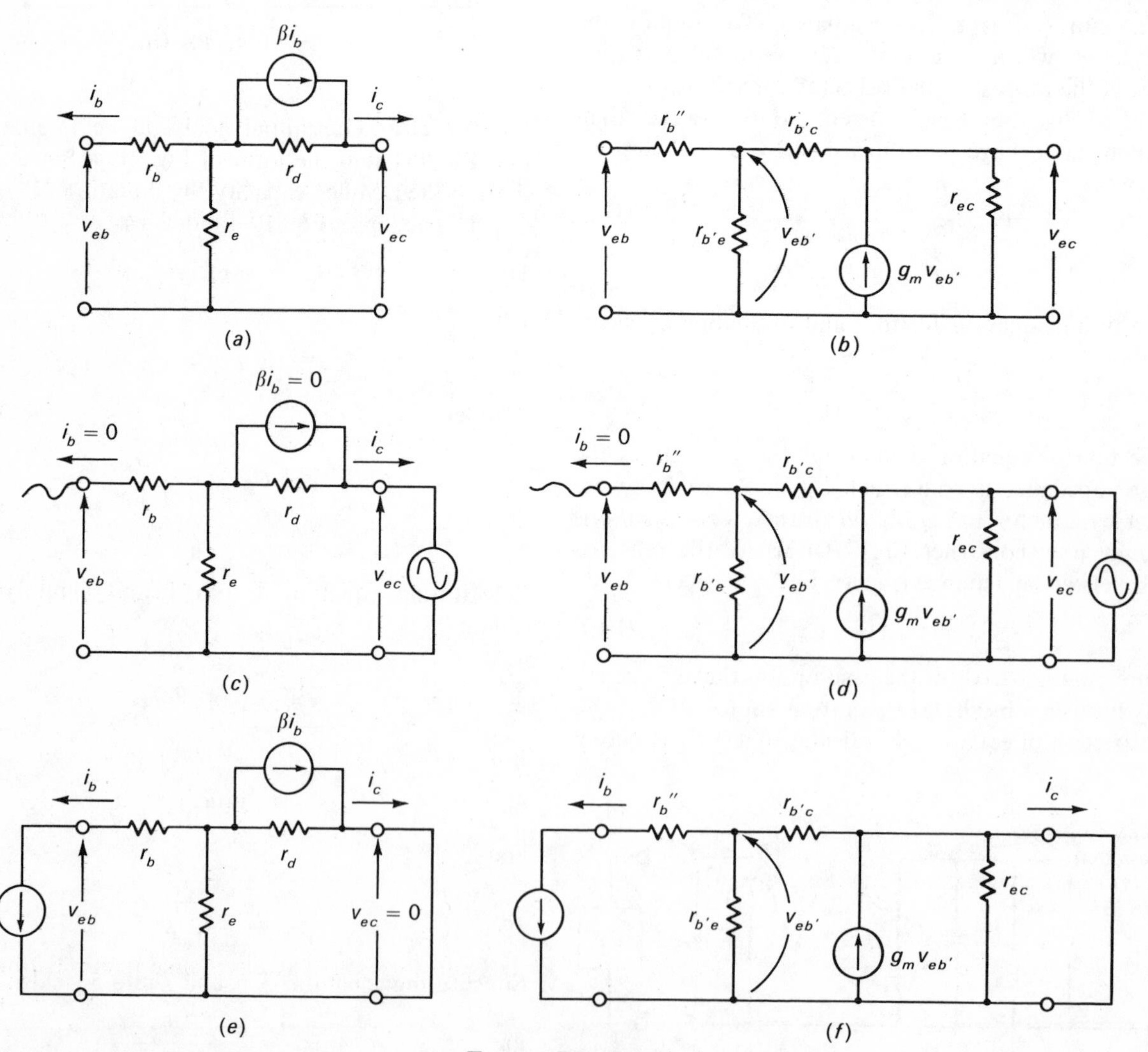

FIGURE PS 10-3

T and hybrid-π cases we can develop relationships between the common-emitter T and hybrid-π parameters.

For example, we can start with the T and hybrid-π models of Fig. PS 10-3a and 10-3b respectively. To determine h_r and h_o we set $i_b = 0$ by opening the input and placing a v_{ec} voltage source across the output, as shown in Fig. PS 10-3c and 10-3d respectively. For the T model of Fig. PS 10-3c, we obtain

1) $$h_{re} = \frac{v_{eb}}{v_{ec}} = \frac{r_e}{r_e + r_d} \approx \frac{r_e}{r_d} \qquad (10\text{-}10)$$

2) $$h_{oe} = \frac{1}{r_e + r_d} \approx \frac{1}{r_d} \qquad (10\text{-}11)$$

For the π model of Fig. PS 10-3d, we obtain $h_{r\pi}$ by noting that

3) $$h_{r\pi} = \frac{v_{eb}}{v_{ec}} = \frac{r_{b'e}}{r_{b'e} + r_{b'c}} \approx \frac{r_{b'e}}{r_{b'c}} \qquad (10\text{-}12)$$

4) $$h_{o\pi} = \frac{i_c}{v_{ec}} = ? \qquad \text{(not very obvious)}$$

As a preliminary step, we may write by superposition

5) $$i_c = g_m v_{eb'} + \frac{v_{ec}}{r_{ec} \parallel (r_{b'c} + r_{b'e})}$$

where

6) $$v_{eb'} = \frac{r_{b'e}}{r_{b'e} + r_{b'c}} v_{ec}$$

Substituting 5 and 6 into 4 yields

7) $$h_{o\pi} = g_m \frac{r_{b'e}}{r_{b'e} + r_{b'c}} + \frac{1}{r_{ec} \parallel (r_{b'c} + r_{b'e})} \qquad (10\text{-}13)$$

8) $$h_{o\pi} \approx g_m \frac{r_{b'e}}{r_{b'c}} + \frac{1}{r_{ec} \parallel r_{b'c}} \qquad (10\text{-}14)$$

To determine h_f and h_i, we drive the input with a current source and short the output as shown in Fig. PS 10-3e and f, respectively. Applying the superposition theorem to Fig. PS 10-3e, we may write

9) $$h_{fe} = \frac{i_c}{i_b} = \frac{\beta r_d}{r_d + r_e} - \frac{r_e}{r_d + r_e}$$

or

10) $$h_{fe} = \frac{\beta r_d - r_e}{r_d + r_e} \approx \beta \qquad (10\text{-}15)$$

Applying superposition to Fig. PS 10-3f, we obtain

11) $$i_c = g_m v_{eb'} - \frac{r_{b'e}}{r_{b'e} + r_{b'c}} i_b$$

but

12) $$v_{eb'} = (r_{b'e} \parallel r_{b'c}) i_b$$

Substituting 12 into 11 yields

13) $$h_{f\pi} = \frac{i_c}{i_b} = g_m(r_{b'e} \parallel r_{b'c}) - \frac{r_{b'e}}{r_{b'e} + r_{b'c}} \qquad (10\text{-}16)$$

or

14) $$h_{f\pi} \approx g_m r_{b'e} - \frac{r_{b'e}}{r_{b'c}} \approx g_m r_{b'e} \approx \beta \qquad (10\text{-}17)$$

The simplification in equation 14 is because $g_m r_{b'e} = r_e'(\beta + 1)/r_e'$ and $r_{b'c} \gg r_{b'e}$.

For Fig. PS 10-3e, we may obtain h_i by superposition if we first note that

15) $$v_{eb} = (r_b + r_e) i_b + (r_e \parallel r_d)\beta i_b$$

Therefore,

16) $$h_{ie} = \frac{v_{eb}}{i_b} = r_b + r_e + (r_e \parallel r_d)\beta \qquad (10\text{-}18)$$
$$\approx r_b + r_e(\beta + 1)$$

For Fig. PS 10-3f, we may write

17) $$h_{i\pi} = \frac{v_{eb}}{i_b} = r_b'' + (r_{b'e} \parallel r_{b'c}) \qquad (10\text{-}19)$$
$$\approx r_b'' + r_{b'e}$$

Now we may equate the respective hybrid common-emitter and hybrid-π parameters. Thus, from equations 1 and 3, 2 and 8, 10 and 14, and 16 and 17, we may write

18) $$\frac{r_e}{r_d} \approx \frac{r_{b'e}}{r_{b'c}} \approx h_{re}$$

19) $$\frac{1}{r_d} \approx g_m \frac{r_{b'e}}{r_{b'c}} + \frac{1}{r_{ec} \parallel r_{b'c}} \approx h_{oe} \qquad (10\text{-}20)$$

20) $$\beta \approx g_m r_{b'e} \approx h_{fe}$$

21) $$r_b + r_e(\beta + 1) \approx r_b'' + r_{b'e} \approx h_{ie}$$

Normally r_b'' is specified. In that case, from 21 we have

22) $$r_{b'e} \approx h_{ie} - r_b'' \qquad (10\text{-}21)$$

Substituting 22 into 20 yields

23) $$g_m \approx \frac{h_{fe}}{h_{ie} - r_b''} \qquad (10\text{-}22)$$

Substituting 22 into 18 yields

24) $$r_{b'c} \approx \frac{h_{ie} - r_b''}{h_{re}} \qquad (10\text{-}23)$$

Solving 19 for $r_{b'c}$ yields, after some manipulation,

25) $$r_{ec} \approx \frac{h_{ie} - r_b''}{(h_{ie} - r_b'')h_{oe} - h_{re}h_{fe}} \qquad (10\text{-}24)$$

PROBLEMS WITH ANSWERS

PA 10-1 At a Q point of $V_{EC} = 6$ volts and $I_C = 0.5$ ma, the h parameters of a transistor are $h_{ie} = 3.57$ kilohms, $h_{re} = 0.944 \times 10^{-3}$, $h_{fe} = 65$, $h_{oe} = 25$ μmhos. Also, $f_T = 0.85$ megahertz at this Q point, $C_{b'c} = 36$ $\mu\mu$farads, $r_b'' = 190$ ohms. Determine the hybrid-π parameters.

ANSWER

$r_{b'e} = 3.38$ kilohms
$g_m = 19.2$ mmhos
$r_{b'c} = 3.58$ megohms
$r_{ec} = 0.146$ megohm
$C_{b'e} = 3600$ $\mu\mu$farads

PA 10-2 Estimate the low-frequency common-emitter T parameters for the transistor of the previous problem.

ANSWER

$r_b \approx 140$ ohms or more
$r_e \approx 52$ ohms or less
$r_d \approx 40$ kilohms
$\beta = 65$

PROBLEMS WITHOUT ANSWERS

P 10-1 Estimate $r_{b'e}$ for a transistor at a Q point $V_{EC} = 9$ volts, $I_C = 2$ ma, and if $h_{fe} = 80$, $r_b'' = 0.2$ kilohm.

P 10-2 Estimate h_{ie} for the transistor in P 10-1.

P 10-3 Determine $C_{b'e}$ if $f_T = 1.6$ megahertz for the transistor in P 10-1.

P 10-4 Determine g_m for the transistor of P 10-1.

P 10-5 A transistor is operated at a Q point of $V_{EC} =$ 4 volts, $I_C = 0.7$ ma. The common-emitter short-circuit gain is down 3 db at 700 kilohertz, r_b'' is measured as 240 ohms, and $C_{ob} = 40$ $\mu\mu$farads. The low-frequency h parameters are $h_{ie} = 2.72$ kilohms, $h_{re} = 0.323 \times 10^{-3}$, $h_{fe} = 55$, $h_{oe} = 14$ $\mu\mu$mhos. Estimate the hybrid-π parameters at this Q point.

P 10-6 A common-emitter amplifier has a voltage gain of $20\ \underline{/0°}$. If $C_{b'c} \approx C_{ob} = 35$ $\mu\mu$farads, what is the reflected Miller-effect capacitance approximately equal to?

11 basic transistor amplifiers

In this chapter we will learn how to recognize and analyze some basic transistor amplifier configurations. Generally, we will use approximate transistor models in the analysis since the labor is considerably simplified and the approximations are usually justified. If you understand how the models were developed, you are aware of the approximations made and therefore can readily recognize situations in which they are not applicable.

The three basic transistor amplifier configurations are the *common-base* (CB), *common-emitter* (CE), and *common-collector* (CC) circuits as shown in Fig. 11-1*a*,

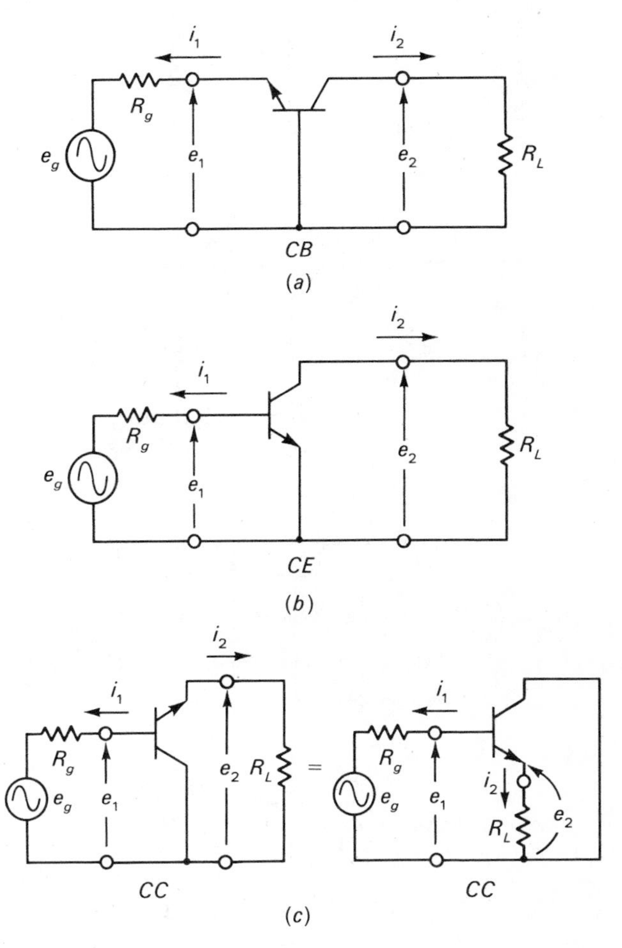

FIGURE 11-1

11-1*b*, and 11-1*c*, respectively. The common-collector circuit is usually referred to as an *emitter follower* (EF). The CC circuit is shown in the two different orientations in which it frequently appears to emphasize that the circuits are identical and the load, R_L, is in the emitter lead. The implication behind the term *common* in these (or other device) amplifiers is that the terminal in question is common to both the input and output sides of the circuit. In more advanced circuits it is not as obvious which configuration is used because normally the common terminal is not grounded. However, inspection of the input and output terminals will usually reveal the basic nature of the circuit.

11-1 The common-base amplifier

Figure 11-2*a* illustrates a typical CB amplifier stage. The V_{EE} supply and R_E forward-bias the emitter diode and establish the dc emitter current. Voltage e_s is the *signal voltage* and R_S is the *signal source resistance.*

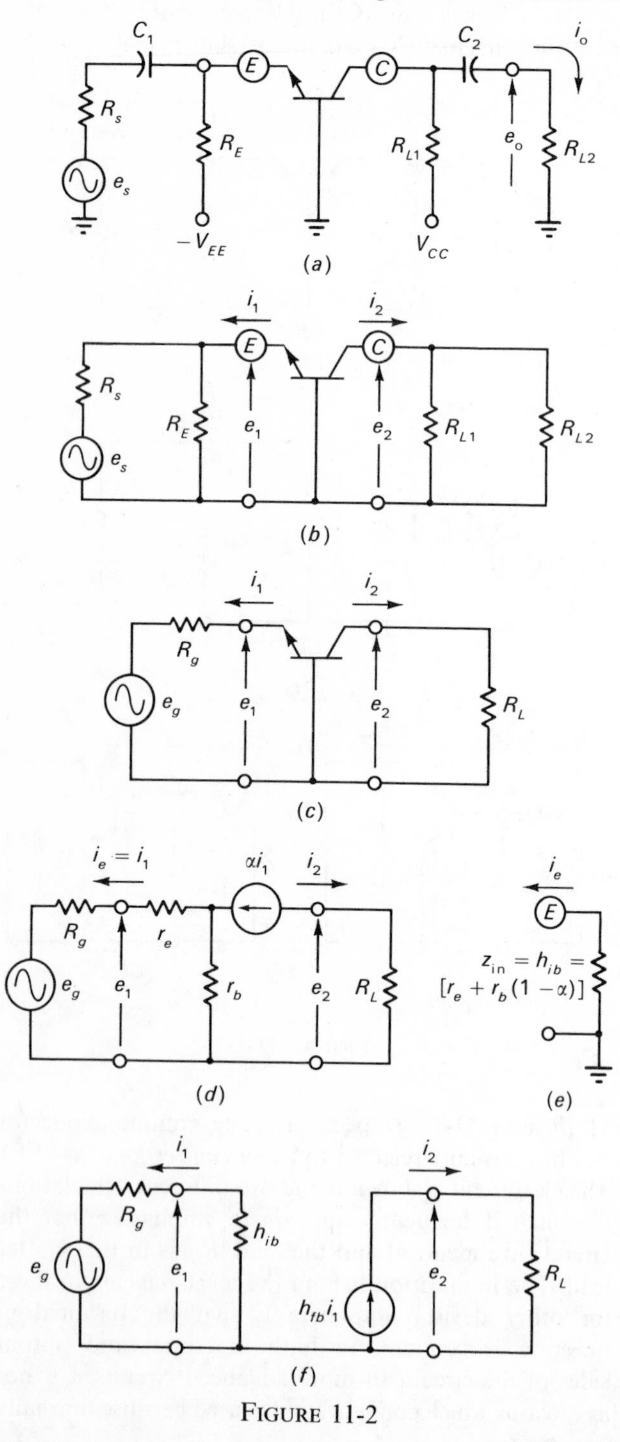

FIGURE 11-2

Capacitor C_1 is used to couple the signal into the emitter while simultaneously blocking any dc current from e_s. In the normal frequency range of the amplifier, the reactance of C_1 is negligible so that it may be assumed a short circuit.

Resistor R_{L1} and coupling capacitor C_2 form a coupling network which forces most of the ac signal component of collector current to flow through the desired load R_{L2}. The reactance of C_2 should be negligible in the passband of the amplifier.

Since we have already learned how to determine the dc or bias conditions in previous lessons, our present goal is to study the ac or signal response of the amplifier. This means we will in effect apply the superposition theorem and study the ac response by constructing the equivalent circuit of Fig. 11-2*b* from Fig. 11-2*a*. This equivalent circuit is derived by replacing all the coupling capacitors and power supplies with short circuits. For purposes of analysis, Fig. 11-2*b* may be simplified to Fig. 11-2*c* by thevenizing to the left of the emitter and to the right of the collector in Fig. 11-2*b*. This results in an open-circuit voltage and Thévenin's equivalent resistance represented by e_g and R_g on the emitter side. On the collector side, R_L is simply the parallel combination of R_{L1} and R_{L2}.

Figure 11-2*c* is the basic CB amplifier. To analyze it we can replace the transistor with the simplified *T* parameter model of Fig. 11-2*d*. The currents i_1 and i_2 are, of course, analogous to the emitter and collector currents. The amplifier characteristics we are interested in are voltage gain, A_v, current gain, A_i, power gain, A_p, and the input and output impedances z_{in} and z_o, respectively.

Since $A_p = A_v A_i$, we shall not bother with A_p. For Fig. 11-2*d* we may write

1) $$e_g = (R_g + r_e + r_b)\, i_1 - r_b \alpha i_1$$

or

2) $$e_g = [R_g + r_e + r_b(1 - \alpha)]\, i_1$$

Note that any external impedance in the base lead is in series with r_b and would, therefore, also be multiplied by $1 - \alpha$. Recalling that $h_{ib} \approx r_e + r_b(1 - \alpha)$

3) $$i_1 = \frac{e_g}{R_g + r_e + r_b(1 - \alpha)} = \frac{e_g}{R_g + h_{ib}}$$

Now

4) $$i_2 = -\alpha i_1$$

and

5) $$e_2 = -R_L i_2$$

Substituting equations 3 and 4 into 5 we obtain

6) $$A_v = \frac{e_2}{e_g} = \frac{\alpha R_L}{R_g + r_e + r_b(1 - \alpha)}$$

$$= \frac{\alpha R_L}{R_g + h_{ib}} \quad (11\text{-}1)$$

The output e_2 is in phase with the input e_g. From equation 4

7) $$A_i = \frac{i_2}{i_1} = -\alpha = h_{fb} \quad (11\text{-}2)$$

And from equation 3, by setting $R_g = 0$ so that $e_g = e_1$ and we are looking directly into the input terminals,

8) $$z_{\text{in}} = \frac{e_1}{i_1} = r_e + r_b(1 - \alpha) = h_{ib} \quad (11\text{-}3)$$

By inspection of Fig. 12-2*d* and with R_L removed,

9) $$z_o = \frac{e_2}{i_2} = \infty \quad (11\text{-}4)$$

since the αi_1 current source is assumed to be ideal.

Figure 11-2*e* shows a circuit which represents the active-region input impedance of the CB amplifier based upon equation (11-3). This result should be memorized as it considerably simplifies transistor circuit analysis. Fig. 11-2*f* illustrates a simplified *h* parameter model of the common-base amplifier derived from Fig. 11-2*d*. Note how simple the circuit is.

It is suggested that at this time you work through problems with solutions PS 11-1 and PS 11-2 and then continue with the lesson.

11-2 The common-emitter amplifier

Figure 11-3*a* illustrates a typical common-emitter amplifier stage. The R_1, R_2 voltage divider, R_E, and the V_{CC} supply essentially establish the *Q* point as discussed in the section on biasing. Capacitors C_1 and C_2 are coupling (blocking) capacitors, while C_E is a bypass capacitor to permit the ac signal a low-impedance path around R_E to ground. If C_E were removed, the amplifier gain would be considerably reduced.

Figure 11-3*b* may be derived from Fig. 11-3*a* by

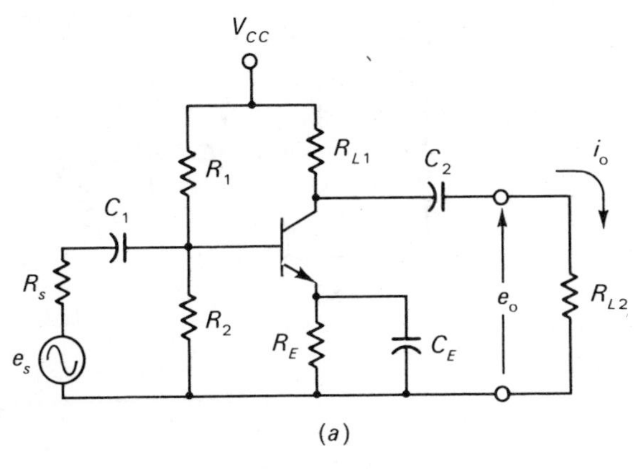

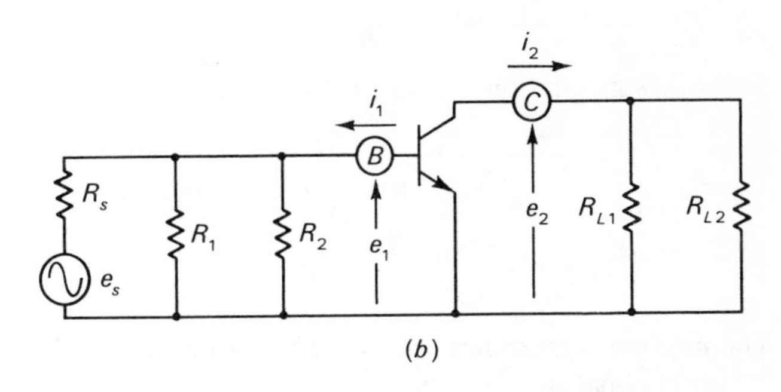

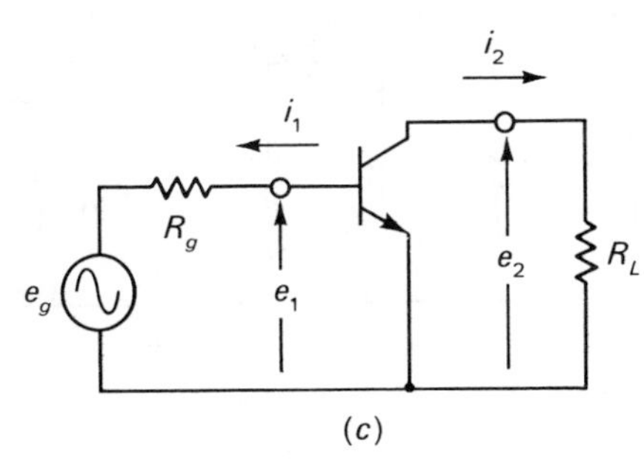

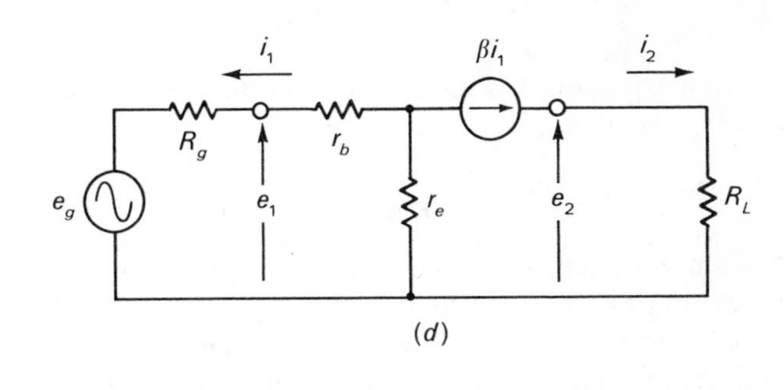

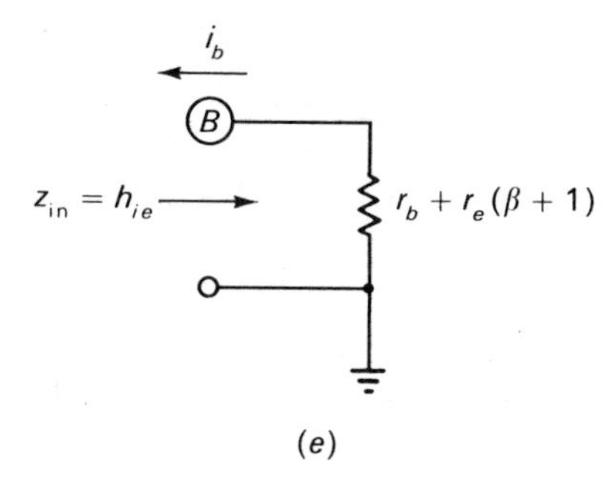

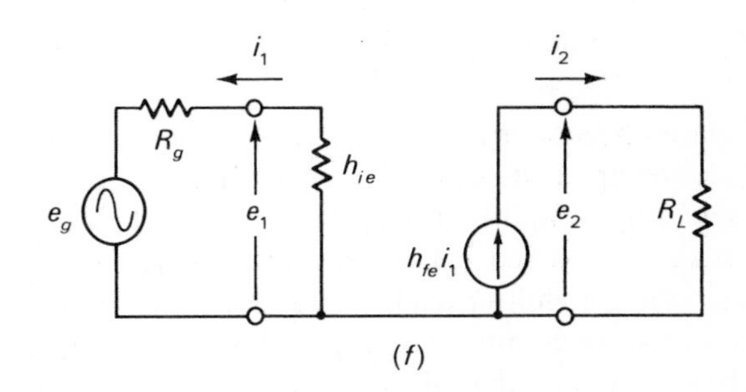

FIGURE 11-3

considering the ac signal paths only. Thus the V_{CC} supply and all the blocking and bypass capacitors become short circuits. Applying Thévenin's theorem to Fig. 11-3*b* yields Fig. 11-3*c* where R_L is the parallel combination of $R_{L1} \parallel R_{L2}$. Figure 11-3*c* is the basic CE amplifier and it may readily be analyzed with the approximate equivalent circuit of Fig. 11-3*d* from which we may write

1) $$e_g = (R_g + r_b + r_e)\, i_1 + r_e i_2$$

but

2) $$i_2 = \beta i_1 = h_{fe} i_1$$

Substituting 2 into 1 and recalling that

3) $$h_{ie} = r_b + r_e\,(\beta + 1)$$

we obtain

4) $$i_1 = \frac{e_g}{R_g + r_b + r_e\,(\beta + 1)} = \frac{e_g}{R_g + h_{ie}}$$

Clearly

5) $$e_2 = -R_L i_2$$

Substituting equations 4 and 2 into 5 yields

6) $$A_v = \frac{e_2}{e_g} = \frac{-R_L \beta}{R_g + r_b + r_e\,(\beta + 1)} = \frac{-R_L h_{fe}}{R_g + h_{ie}} \qquad (11\text{-}5)$$

The negative sign indicates e_o is 180° out of phase with e_g. From equation 2 we obtain

7) $$A_i = \frac{i_2}{i_1} = \beta = h_{fe} \qquad (11\text{-}6)$$

From equation 4, with R_g set equal to zero so that $e_g = e_1$, we obtain

8) $$z_{in} = \frac{e_1}{i_1} = r_b + r_e\,(\beta + 1) = h_{ie} \qquad (11\text{-}7)$$

And since the dependent current source βi_1 is assumed to be ideal in Fig. 11-3*d*, it follows that any voltage source e_2 connected across the output terminals, with R_L removed, would be unable to force any i_1 or resultant βi_1. Therefore

9) $$z_o = \frac{e_2}{i_2} = \infty \qquad (11\text{-}8)$$

Figure 11-3*e* illustrates the active-region input impedance based upon equation 11-7. Memorize it, as this will considerably simplify future work. Note that if an external, unbypassed emitter resistor, R_E, is added in series with the emitter lead, the same i_e flows through R_E as r_e. Since r_e is multiplied by $\beta + 1$ when looking into the base, it follows that any external R_E raises the input impedance by an amount $R_E(\beta + 1)$.

Figure 11-3*f* illustrates a simplified but most useful *h* parameter model of the CE amplifier as it would be derived from Fig. 11-3*c* or Fig. 11-3*d*.

11-3 The emitter-follower amplifier

Figure 11-4*a* illustrates a typical common-collector (CC) or emitter-follower (EF) amplifier as it is more popularly called. The V_{CC} supply, R_1, R_2, and R_{L1}, establish the *Q* point. Capacitors C_1 and C_2 are coupling (blocking) capacitors. Note the load is in the emitter circuit.

The ac equivalent circuit of Fig. 11-4*a* reduces to Fig. 11-4*b* if operation is assumed to be in a frequency range where the capacitive reactances are negligible. Applying Thévenin's theorem to Fig. 11-4*b* and letting $R_L = R_{L1} \parallel R_{L2}$, we obtain Fig. 11-4*c* which is the basic

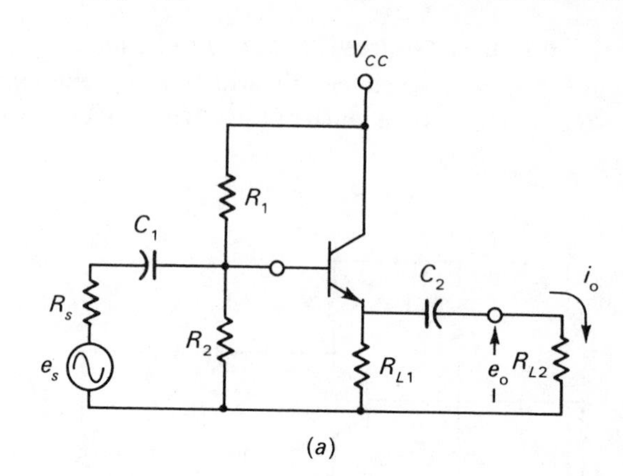

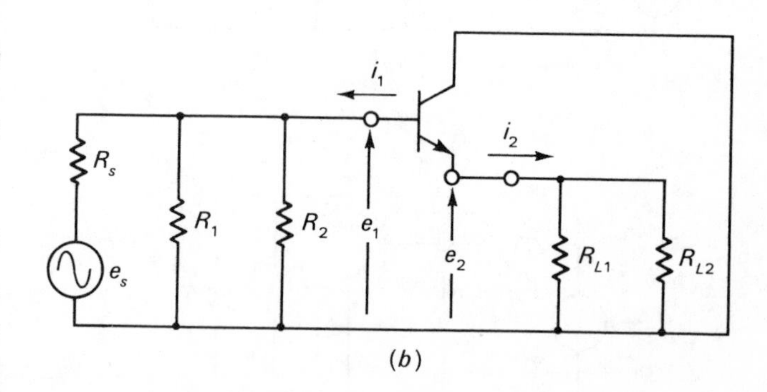

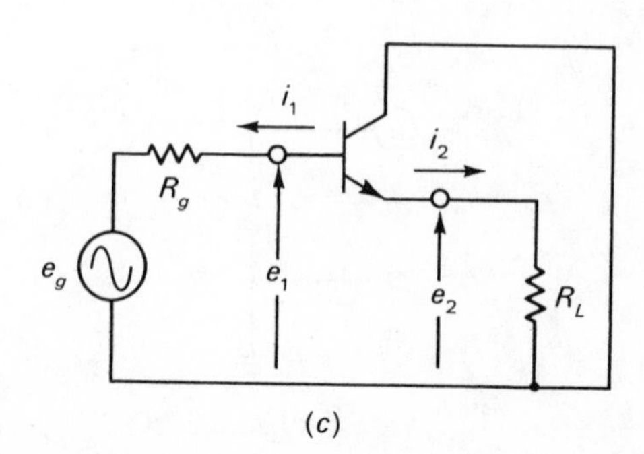

FIGURE 11-4

EF. The T parameter equivalent circuit of Fig. 11-4c would appear as shown in Fig. 11-4d, and the h parameter equivalent circuit would appear as shown in Fig. 11-4e.

The circuit of Fig. 11-4c may actually be analyzed by inspection if we apply our prior knowledge. The most important point to recall is that when looking into the base of a transistor operating in its active region, we see h_{ie} in series with any external emitter resistance looking as if it was $(h_{fe} + 1)$ or $(\beta + 1)$ times greater. This is illustrated by Fig. 11-4f. Thus we may write

1) $$A_v = \frac{e_2}{e_g} = \frac{-R_L i_2}{e_g} = \frac{-R_L}{e_g}[-i_1(\beta + 1)]$$

Remember, the emitter current i_2 is equal to $\beta + 1$ times the base current. Now

2) $$i_1 = \frac{e_g}{R_g + h_{ie} + R_L(\beta + 1)}$$

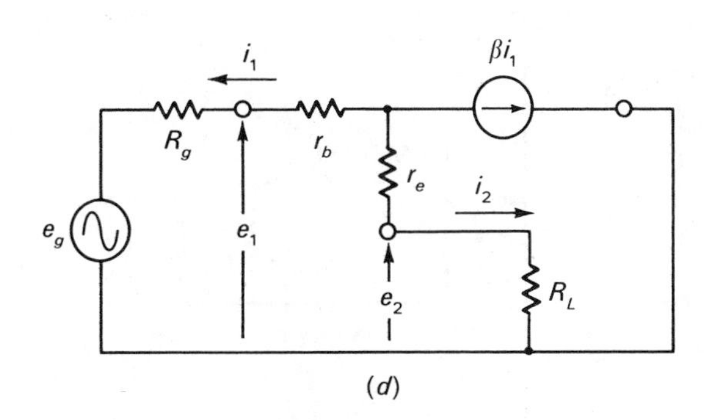

(d)

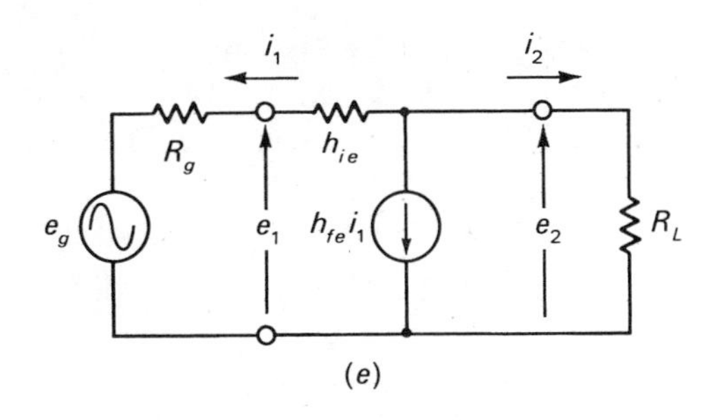

(e)

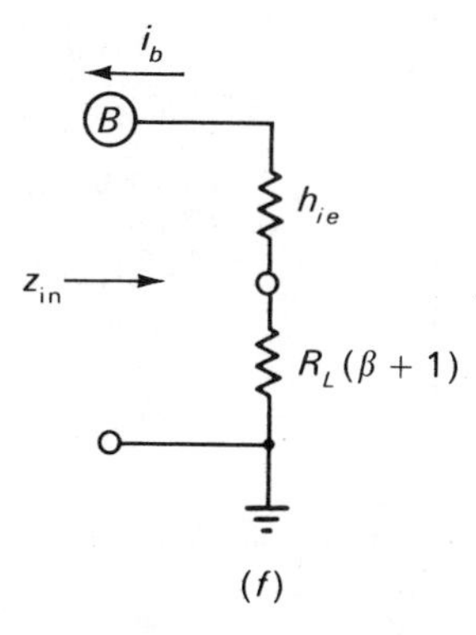

(f)

FIGURE 11-4 (*Continued*)

Substituting 2 into 1 yields

3) $$A_v = \frac{e_2}{e_1} = \frac{R_L(h_{fe} + 1)}{R_g + h_{ie} + R_L(h_{fe} + 1)} \quad (11\text{-}9)$$

or in terms of the T parameters

4) $$A_v = \frac{e_2}{e_1}$$
$$= \frac{R_L(\beta + 1)}{R_g + r_b + r_e(\beta + 1) + R_L(\beta + 1)} \quad (11\text{-}10)$$

Note that if $R_L(h_{fe} + 1) \gg R_g + h_{ie}$, the voltage gain approaches 1.

The current gain with the reference directions shown is

5) $$A_i = \frac{i_2}{i_1} = \frac{-(\beta + 1)i_1}{i_1} = -(\beta + 1)$$
$$= -(h_{fe} + 1) \quad (11\text{-}11)$$

The input impedance by Fig. 11-4f is

6) $$z_{in} = \frac{e_1}{i_1} = h_{ie} + R_L(h_{fe} + 1)$$
$$= r_b + (r_e + R_L)(\beta + 1) \quad (11\text{-}12)$$

To intuitively determine the output impedance, imagine we are looking into the emitter of the transistor of Fig. 11-4c with R_L removed. Also imagine R_g and e_g equal to zero so that the base is grounded. Couldn't we then say that looking into the output terminals of an EF is the same as looking into the input terminals of a CB amplifier? Now the z_{in} for the CB amplifier was given by equation 11-3 as $z_{in} = r_e + r_b(1 - \alpha)$. Therefore, doesn't it follow that any external base resistance in series with r_b would also be multiplied by $1 - \alpha$? Since R_g is an external base resistance, we may now write

7) $$z_o = \frac{e_2}{i_2} = r_e + (r_b + R_g)(1 - \alpha)$$
$$= r_e + \frac{r_b + R_g}{\beta + 1} \quad (11\text{-}13)$$

or in terms of the CB's h parameters and recalling that $h_{fb} = -\alpha$

8) $$z_o = \frac{e_2}{i_2} = h_{ib} + R_g(1 + h_{fb}) \quad (11\text{-}14)$$

or, in terms of the CE's h parameters,

9) $$z_o = \frac{e_2}{i_2} = \frac{h_{ie} + R_g}{h_{fe} + 1} \quad (11\text{-}15)$$

Note that for h_{fe} large, z_o may be quite small.

A formal rather than intuitive proof of all the above relations may be easily derived from Fig. 11-4e.

At this point you should work through problems PS 11-6 and PS 11-7.

11-4 Summary of basic amplifier characteristics

The general properties of the basic amplifier configurations may be tabulated as follows:

Table 11-1 Basic amplifier properties.

	A_v	PHASE OF A_v	A_i	A_p	Z_{in}	Z_o
CB	largest	0°	<1	large	lowest	highest
CE	large	180°	large	largest	low	high
CC	<1	0°	largest	large	highest	low

Table 11-2 The *h* to *T* conversions.

	EXACT	APPROXIMATE	EQUIVALENT
r_e	$h_{ib} - \dfrac{h_{rb}(1 + h_{fb})}{h_{ob}}$	$h_{ib} - \dfrac{h_{rb}(1 + h_{fb})}{h_{ob}}$	
r_b	$\dfrac{h_{rb}}{h_{ob}}$	$\dfrac{h_{rb}}{h_{ob}}$	
r_c	$\dfrac{1 - h_{rb}}{h_{ob}}$	$\dfrac{1}{h_{ob}}$	
α	$\dfrac{-(h_{fb} + h_{rb})}{1 - h_{rb}}$	$-h_{fb}$	$\dfrac{h_{fe}}{1 + h_{fe}}$
β	h_{fe}	$\dfrac{-h_{fb}}{1 + h_{fb}}$	

Table 11-3

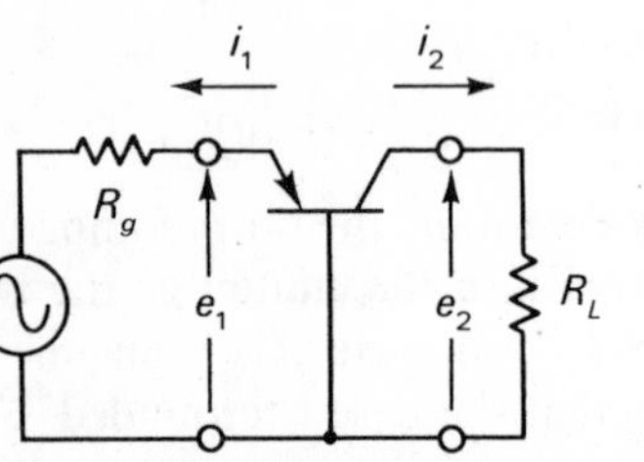

	CB	CE	CC
	i_1, i_2, R_g, e_g, e_1, e_2, R_L	i_2, R_g, i_1, e_g, e_1, e_2, R_L	R_g, i_1, e_g, e_1, R_L, i_2, e_2
A_i	$= -\dfrac{r_b + \alpha r_c}{r_b + r_c + R_L}$ $\approx -\alpha$ if $r_c \gg r_b, r_c \gg R_L$	$= \dfrac{\beta r_d - r_e}{r_e + r_d + R_L}$ $\approx \beta$ if $r_d \gg r_e, r_d \gg R_L$	$= \dfrac{-r_d(\beta + 1)}{r_e + r_d + R_L}$ $\approx -(\beta + 1)$ if $r_d \gg R_L$
A_v	$= \dfrac{(r_b + \alpha r_c)R_L}{r_b[r_c(1 - \alpha) + R_L] + r_e(r_b + r_c + R_L)}$ $\approx R_L/r_e$ if $r_d \gg R_L, r_e \gg r_b$	$= \dfrac{-(\beta r_d - r_e)R_L}{(r_e + r_b)(r_d + R_L) + r_e(r_b + \beta r_d)}$ $\approx -\dfrac{R_L}{r_e + r_b(1 - \alpha)}$ if $r_d \gg R_L$	$= \dfrac{r_d(\beta + 1)R_L}{r_b(r_e + r_d + R_L) + r_d(\beta + 1)(r_e + R_L)}$ ≈ 1 if $r_d \gg R_L, R_L \gg r_e + r_b(1 - \alpha)$
z_{in}	$= \dfrac{r_b[r_c(1 - \alpha) + R_L] + r_e(r_b + r_c + R_L)}{r_b + r + R_L}$ $\approx r_e + r_b(1 - \alpha)$ if $r_c \gg r_b, r_d \gg R_L$	$= \dfrac{(r_b + r_e)(r_d + R_L) + r_e(r_b + \beta r_d)}{r_e + r_d + R_L}$ $\approx r_b + r_e(\beta + 1)$ if $r_d \gg R_L, r_d \gg r_e$	$= r_b + \dfrac{r_d(\beta + 1)(r_e + R_L)}{r_e + r_d + R_L}$ $\approx R_L(\beta + 1)$ if $r_d \gg R_L, R_L \gg r_b, R_L \gg r_e$
z_o	$= \dfrac{r_b[r_e + r_c(1 - \alpha)] + r_e r_c}{r_e + r_b}$ $\approx r_c\left[\dfrac{r_e + r_b(1 - \alpha)}{r_e + r_b}\right]$ if $r_d \gg r_e$	$= \dfrac{r_d[r_e(\beta + 1) + r_b] + r_e r_b}{r_e + r_b}$	$= \dfrac{r_b(r_e + r_d) + r_e r_d(\beta + 1)}{r_b + r_d(\beta + 1)}$ $\approx r_e + r_b(1 - \alpha)$

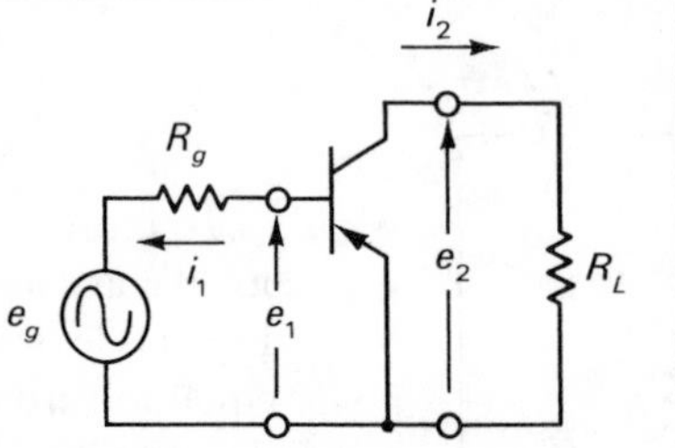

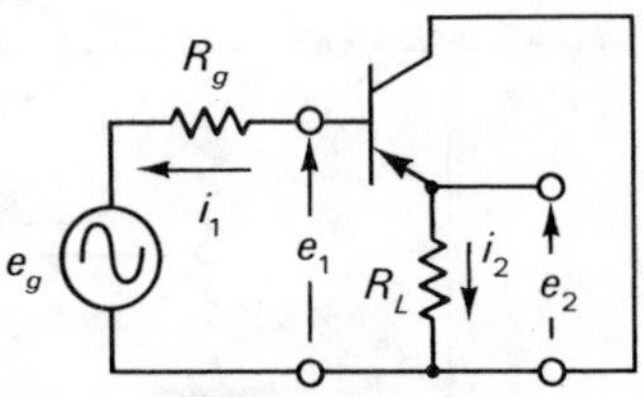

* $r_d = r_c(1 - \alpha)$ $\quad A_i = \dfrac{i_2}{i_1}, A_v = \dfrac{e_2}{e_1}, z_{in} = \left.\dfrac{e_1}{i_1}\right|_{R_g = \infty}, z_o = \left.\dfrac{e_2}{i_2}\right|_{R_L = \infty}$

PROBLEMS WITH SOLUTIONS

PS 11-1 Estimate the quiescent (V_{BC}, I_C) operating point of the silicon transistor and the signal output voltage e_0 in the circuit of Fig. PS 11-1*a*.

SOLUTION The *Q* point may be determined with the aid of the dc model in Fig. PS 11-1*b*. I_{CBO} is not specified so it may be neglected since the transistor is silicon. Inspection of Fig. PS 11-1*b* indicates that the emitter will be forward-biased and the collector most likely reverse-biased. With these assumptions, Fig. PS 11-1*c* may be constructed. Therefore,

1) $$I_E = \frac{6 \text{ volts} - 0.6 \text{ volt}}{6 \text{ kilohms}} = 0.9 \text{ ma}$$

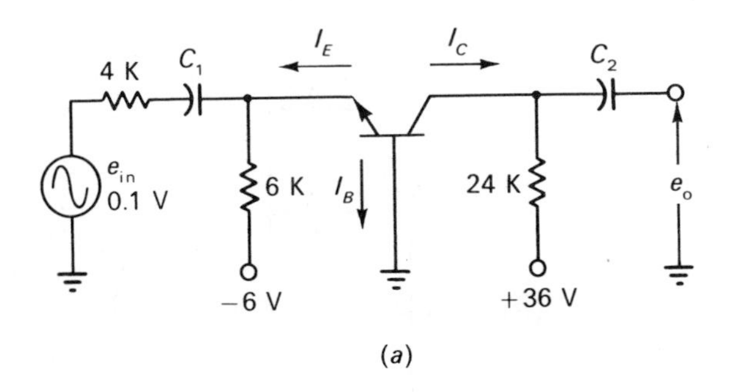

(*a*)

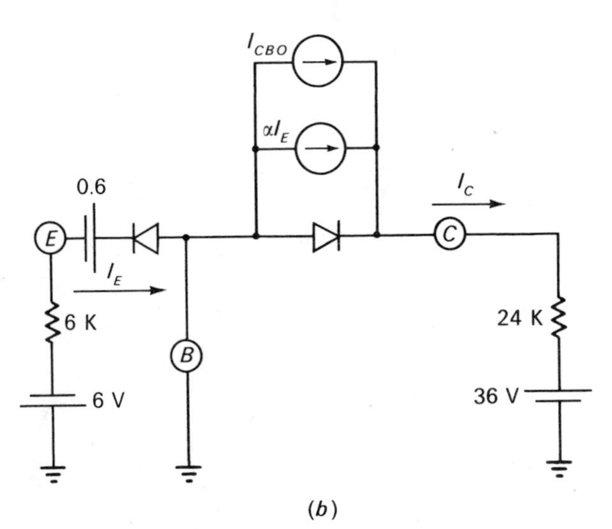

(*b*)

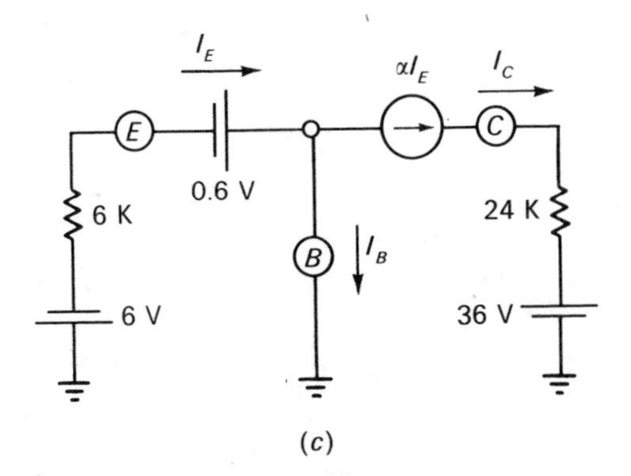

(*c*)

FIGURE PS 11-1

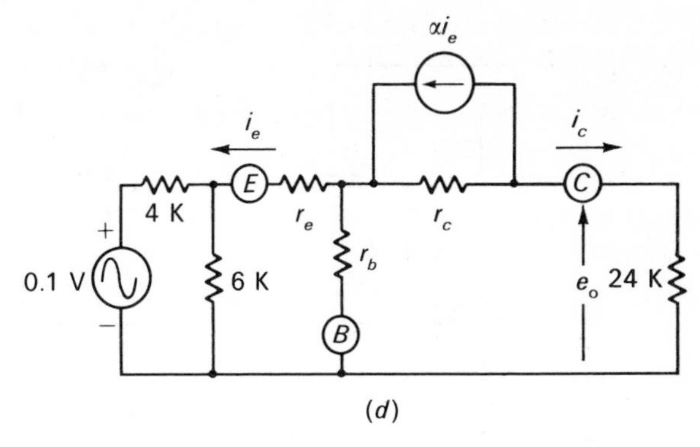

(*d*)

FIGURE PS 11-1 (*Continued*)

Since $\alpha \approx 1$

2) $$I_C \approx I_E = 0.9 \text{ ma}$$

3) $V_{BC} = 36 \text{ volts} - 24 \text{ kilohms } (0.9 \text{ ma}) = 14.4 \text{ volts}$

Note that the collector is reverse-biased as initially assumed.

For the ac analysis, we derive Fig. PS 11-1*d* from Fig. PS 11-1*a*. The Thévenin's equivalent resistance and open-circuit voltage the emitter sees is

4) $z_{th} = 4 \text{ kilohms} \parallel 6 \text{ kilohms} = 2.4 \text{ kilohms} = R_g$

5) $$v_{oc} = \frac{6 \text{ kilohms}}{4 \text{ kilohms} + 6 \text{ kilohms}} (0.1 \text{ volt})$$
$$= 0.06 \text{ volt} = e_g$$

Since r_e isn't specified, we may estimate it as

6) $r_e = 26 \text{ mv}/I_E = 26 \text{ mv}/0.9 \text{ ma} = 28.8 \text{ ohms}$

Now r_e, plus r_b looking as if it is divided by $\beta + 1$, will most surely be much less than R_g. Hence, for all practical purposes,

7) $$i_e = \frac{0.06 \text{ volt}}{2.4 \text{ kilohms}} = 0.025 \text{ ma}$$

Since r_c will probably exceed a megohm at the *Q* point calculated (r_c tends to increase as the collector becomes increasingly reverse-biased and the emitter current decreases), we may safely assume that most of the αi_e current source will exit through the load. Thus,

8) $$i_c = -\alpha i_e \approx -i_e = -0.025 \text{ ma}$$

and

9) $$e_o = -R_L i_c = -(24 \text{ kilohms})(-0.025 \text{ ma})$$
$$= 0.6 \text{ volt}$$

PS 11-2 Assuming the transistor in Fig. PS 11-2*a* is biased into its active region, determine the input impedance, z_{in}, presented to the flow of emitter current.

SOLUTION When looking into the emitter, any

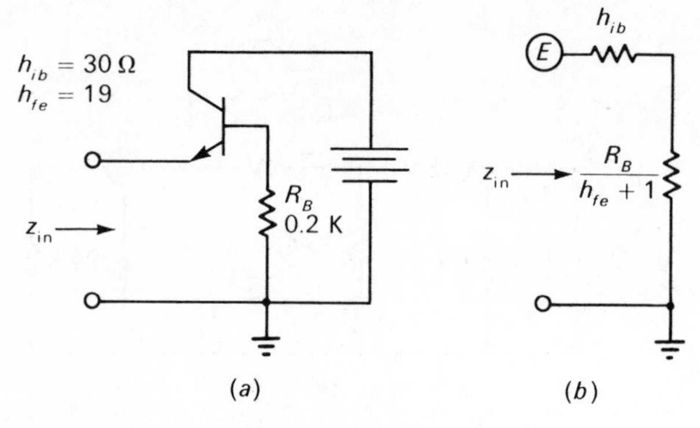

FIGURE PS 11-2

impedance in the base lead is divided by $\beta + 1$. Thus z_{in} as determined by Fig. PS 11-2*b* is given by

$$z_{in} = h_{ib} + \frac{R_B}{h_{fe} + 1} = 30 + \frac{200}{19 + 1} = 40 \text{ ohms}$$

PS 11-3 Assuming the transistor of Fig. 11-3*a* is biased into its active region, determine the input impedance z_{in} presented to the flow of base current.

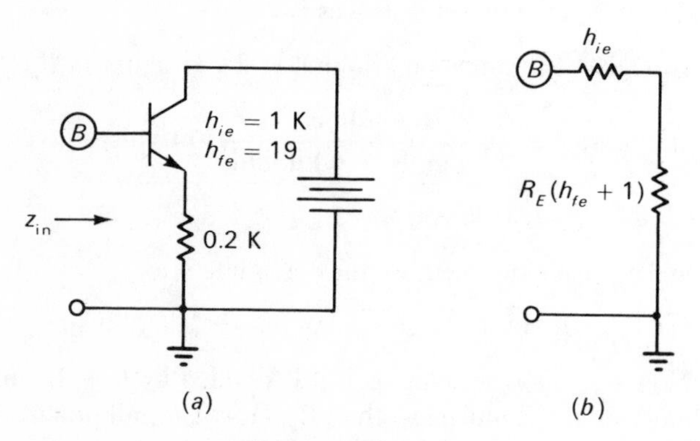

FIGURE PS 11-3

SOLUTION When looking into the base, any impedance in the emitter lead is multiplied by $\beta + 1$. Thus z_{in} as determined from Fig. PS 11-3*b* is given by

$$z_{in} = h_{ie} + R_E(h_{fe} + 1)$$
$$= 1 \text{ kilohm} + 0.2 \text{ kilohm } (19 + 1) = 5 \text{ kilohms}$$

PS 11-4 Estimate V_{EC} and e_o in the circuit of Fig. PS 11-4*a*. Assume that the dc β, also referred to as h_{FE}, equals 49, whereas the ac β, also referred to as h_{fe}, equals 58. Also assume a silicon transistor and $h_{ob} = 5\ \mu$mho.

SOLUTION An approximate dc equivalent circuit is shown in Fig. PS 11-4*b*. Since the transistor is silicon and no I_{CBO} value is specified, we may reasonably assume I_{CBO}, and most likely I_{CEO}, is negligible. However, the latter assumption becomes increasingly poorer as h_{FE} and temperature increase and as the base is driven by a constant-current source. We will, however, assume

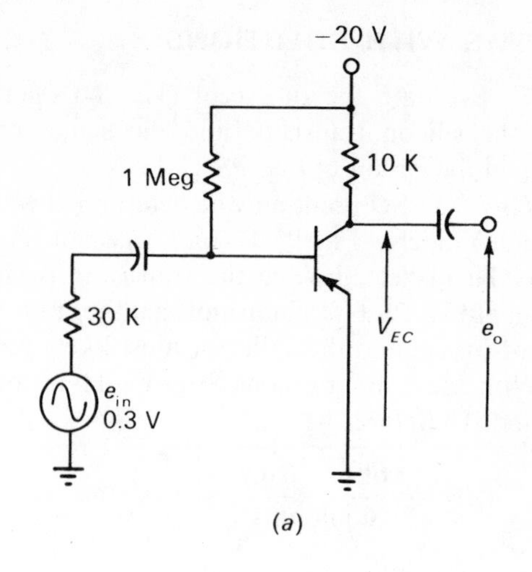

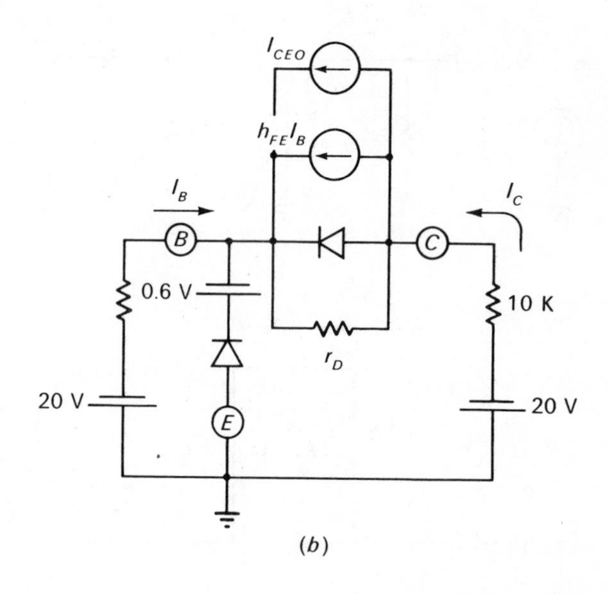

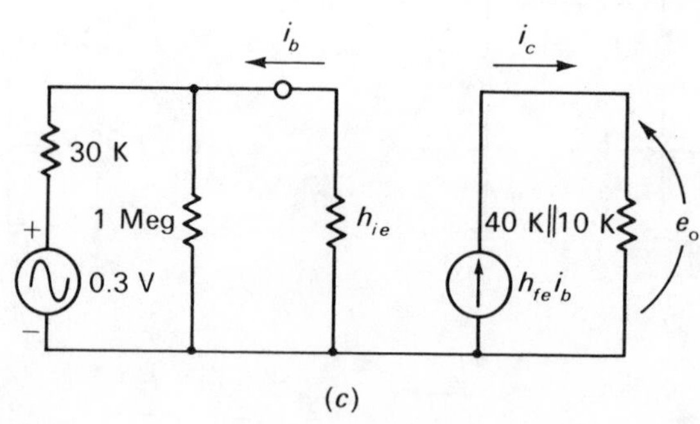

FIGURE PS 11-4

I_{CEO} is negligible. We may approximate $r_D = r_d$ and from

$$r_c \approx \frac{1}{h_{ob}} = \frac{1}{0.5\ \mu\text{mho}} = 2 \text{ megohms}$$

$$r_d \approx \frac{r_c}{h_{fe} + 1} = \frac{2 \text{ megohms}}{50} = 40 \text{ kilohms}$$

Now an $r_d = 40$ kilohms isn't exactly negligible when the load resistor is as large as 10 kilohms, and consequently it must be taken into account.

$$I_B = \frac{10 \text{ volts} - 0.6 \text{ volt}}{1 \text{ megohm}} = 9.4\ \mu\text{a}$$

By superposition,

$$I_C = (h_{FE}I_B + I_{CEO})\frac{r_d}{r_d + R_L} + \frac{V_{CC} - V_{EB'}}{r_d + R_L}$$

$$= [49(9.4\ \mu\text{a}) + 0]\frac{40 \text{ kilohms}}{50 \text{ kilohms}} + \frac{20 \text{ volts} - 0.6 \text{ volt}}{50 \text{ kilohms}}$$

$$= 0.368 \text{ ma} + 0.388 \text{ ma}$$

$$I_C = 0.756 \text{ ma}$$

$$V_{EC} = -20 \text{ volts} + 0.756 \text{ ma } (10 \text{ kilohms})$$

$$= -12.4 \text{ volts}$$

An approximate ac equivalent circuit is shown in Fig. PS 11-4*c*. Since h_{ie} is not specified, it may be estimated by first determining

$$r_e \approx \frac{26 \text{ mv}}{0.756 \text{ ma}} = 34.4 \text{ ohms}$$

Therefore, with say 200 ohms assumed for r_b,

$$h_{ie} = r_b + r_e(\beta + 1)$$

$$= 200 + 34.4\,(58 + 1) = 2.23 \text{ kilohms}$$

Thus

$$i_b \approx \frac{0.3 \text{ volt}}{30 \text{ kilohms} + 2.23 \text{ kilohms}} = 9.3\ \mu\text{a}$$

$$h_{fe}i_b = 9.3\ \mu\text{a}\,(58) = 0.540 \text{ ma}$$

$$e_o = -(r_d \parallel R_L)h_{fe}i_b = -8 \text{ kilohms } (0.540 \text{ ma})$$

$$= 4.32 \text{ volts}$$

PS 11-5 Estimate I_C and e_o in the circuit of Fig. PS 11-5*a*. Assume a silicon transistor is used with a $\beta \approx 50$.
SOLUTION Since the β given was not specified as dc or ac and there is no information given to clarify this point, it may be assumed that $\beta_{dc} \approx \beta_{ac}$. A dc equivalent circuit is shown in Fig. PS 11-5*b*. Thus

$$I_E \approx \frac{12 \text{ volts} - 0.6 \text{ volt}}{12.01 \text{ kilohms} + (60 \text{ kilohms}/50)}$$

$$= \frac{11.4 \text{ volts}}{13.2 \text{ kilohms}} = 0.863 \text{ ma}$$

changing β to α yields

$$I_C = \alpha I_E = 0.846 \text{ ma}$$

For the ac solution, consider Fig. PS 11-5*c*. By now you should be able to work this type of equivalent

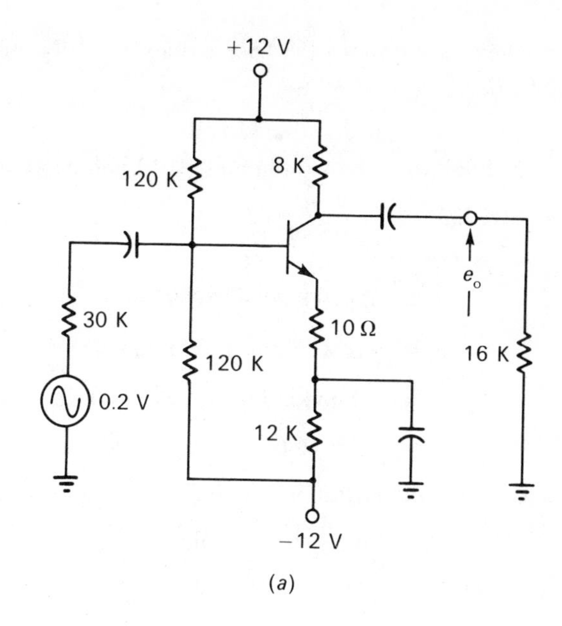

(*a*)

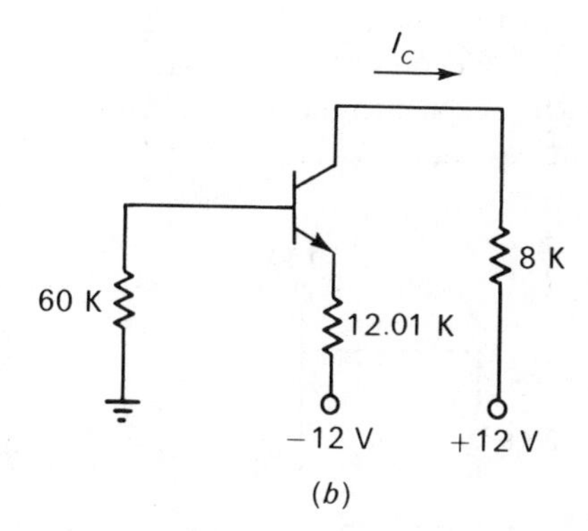

(*b*)

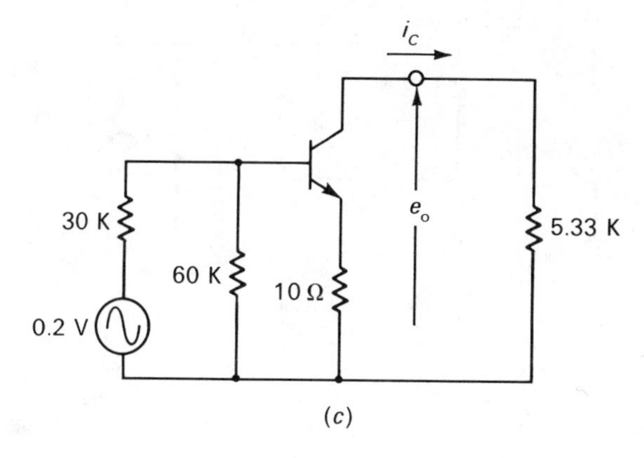

(*c*)

FIGURE PS 11-5

circuit and not have to model the transistor. Therefore, if the base circuit is thevenized, we obtain a $v_{oc} = 0.1333$ volts and $z_{th} = 20$ kilohms. Since r_e or h_{ie} isn't specified and the external emitter resistance R_E is only 10 ohms, the internal r_e should be estimated. Thus $r_e \approx 26 \text{ mv}/I_E = 30.1$ ohms, and assuming an r_b of, say 200 ohms, an educated guess for a low-power transistor, we obtain $h_{ie} = 200 + 30.1\ (50 + 1) = 1.74$ kilohms. Recalling that when looking into the base any external R_E looks like $R_E(\beta + 1)$, we have

$$i_b = \frac{0.1333 \text{ volt}}{20 \text{ kilohms} + 1.74 \text{ kilohms} + 0.01 \text{ kilohm} (50 + 1)}$$

$$= \frac{0.1333 \text{ volt}}{22.25 \text{ kilohms}} = 5.99\ \mu a$$

$$i_c = 50\,(5.99\ \mu a) = 0.2995 \text{ ma}$$

$$e_o = -(8 \text{ kilohms} \parallel 16 \text{ kilohms})\, i_c$$

$$= -5.33 \text{ kilohms}\,(0.2995 \text{ ma})$$

$$= -1.6 \text{ volts}$$

The overall voltage gain is

$$A_V = \frac{e_o}{e_{in}} = \frac{-1.6 \text{ volts}}{0.2 \text{ volt}} = -8$$

PS 11-6 Determine e_o in the circuit of Fig. PS 11-6*a*.
SOLUTION We must first estimate r_e to see if it is negligible with respect to the external emitter resistance. From the dc equivalent circuit of Fig. PS 11-6*b*,

$$I_E = \frac{12 \text{ volts}}{1.2 \text{ kilohms} + 60 \text{ kilohms}/(\beta + 1)} = 6.7 \text{ ma}$$

$$r_e \approx \frac{26 \text{ mv}}{6.7 \text{ ma}} = 3.88 \text{ ohms}$$

Therefore, with r_b assumed as 200 ohms

$$h_{ie} = r_b + r_e(\beta + 1)$$

$$= 200 + 3.88\,(101) = 0.588 \text{ kilohm}$$

Noting that the net ac load in the emitter is 1.2 kilohms $\parallel$ 2.4 kilohms, we obtain Fig. PS 11-6*c* which in turn leads to Fig. PS 11-6*d* if we try to determine the equivalent circuit presented to the flow of base current. $R_L(\beta + 1)$ allows for the fact that R_L not only carries i_b, but $i_c = \beta i_b$. Also, we may recognize the equivalent voltage divider and determine

$$e_o = \frac{80.8 \text{ kilohms}\,(0.6 \text{ volt})}{24 \text{ kilohms} + 0.588 \text{ kilohm} + 80.8 \text{ kilohms}}$$

$$= 0.46 \text{ volt}$$

Note that the voltage gain from the 1-volt input to the

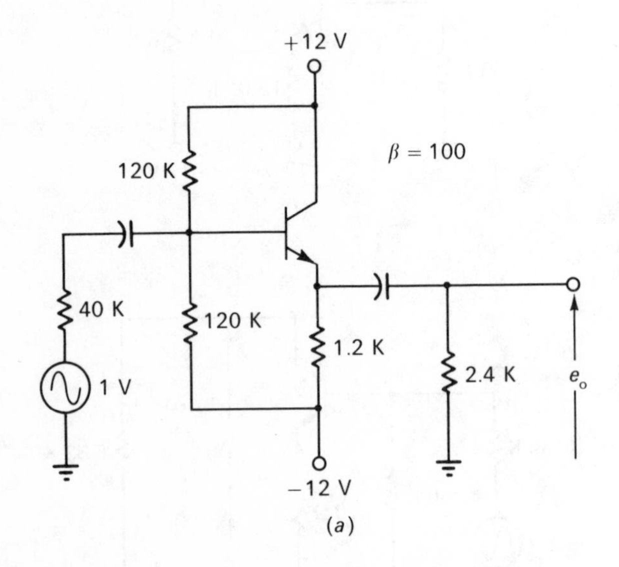

(*a*)

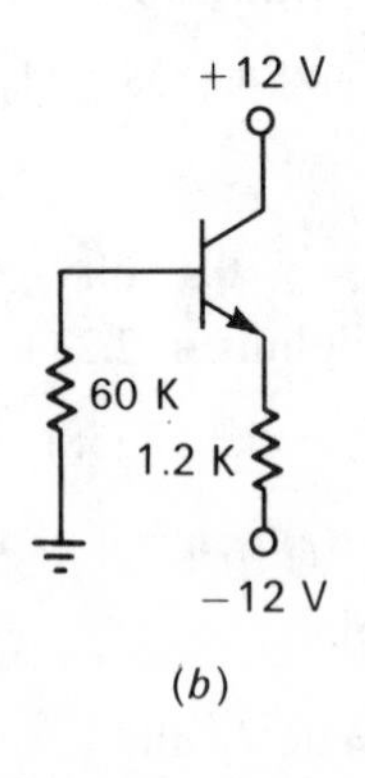

(*b*)

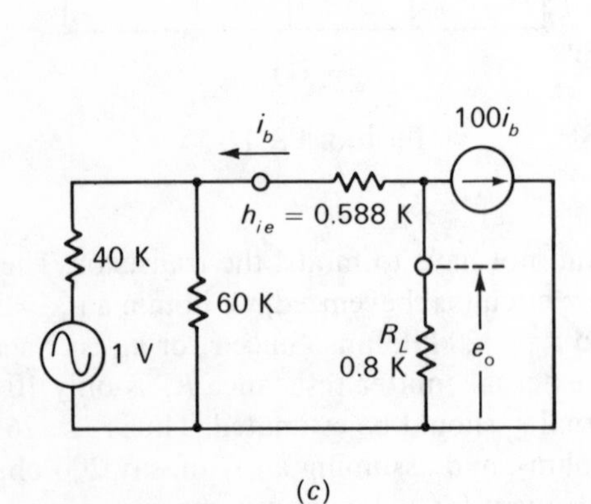

(*c*)

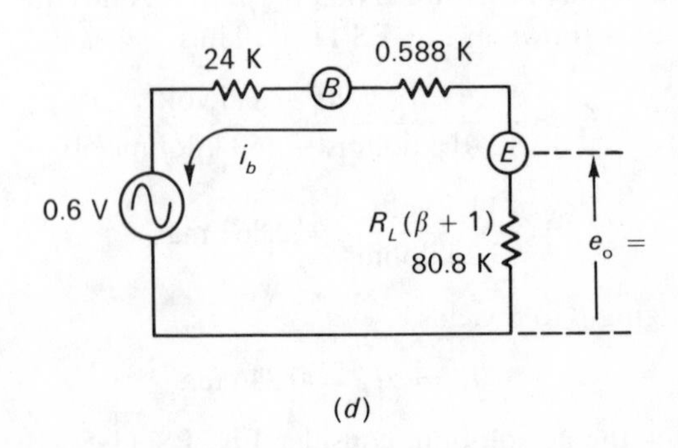

(*d*)

FIGURE PS 11-6

device's output is considerably less than one. This is to be expected, despite the popular notion that an EF has a gain of 1, because the 40-kilohm source impedance is severely loaded by the base bleeder resistors (120 kilohms ∥ 120 kilohms). Furthermore, the loading effect seen directly at the base terminal is $z_{in} = h_{ie} + R_L (\beta + 1) = 81.4$ kilohms which also is not high enough to be ignored.

PS 11-7 Since the input impedance of the EF is raised by increasing R_L, it might be interesting to determine the maximum possible input impedance by letting R_L approach infinity in the circuit of Fig. PS 11-7*a*. Determine $z_{in(max)}$.

SOLUTION A simple, approximate solution may be obtained by drawing the *h* parameter equivalent circuit of Fig. PS 11-7*b* with the $h_{re}v_{ec}$ source neglected. Since $1/h_{oe}$ is the effective resistance in the emitter, it follows that, insofar as looking into the base is concerned, $1/h_{oe}$ will appear to be $h_{fe} + 1$ times larger. Thus we obtain Fig. PS 11-7*c*. The same result could, of course, be obtained by formal circuit analysis procedures.

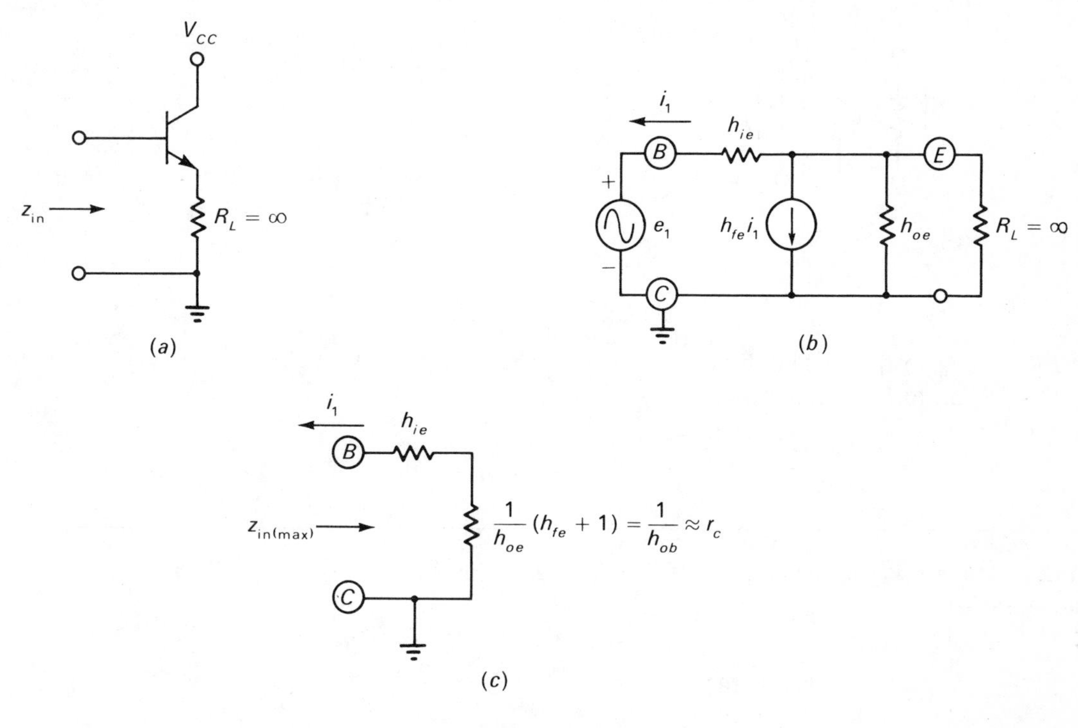

FIGURE PS 11-7

PROBLEMS WITH ANSWERS

PA 11-1 What is the impedance presented to the flow of: (*a*) base current, I_b, and (*b*) emitter current, I_e, in the circuit of Fig. PA 11-1? Assume the transistor is biased into its active region and $r_e = 40$ ohms, $r_b = 200$ ohms, $r_c = \infty$, $\alpha = 0.98$.

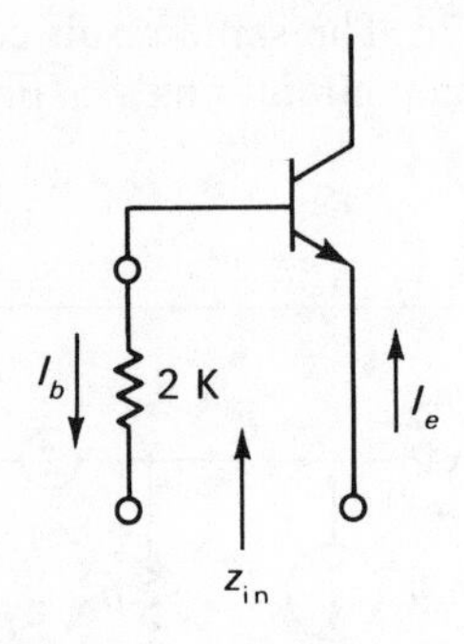

ANSWER z_{in} to I_b = 4.2 kilohms
z_{in} to I_e = 85 ohms

PA 11-2 Estimate V_{EC} in the circuit of Fig. PA 11-2. Estimate e_o.

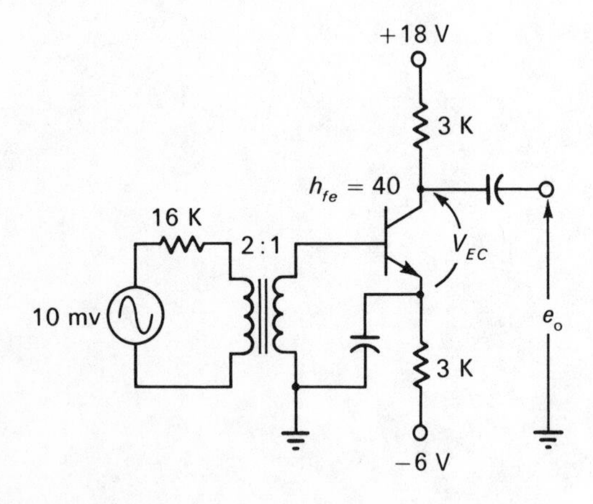

ANSWER V_{EC} = 12 volts
e_o = 0.12 volt

PA 11-3 Estimate the V_{in} that will just saturate the silicon transistor in Fig. PA 11-3.

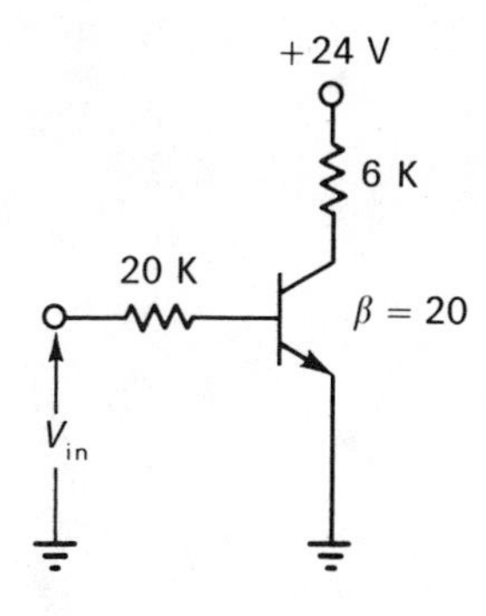

ANSWER 4.6 volts

PA 11-4 Assume a silicon transistor in Fig. PA 11-4. Estimate V_{EC} and e_o.

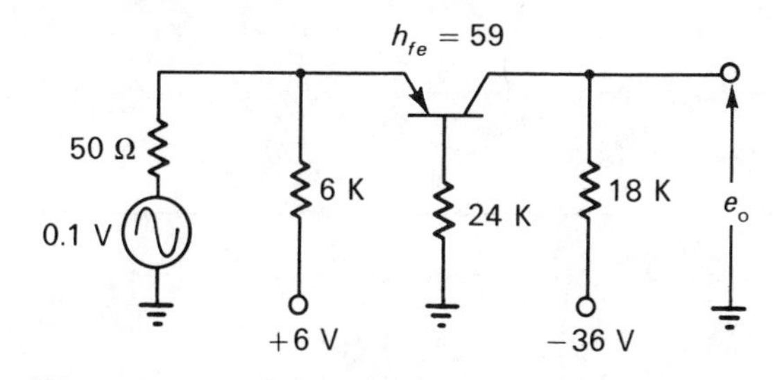

ANSWER

$$V_{EC} = -22 \text{ volts}$$
$$e_o = 3.63 \text{ volts}$$

PA 11-5 Estimate the value of R_X to decrease e_o to one-half its open-circuit value in Fig. PA 11-5.

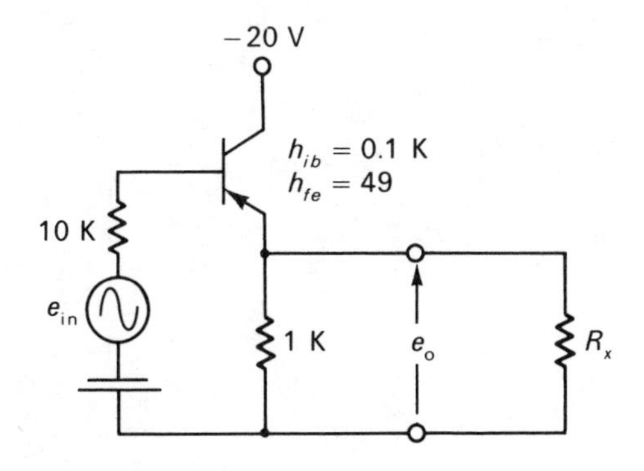

ANSWER 231 ohms

PA 11-6 Estimate the voltage gain in the circuit of Fig. PA 11-6.

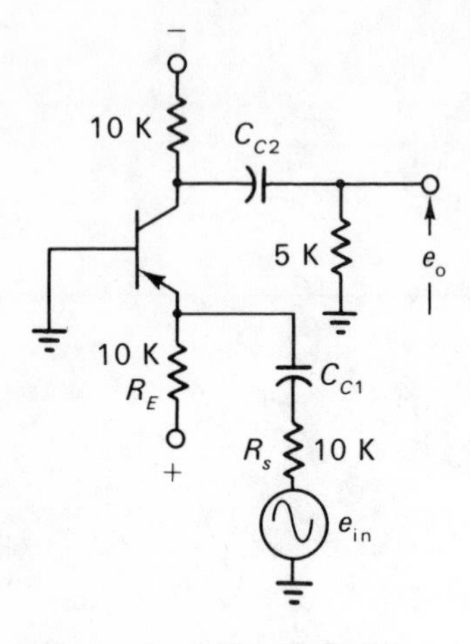

ANSWER $A_V = 1$

PA 11-7 Estimate e_o in Fig. PA 11-7.

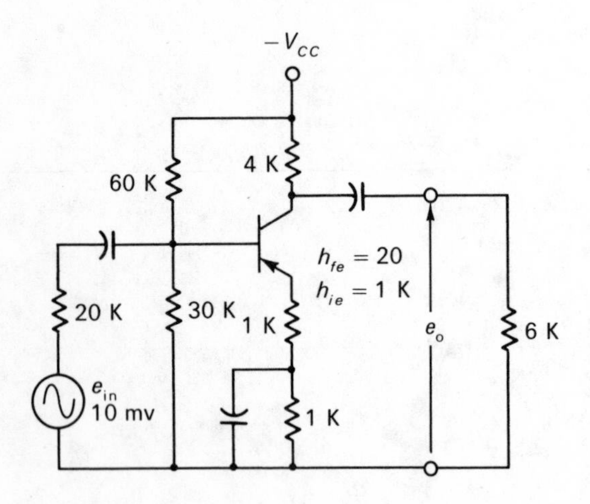

ANSWER $e_o = 7.5$ mv

PA 11-8 Determine z_{in} to e_g with $R_g = 0$ and A_V in the circuit of Fig. PA 11-8. Assume $R_g = 2.4$ kilohms, $R_B = 18$ kilohms, $R_L = 6.8$ kilohms, $h_{ie} = 1$ kilohm, $h_{re} = 0$, $h_{fe} = 70$, $h_{oe} = 0$ mho.

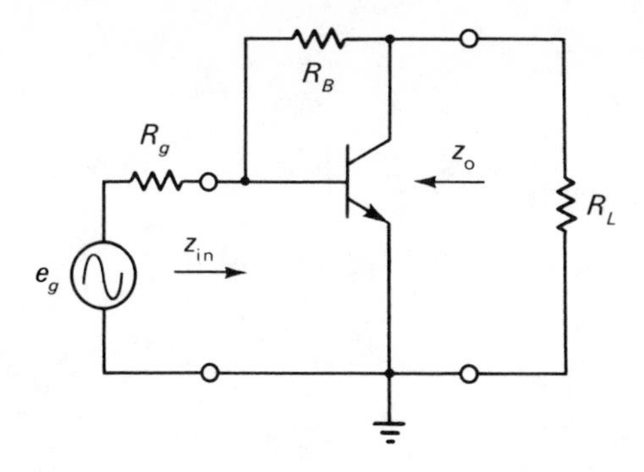

ANSWER $z_{in} = 0.336$ kilohm for $R_g = 0$
$A_V = -6.97$

PROBLEMS WITHOUT ANSWERS

P 11-1 Assuming the collector is reverse-biased and the emitter is forward-biased by some circuitry not shown in Fig. P 11-1, determine the impedance presented to: (*a*) the emitter current, I_e, and (*b*) the base current, I_b.

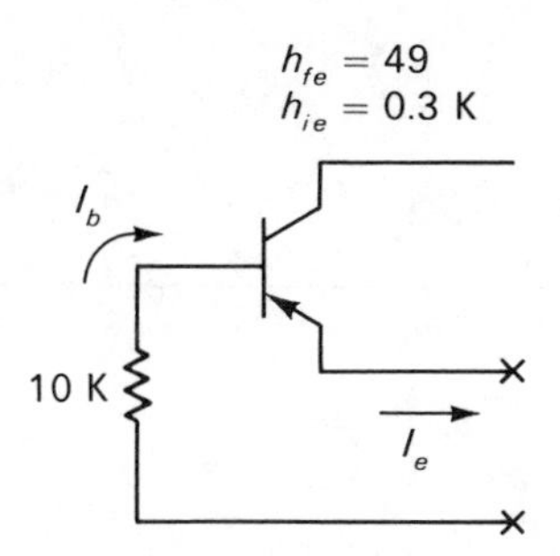

P 11-2 Determine the ac collector current, I_c, in Fig. P 11-2.

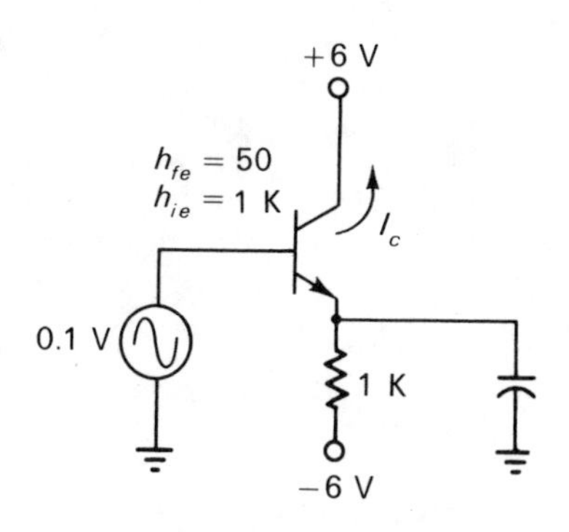

P 11-3 If $\alpha = 0.98$ and $\Delta I_{CBO} = 20\ \mu a$, determine ΔV_C in Fig. P 11-3.

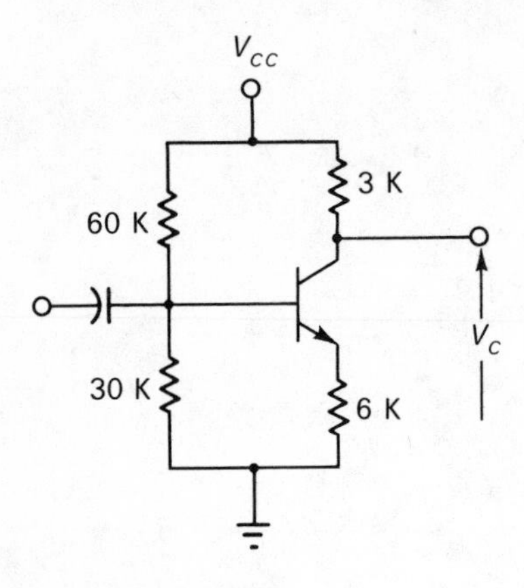

P 11-4 Estimate the output voltage e_o in Fig. P 11-4.

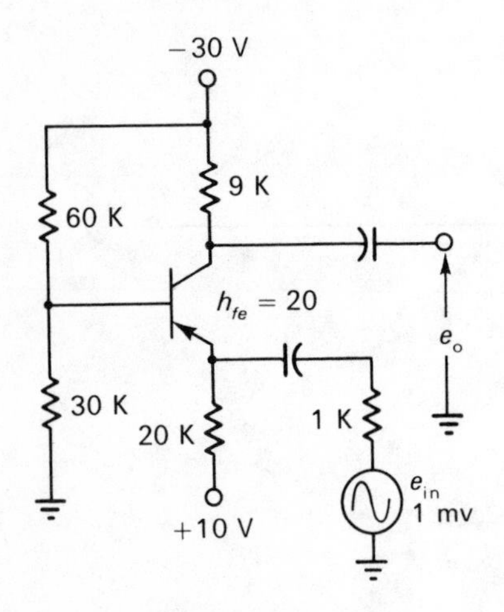

P 11-5 Determine the dc voltage V_E and the ac output in Fig. P 11-5. Assume a silicon transistor.

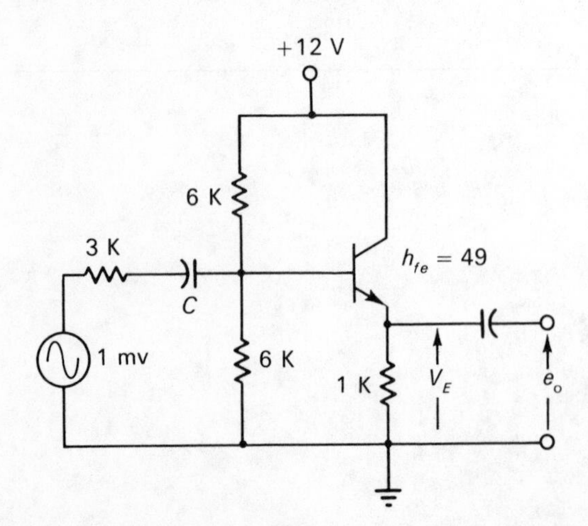

P 11-6 Estimate e_o without any detailed calculations in Fig. P 11-6. Assume a typical transistor with $h_{fe} \geqq 30$.

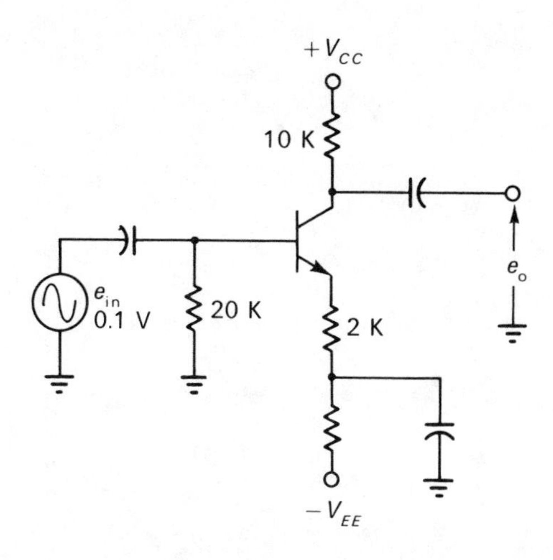

P 11-7 Design a small signal amplifier using a silicon transistor in the circuit configuration of Fig. P 11-7 for a voltage gain of approximately 10, stability factor $\leqq 6$, and base bleeder current equal to ten times the base current. Assume $V_{CC} = 18$ volts and a desired Q point at $V_{EC} = 6$ volts, $I_C = 1$ ma. Also assume $60 < h_{fe} < 100$, $50 < h_{FE} < 80$ and $R_L = 8$ kilohms. If the design is not feasible, what parameters would you preferably vary? Many solutions are possible.

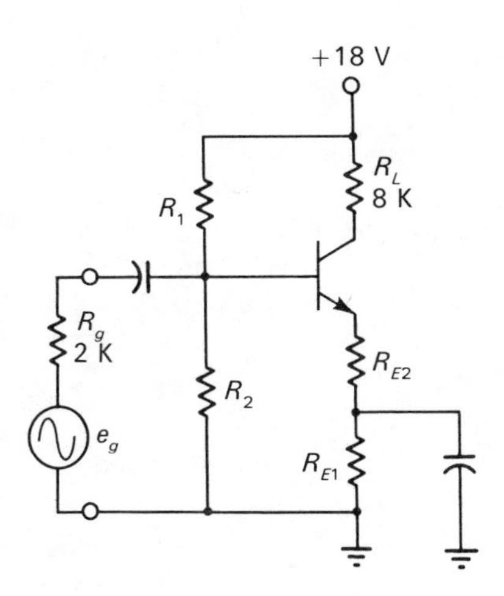

P 11-8 Determine the Thévenin's equivalent circuit seen looking into the output terminals of the amplifier in Fig. P 11-8. Assume $r_e = 18$ ohms, $r_b = 320$ ohms, $\beta = 80$, $r = 1.6$ megohms, $R_B = 10$ kilohms, and $R_E = 60$ ohms.

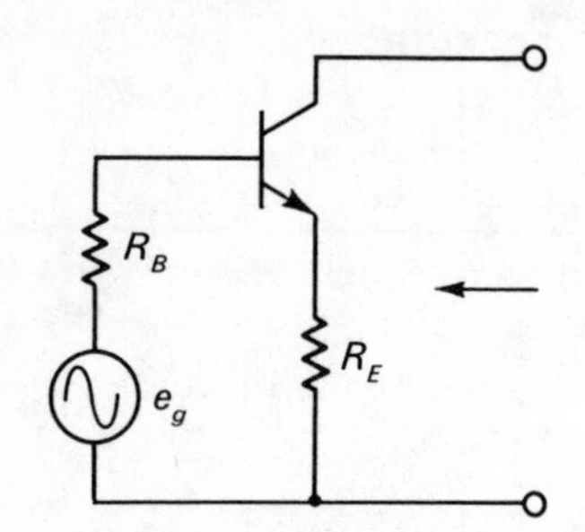

12 the field-effect transistor (fet)

The *field-effect* or *unipolar transistor* (FET), as it is sometimes called, belongs to a class of solid-state amplifying devices which, in particular, are characterized by their high input impedance in contrast to the relatively low input impedance of the junction or bipolar transistor that we previously discussed. Other advantages of the FET may include properties such as low internal noise, inherent reliability, ease of fabrication, and a square-law characteristic in which the current through the device varies with the square of the input voltage. Although this square-law characteristic may be undesirable in certain linear amplifying applications, it can be particularly valuable in certain high-frequency applications.

12-1 FET types

Figure 12-1 illustrates in block form the FET family. The FET family breaks down into two major subclassifications, the JFET, or *junction field-effect transistor,*

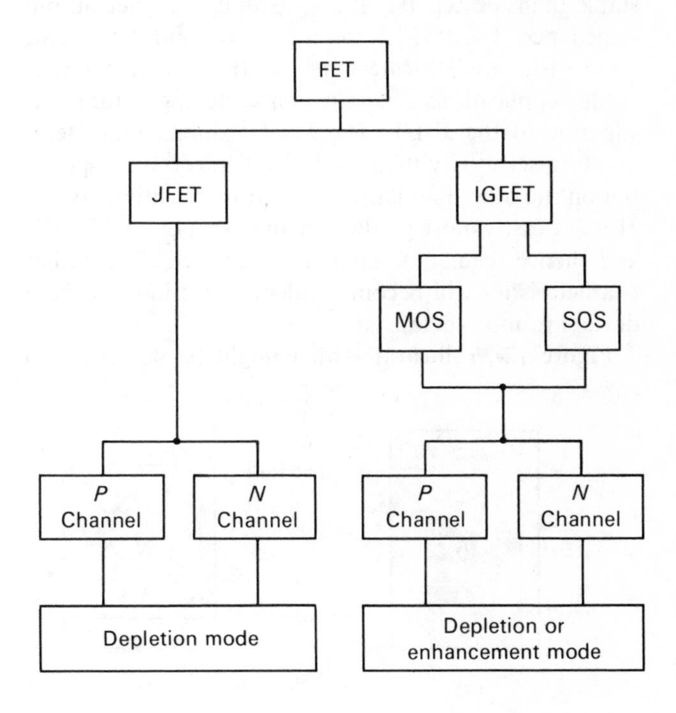

FIGURE 12-1

and the IGFET, or *insulated-gate field-effect transistor.* In the JFET the input signal looks into a reverse-biased *PN* junction which, as we have learned, presents a high input impedance, perhaps on the order of 100 megohms. In the IGFET the input signal actually looks into what amounts to one terminal of a capacitor with an insulating medium for the dielectric, and therefore, in the low-frequency spectrum where capacitive effects may be neglected, the input impedance is essentially that of the dielectric medium, which may be in the neighborhood of 10,000 megohms. Without going into details at this time, IGFET or insulated-gate field-effect transistors break down into two subclassifications which may increase as the technology advances. One type, and the most common, is the MOS or *metal on silicon* type, whereas the other is the SOS or *silicon on sapphire* type.

Both the JFET and the IGFET may be produced in complimentary form, that is, made of *P* material as well as *N* material, just as we have *PNP* and *NPN* transistors. In a field-effect transistor, however, we speak of a conducting or nonconducting channel; and this channel may be made of *P* or *N* material depending on the particular device. The JFET normally operates only in what is called the depletion mode, whereas the IGFET may be made so that one type operates in just the depletion mode, and another in the depletion or enhancement mode. The meaning of these terms will also be clarified as we move along.

In general the JFET tends to be less noisy and more stable than the IGFET. It also exhibits a higher output impedance. The IGFET, on the other hand, is in some ways easier to fabricate than the JFET, and, for particular applications generally of a switching nature, it is superior to the JFET. The JFET exhibits more temperature sensitivity than the IGFET in certain respects; but on the other hand, the temperature sensitivity of the JFET is vastly more predictable than that of the IGFET, and hence can be better compensated for. Other characteristics will become evident as we look at these devices in more detail.

Figure 12-2*a* illustrates what might be the first step

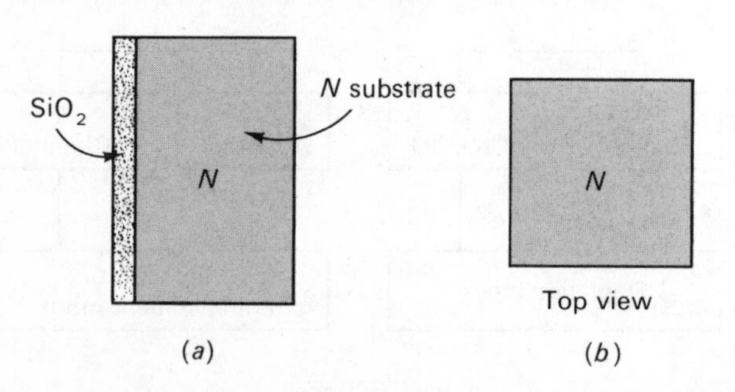

FIGURE 12-2

in the construction of a JFET. A layer of silicon dioxide (SiO_2) is deposited on some *N*-type material which is called the substrate. The silicon dioxide is an insulator. If we could look through the dioxide, a top view of Fig. 12-2*a* would appear as shown in Fig. 12-2*b*. Using photochemical etching methods a portion of the dioxide film is removed and some *P* material is diffused into the *N*-type substrate. The silicon dioxide layer is then reformed so that we now have the structure shown in Fig. 12-3*a* and a top view as shown in Fig. 12.3*b*.

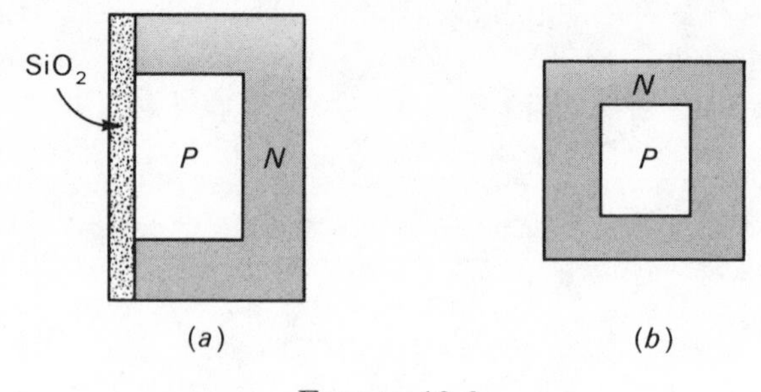

FIGURE 12-3

A window is etched in the dioxide coating and this time some *N*-type material is diffused into the *P* region as shown in Fig. 12-4*a*. The dioxide film is again reformed. A top view of Fig. 12-4*a* is shown in Fig. 12-4*b*. The dioxide coating is then etched again so that ohmic contacts may be made to the *N* and *P* regions. We

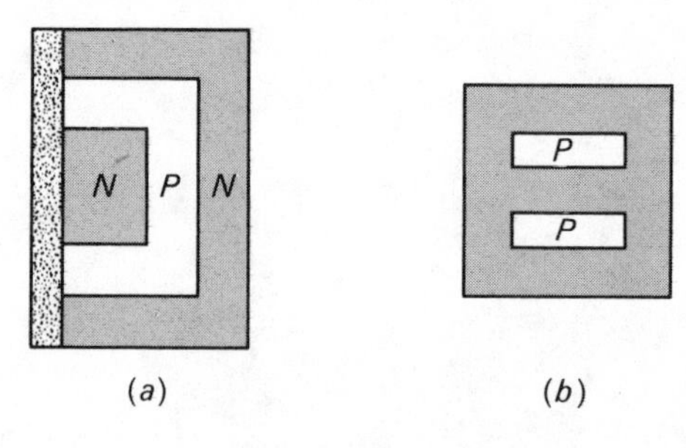

FIGURE 12-4

now have a structure with the cross-sectional view shown in Fig. 12-5*a* and the top view shown in Fig. 12-5*b*. Note that in this case we have produced a channel of *P* material running between two layers of *N* material which are in effect electrically tied together. Since this channel is constructed of *P* material we have what is called a *P*-channel FET. Were the channel constructed of *N* material it would be called an *N*-channel FET. In that case the region surrounding the channel would be made of *P* material rather than *N* material, as in this

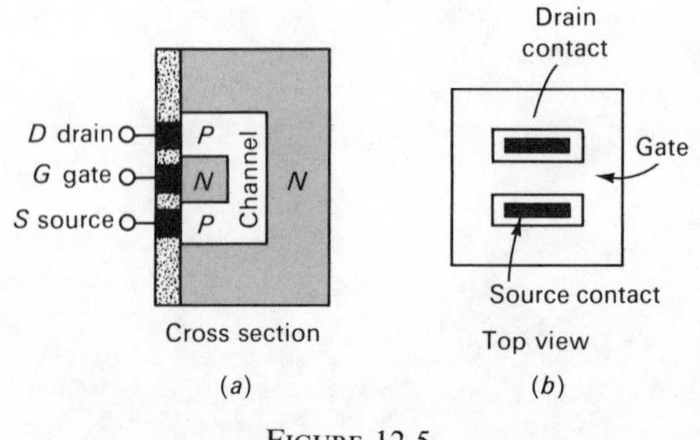

FIGURE 12-5

case. One end of this channel is called the *source* and it is designated by the letter S, and the other end is called the *drain* and it is designated by the letter D. The material surrounding the channel is called the *gate* and it may be designated as terminal G. If the construction of the device is such that there are two separate gates, the gates may be designated as $G1$ and $G2$ or perhaps G and some other symbol (SB) to indicate that the other gate may be a substrate. Irrespective of the construction details, the goal of all the JFET manufacturing techniques is to generate a channel of semiconductor material sandwiched between two layers of opposite-type semiconductor. Now that we have some familiarity with the construction of a JFET, we will direct our attention to the theory of its operation.

12-2 JFET operation

To understand how the JFET operates, direct your attention to Fig. 12-6, which is a simplified representation of the previously developed structure shown in

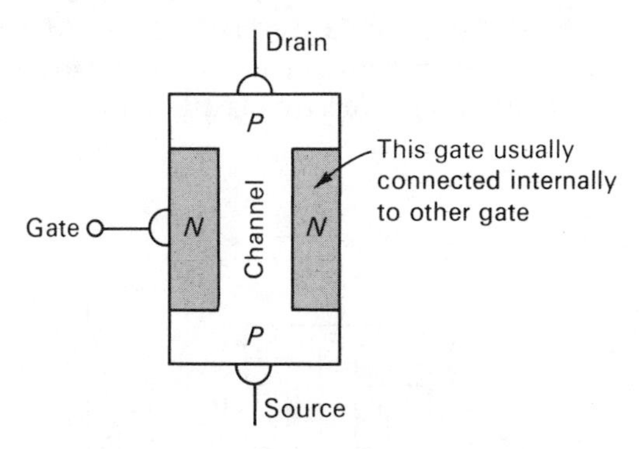

FIGURE 12-6

Fig. 12-5*a*. The bulk of the material consists of a rather lightly doped bar of *P*-type material so that we obtain a *P*-channel JFET. On each side of the channel we have regions of heavily doped *N*-type material, and these regions are the gates of the JFET. The two gates or end regions in this case are actually connected together internally although in some particular FET types the gate connections may be brought out separately. We will assume that the gates are internally connected even though the connection isn't shown in the interest of simplicity. The channel itself is terminated in the source and drain contacts.

In Fig. 12-7 the depletion regions which exist about the *PN* junctions with no applied voltage are indicated by the shaded areas. Since the gate material is doped much more heavily than the channel material, it has much lower resistivity. Now you will recall that the depletion region extends much further into a region of high resistivity than into a region of low resistivity. Hence

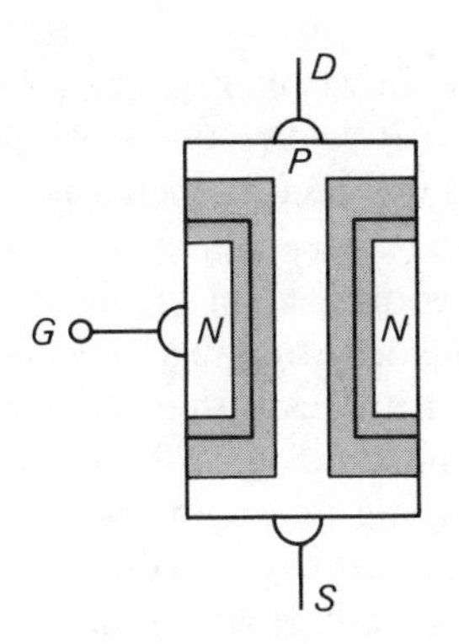

FIGURE 12-7

the depletion region can be better sustained and will extend further into the channel than it does into the gate material. Actually, we are not interested in the depletion region in the gate material at all. It is the behavior of the depletion region in the channel that is of interest.

In Fig. 12-8*a* the source and drain have been shorted together so that the *source-to-drain voltage* $V_{SD} = 0$. An adjustable potential, V_{SG}, is applied between source and gate. Now, when $V_{SG} = 0$ the depletion region will

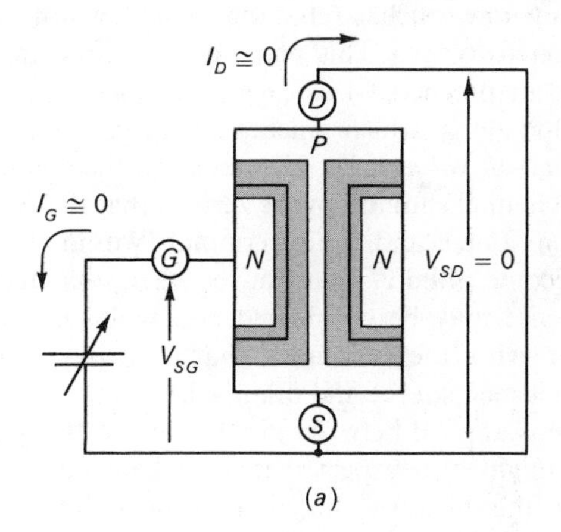

(*a*)

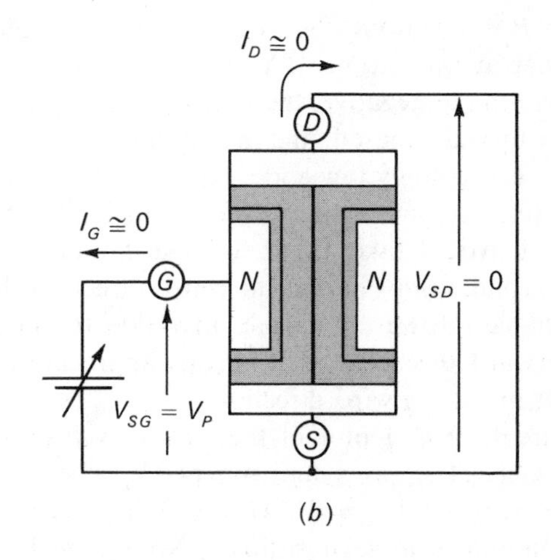

(*b*)

FIGURE 12-8

look exactly as shown in Fig. 12-7. Then if V_{SG} is increased in the positive direction as shown in Fig. 12-8*a*, the *PN* junction will become increasingly reverse-biased, which causes the depletion region to widen; and since the channel is composed of higher resistivity material than the gate, the depletion region will grow further into the channel than it does within the gate region. As you can see, the channel in Fig. 12-8*a* has become somewhat constricted all along its length due to V_{SG}. Be sure you note that V_{SG} is effective in controlling the depletion region *along the full length of the channel.* The *gate current,* I_G, will essentially be equal to zero as well as the *drain current,* I_D, since we effectively just have some reverse-biased diodes here. Actually, some minute leakage currents will flow. However, this is of no interest at the moment. You can appreciate how the input impedance seen looking into the gate and source terminals of a FET will tend to be high since you are looking into a reverse-biased *PN* junction, whereas in the bipolar transistor you are looking into a forward-biased emitter diode which presents a relatively low input impedance.

If we keep increasing V_{SG} in the positive direction, a point is reached where the depletion regions have actually grown throughout the full length of the channel and they now touch, so that the channel width has been constricted to zero. This point, at which the channel is choked or pinched off, is called the *pinch-off point* and the value of V_{SG} which pinches off the channel is called the *pinch-off voltage,* V_P. Theoretically, the conductivity of the channel should now be zero so that the resistance between source and drain terminals within the device has become infinite. You can see, then, that the JFET in a sense may be considered as a voltage-controlled resistor with the resistance that is being controlled lying between source and drain, whereas the controlling voltage is applied between the source and the gate.

We might also expect that if we kept on increasing V_{SG} in the positive direction, a breakdown voltage point would be reached as is characteristic of any reverse-biased *PN* junction. We will not concern ourselves with that at this moment. You might also note that if V_{SG} were made negative, the gate diodes would tend to become forward-biased; and the input impedance would become exceedingly low and a large gate current would flow. Actually, since the JFET is normally made of silicon, it would take a V_{SG} equal to about -0.6 volt before conduction in the forward direction became appreciable. However, it isn't advisable to operate in this forward direction as z_{in} drops appreciably below 0.4 volt in the forward direction.

Figure 12-9 is a plot of the *source-to-drain channel conductance,* G_{SD}, as a function of V_{SG} when $V_{SD} = 0$. Note that when $V_{SG} = V_P$, $G_{SD} = 0$. This corresponds to an infinite channel resistance. Note also that when V_{SG} swings negative, which corresponds to a forward-

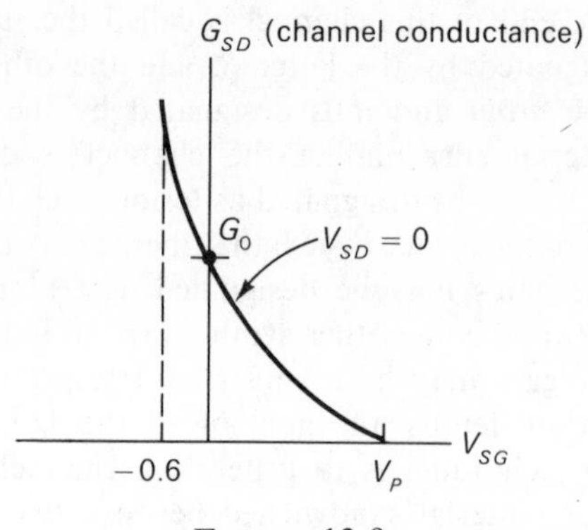

FIGURE 12-9

biased condition of the gate diode, the channel conductance increases markedly.

Now consider Fig. 12-10. Here we set V_{SG} equal to zero and are going to study the effect of V_{SD} upon the drain current I_D. When $V_{SD} = 0$, I_D clearly is equal to zero. As we increase V_{SD} in the negative direction, I_D starts to increase in magnitude. Algebraically I_D will be a negative quantity, since V_{SD} must force a current in the direction opposite to the assumed reference direction of I_D. V_{SD} must be a negative quantity in order to guarantee the gate junction is reverse-biased.

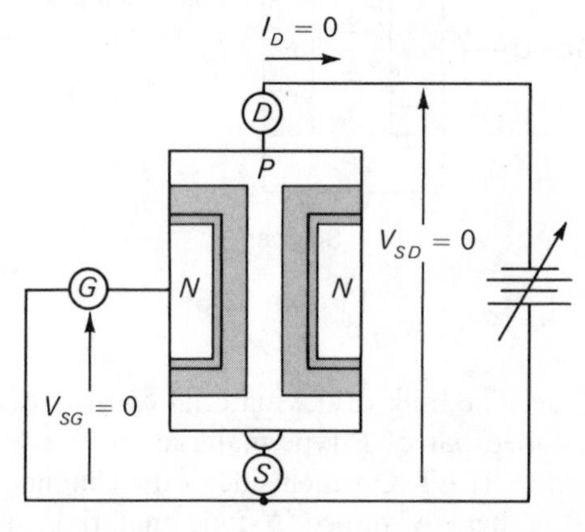

FIGURE 12-10

As we increase the magnitude of V_{SD} the reverse bias tends to increase and the depletion region will widen. However, it does not widen uniformly, as shown in Fig. 12-11*a*. The depletion region is actually wider at the drain end of the channel than at the source end, which means the channel is more constricted at the drain end than at the source. This effect is caused by the voltage drop along the length of the channel due to the flow of drain current which causes the drain end of the gate diode to be more reverse-biased than the source end. We will take a closer look at this later. For the moment, just accept this nonuniform growth of the depletion region due to V_{SD}. As long as the channel remains

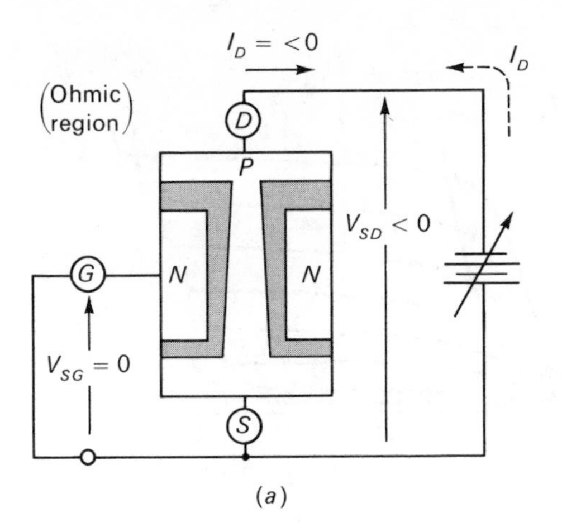

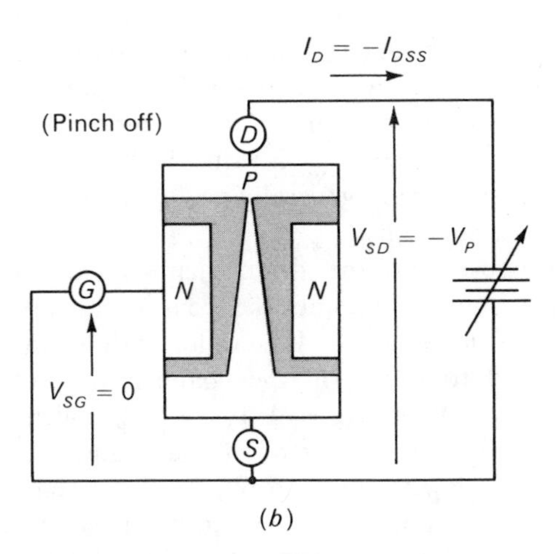

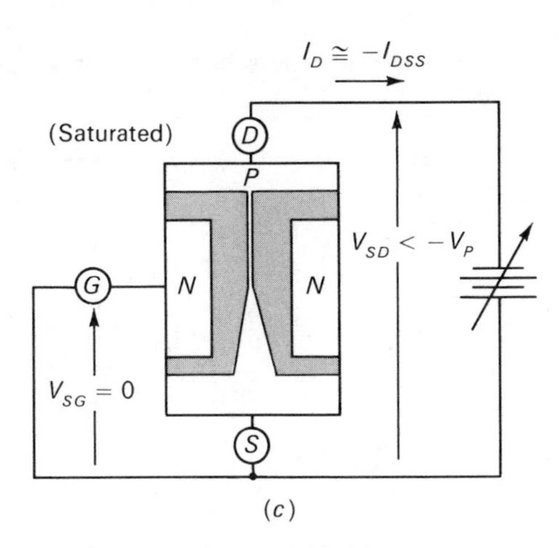

FIGURE 12-11

appreciably open the drain current is essentially proportional to the voltage V_{SD}. This is sometimes called the ohmic operating region of the JFET.

If we keep increasing the magnitude of V_{SD}, the magnitude of the drain current also increases until a point is reached where the depletion regions almost touch at the drain end of the JFET. This situation is illustrated in Fig. 12-11*b*. One might suspect that if V_{SD} is increased still further, the channel would become pinched off due to the depletion regions touching. Actually this effect does not take place. What does happen is that when the depletion regions *almost* touch, further increases in V_{SD} do not bring the depletion regions closer together at the drain end but instead the depletion region tends to grow closer together along the length of the channel as shown in Fig. 12-11*c*. Go back to Fig. 12-11*b* for the moment. At the point where the depletion regions almost meet, the source-to-drain voltage, V_{SD}, is equal to the negative of the pinch-off voltage, V_P, that was previously applied as the V_{SG} required to just choke off the channel. The value of drain current at the point where the depletion regions almost touch ($V_{SD} = -V_P$) is labeled I_{DSS} or $I_{D(\text{on})}$ as it is sometimes called.[1] Now if the magnitude of V_{SD} is increased beyond V_P, we have the situation shown in Fig. 12-11*c*. Note that the drain current remains substantially constant at the value I_{DSS}.

At this point a valid question is, "Why does I_D tend to remain essentially constant once V_{SD} increases beyond V_P in magnitude?" The explanation is that for every unit increase in V_{SD}, the depletion regions tend to grow closer together along the length of the channel which increases the channel resistance. It turns out that for every unit increase in V_{SD} the channel resistance increases almost in proportion. Consequently, the drain current will tend to remain almost constant and increase only slightly in magnitude with an increase in V_{SD} beyond $|V_{SD}| = |V_P|$. This constant drain-current region of the JFET is commonly referred to as the saturated or constant-current region.

In case the wedge shape of the depletion region due to the effect of V_{SD} isn't clear, consider Fig. 12-12. Diodes

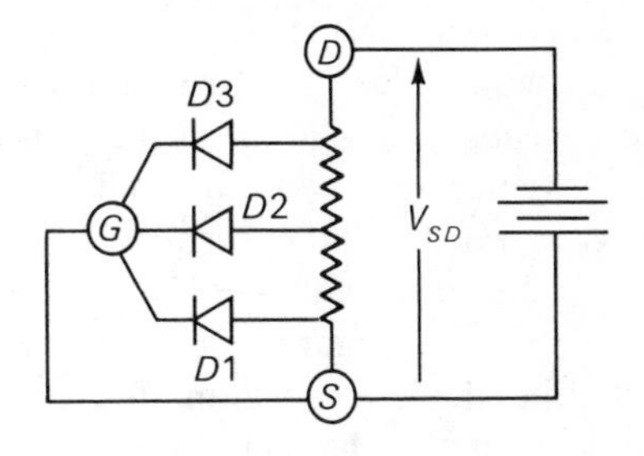

FIGURE 12-12

*D*1, *D*2, and *D*3 represent the *PN* junction at the source, middle, and drain ends of the channel. The resistance corresponds to channel resistance and the anodes of the diodes represent the *P* material of the channel. As you can see, the flow of drain current through the channel

[1] I_D (on) and I_{DSS} are not necessarily identical but usually close enough to be used interchangeably.

resistance creates a voltage drop along the channel such that $D3$ sees more reverse bias than $D2$, and $D2$ is more reverse-biased than $D1$. Since the depletion region is a function of the reverse bias, it should be clear that the depletion region will be wider at the drain end than at the source end of the channel.

Figure 12-13 is a plot of the drain current versus the

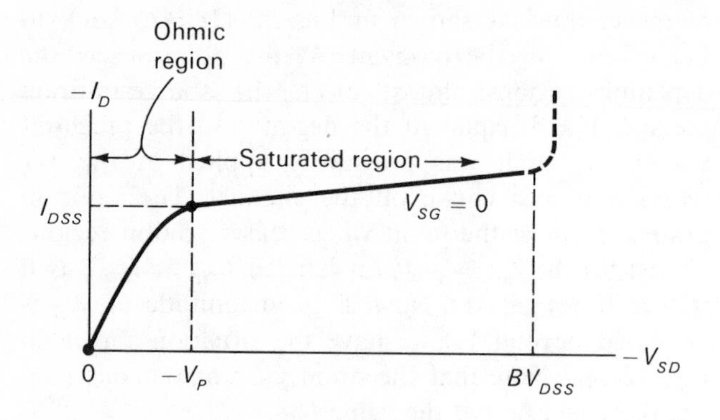

FIGURE 12-13

source-to-drain voltage for a P-channel FET.[1] Note that when the magnitude of V_{SD} is equal to the magnitude of V_P, the drain current tends to remain substantially constant at the value I_{DSS}. This current is plotted for $V_{SG} = 0$. If V_{SD} is increased in magnitude we reach a point where the reverse-biased diode at the drain end of the channel will break down and the drain current will increase considerably and damage the device unless it is limited by some external resistance to a safe value. This breakdown voltage when $V_{SG} = 0$ is sometimes designated as BV_{DSS}, and typically may range from 20 to perhaps 250 volts depending on the particular FET in question.

Figure 12-14 illustrates the P-channel JFET volt-ampere characteristics for the drain current, I_D, as a function of V_{SD} with different values of V_{SG}. This plot is called the *common-source drain characteristic* and it illustrates, in essence, the combined effect or superposition of V_{SG} and V_{SD} in controlling I_D. For this particular FET the drain voltage V_{SD} for pinch-off is given as -8 volts. This figure also means that when $V_{SG} = 8$ volts the depletion region due to V_{SG} only will have choked off the channel. Note that I_D is very small at $V_{SG} = 7$ volts and it may be assumed negligible at $V_{SG} = 8$ volts, in most cases. Just remember the pinch-off voltage is the voltage that must be applied between source and gate to cut off the drain current.

In practice it's quite difficult to measure V_P since the

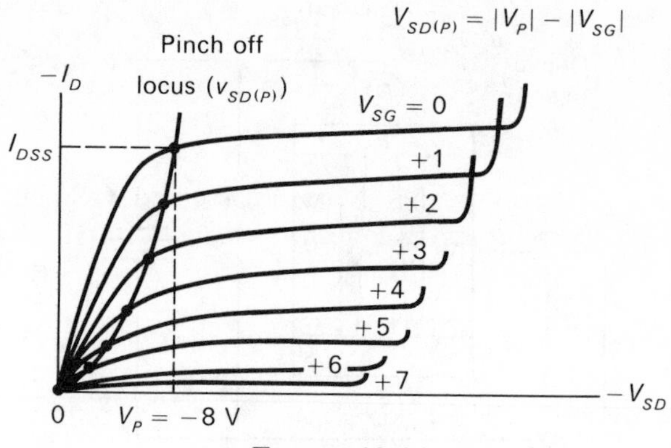

FIGURE 12-14

transition point from the ohmic to the saturated region, or the point at which V_{SG} reduces I_D to approximately zero, is not sharply defined. Hence some JFET data sheets define a $V_{SG} = V_{SG(\text{off})}$ value at which the drain current is some small (generally negligible) value which is sometimes called $I_{D(\text{off})}$.

Superimposed on the drain characteristic of Fig. 12-14 are a number of points, the locus of which is called the pinch-off locus, $V_{SD(P)}$. It is a plot of those values of V_{SD} required to pinch off the channel with different values of V_{SG}. When $V_{SG} = 0$, $V_{SD} = |V_P|$, which is the V_{SD} value required to pinch off the channel. As V_{SG} increases, the V_{SD} required to pinch off the channel or move the operating point into the saturated region decreases since V_{SG} is aiding V_{SD} in pinching off the channel.

Figure 12-15 illustrates the schematic representation

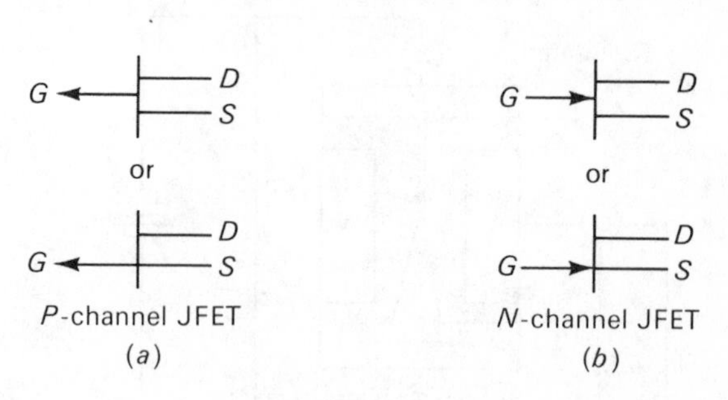

FIGURE 12-15

for the P-channel JFET, and Fig. 12-15*b* illustrates the N-channel JFET. The arrowhead denotes the type of semiconductor used for the gate. A gate arrow directed out of the device indicates an N-type gate, whereas a gate arrow directed into the device indicates a P-type gate. The situation is analogous to the emitter lead of a junction transistor. The channel of the JFET will be composed of a material opposite to that of the gate. Thus in Fig. 12-15*a* we have an N-type gate which implies a P-channel JFET, and in Fig. 12-15*b* we

[1] Most data sheets use V_{DS} and V_{GS} rather than V_{SD} and V_{SG}. This should create no problem if the voltage notation is understood. In our notation, the tail of the arrow (or the first subscript) represents the reference point to which the potential at the head of the arrow (or second subscript) is being compared.

have a P-type gate which implies an N-channel JFET. Memorize these symbols.

12-3 IGFET operation

Figure 12-16*a* illustrates the depletion-mode IGFET, and Fig. 12-16*b* the enhancement-mode IGFET. As

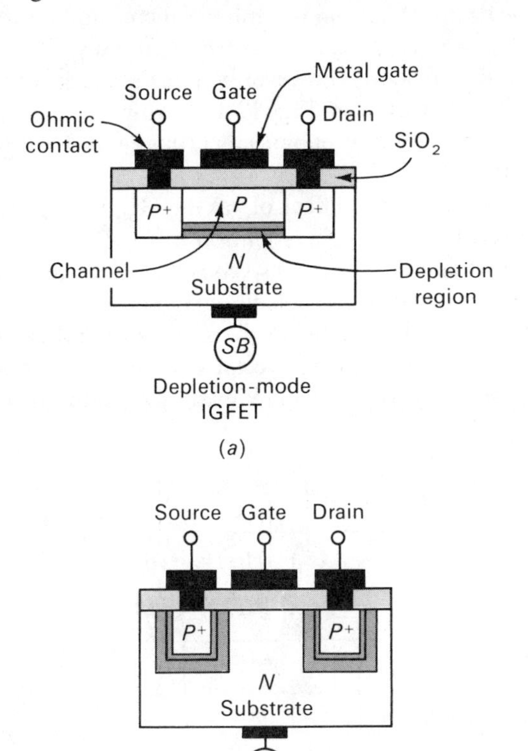

FIGURE 12-16

with the JFET we have three terminals: source, gate, and drain, and the functions they perform are similar to the JFET. A fourth terminal which may be used as another gate goes to the substrate which represents the bulk of the material from which the device is fabricated. The mode of control, however, is somewhat different in the IGFET than in the JFET. In the JFET we used a reverse-biased semiconductor junction to control the width and the channel. In the IGFET a metal gate is isolated from the semiconductor channel by a thin insulating layer of silicon dioxide (SiO_2). Because of this insulated gate the input resistance of the IGFET is extremely high, and it is not affected by the polarity of V_{SG} since we do not have to worry about forward- or reverse-biasing a PN junction. Furthermore, the IGFET gate leakage current is not as thermally sensitive as the leakage current of the reverse-biased gate-to-channel junction in the JFET. The IGFET, however, is not as stable with respect to time and temperature as the JFET; and it is noisier and more sensitive to radiation effects. However, the IGFET is very useful in switching applications; and, because of certain advantages in fabricating techniques, it is finding a wide use in high-frequency and digital type circuits. Inspection of Fig. 12-16*a* clearly indicates that the depletion-mode IGFET consists of a heavily doped source and drain region connected by a moderately doped channel made of the same material as the source and drain. This means the depletion-mode IGFET contains a normally conducting channel with zero voltage applied to the device. On the other hand, the enhancement-mode IGFET of Fig. 12-16*a* clearly indicates that the source and drain are electrically isolated by the substrate which is of the opposite type semiconductor material. Hence we have a depletion region about the source and about the drain.

In Fig. 12-17*a* we now consider the effect of V_{SD} upon the drain current with the condition that the source is tied to the substrate, as is usually the case, and the gate is tied to the source so that $V_{SG} = 0$. The drain voltage, V_{SD}, is maintained negative in this P-channel IGFET so that the PN junctions in the device are all reverse-biased or at least zero-biased as at the

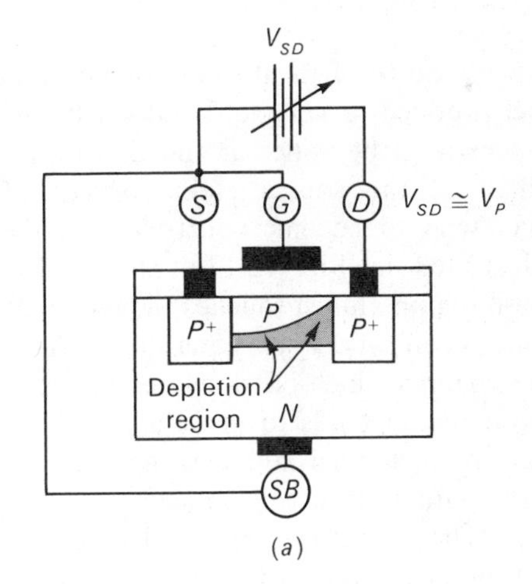

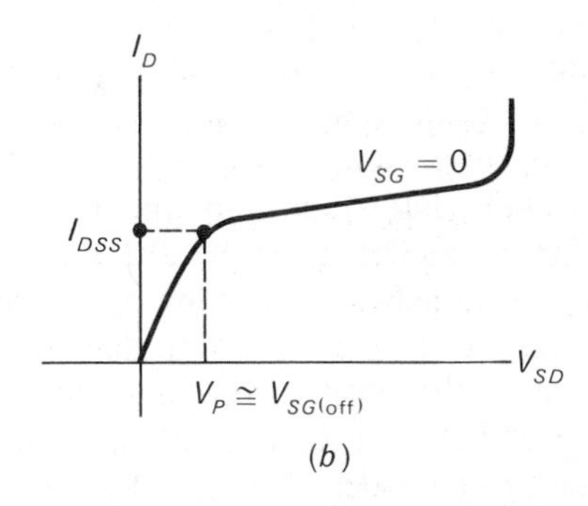

FIGURE 12-17

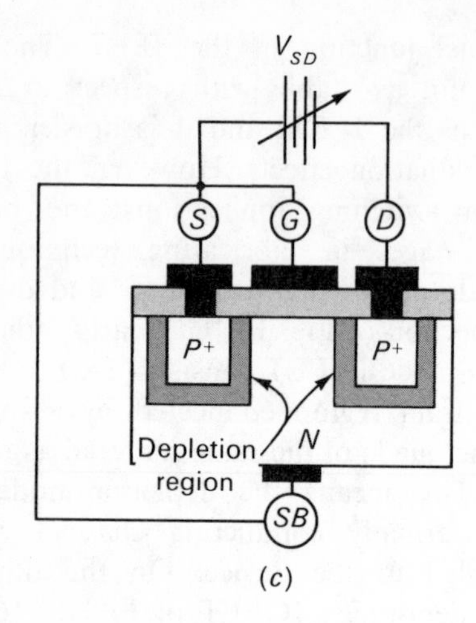

(c)

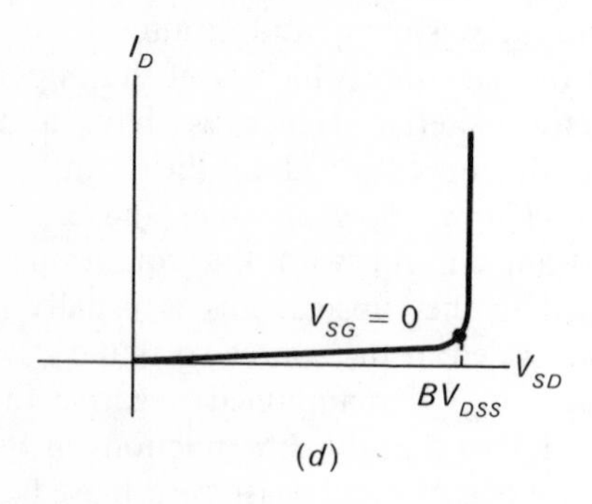

(d)

FIGURE 12-17 (*Continued*)

source. As with the JFET the flow of drain current along the channel produces a voltage drop which causes the depletion region to be wider at the drain end of the channel than at the source end. Consequently the channel will tend to become constricted at the drain end as illustrated in Fig. 12-17*a*. Specifically, when the depletion region almost touches the oxide layer, the channel pinches off; and V_{SD} is equal to V_P. As with the JFET, true channel pinch-off does not occur. Instead, the depletion region tends to grow along the channel with some small channel still existing. The resultant volt-ampere characteristic will appear as illustrated in Fig. 12-17*b*. The curve is very similar to that of a JFET, and in effect the JFET also operates in the depletion mode. Note that increasing V_{SD} sufficiently causes voltage breakdown to occur. Once the channel pinches off, the drain characteristic tends to approach that of a constant-current device. It turns out that the slope of the drain characteristic for the depletion-mode IGFET is larger than that for the JFET. This implies that the output impedance seen looking between drain and source for the IGFET is lower than that for the JFET. Typically, the output impedance for the JFET may be several hundred kilohms or more, whereas it may only be on the order of 20 to 50 kilohms for the IGFET.

In Fig. 12-17*c* we now consider the enhancement-mode IGFET with the substrate tied to the source, and the gate also tied to the source so that $V_{SG} = 0$. In this case, varying V_{SD} causes no current to flow because there is no channel in existence between the source and the drain; and all we do by making V_{SD} increasingly negative is simply widen the depletion region around the source and the drain. Hence, in the enhancement-mode IGFET with $V_{SG} = 0$ there can be no current flow. If V_{SD} is increased sufficiently, however, voltage breakdown may occur as illustrated in Fig. 12-17*d*, which is the drain characteristic for the enhancement-mode IGFET with $V_{SG} = 0$.

Now, to study the effect of varying V_{SG} in depletion-mode IGFET, turn your attention to Fig. 12-18. In the interest of simplicity we may consider the gate as one plate of a capacitor, the silicon dioxide as the dielectric, and the channel underneath the gate as the other plate of the capacitor. You will recall that in a capacitor, a charge on one plate is accompanied by an induced equal

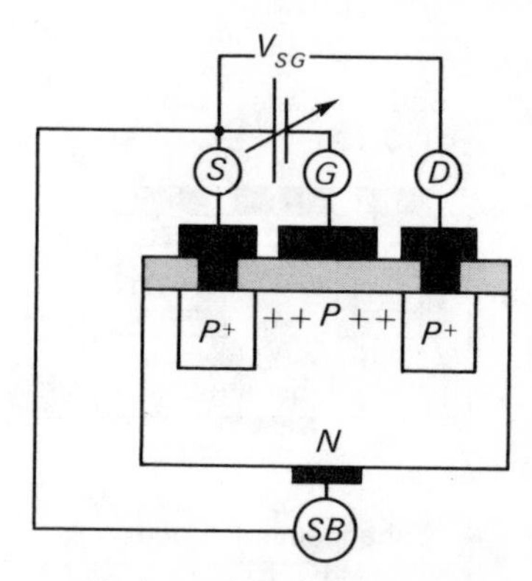

P-channel depletion type being enhanced

(*a*)

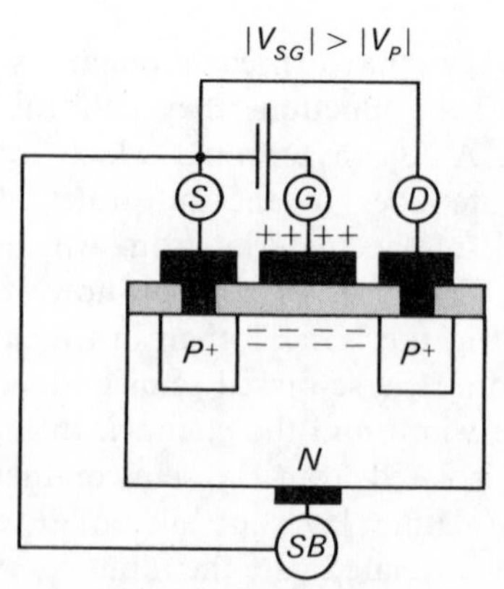

P-channel depletion type being depleted

(*b*)

FIGURE 12-18

and opposite charge on the other plate. Hence if we charge (bias) the gate negative with respect to the source as shown in Fig. 12-18, the negative charge on the gate will induce an equal and opposite positive charge in the channel underneath. Since this channel has normally *P*-type material, inducing further positive charges (holes) to migrate into this region increases the conductivity of the channel. Hence the conduction of drain current would be enhanced.

On the other hand, if the gate is biased positive with respect to the source as shown in Fig. 12-18*b*, the channel acquires an equal and opposite negative charge. If the number of induced negative charges in the channel is just equal to the number of positive charges due to the *P* material that the channel was originally composed of, the channel conductivity should theoretically become zero as there would be no more available carriers. At this point V_{SG} is said to be equal to the pinch-off voltage V_P. In Fig. 12-18*b* V_{SG} is greater than V_P in magnitude since the channel has been changed from *P*- to *N*-type as evidenced by the negative signs.

Figure 12-19 illustrates the drain characteristic for a

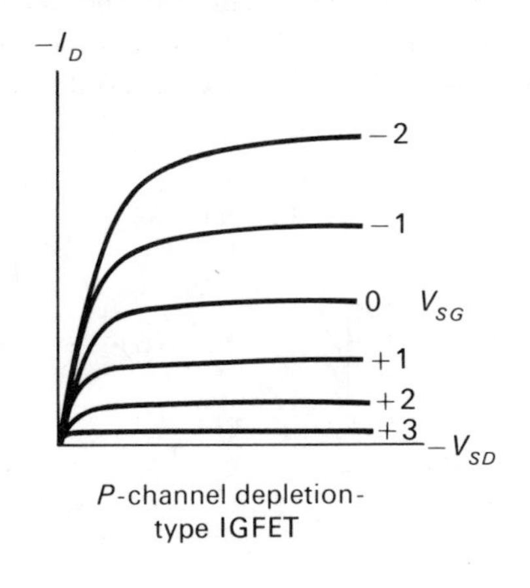

P-channel depletion-type IGFET

FIGURE 12-19

P-channel depletion-type IGFET. Note that V_{SG} can swing positive or negative and that in one case the magnitude of the drain current increases and in the other case it decreases. Thus the drain current is controllable for both polarities of V_{SG} which is not true for a JFET since the *PN* junction at the gate can become forward-biased. Figure 12-20 illustrates what the transfer characteristic or plot of drain current versus V_{SG} for any given V_{SD} might look like for a depletion-type IGFET. The only restriction in Fig. 12-20 is that the magnitude of V_{SD} be greater than the pinch-off voltage so that operation is in the saturated region. Note that the drain current can increase beyond the value of I_{DSS} which is pretty close to the limiting value in the JFET.

Now let us consider the effects of V_{SG} in the enhance-

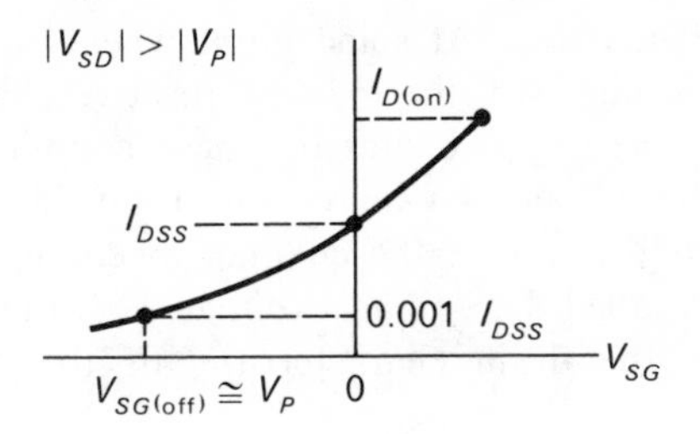

FIGURE 12-20

ment-mode IGFET. In Fig. 12-21, V_{SG} is made positive. This means that the region under the gate will acquire a negative charge. The negative charges do not afford a conducting path between the *P*-type source and drain regions. Consequently, the drain current is negligibly small, and, if we keep making V_{SG} increasingly positive, the drain current still remains virtually negligible.

On the other hand, if V_{SG} is made negative as shown in Fig. 12-21*b*, the negatively charged gate induces positive charges in the *N*-type channel underneath. This has the effect of converting the *N* material directly below the

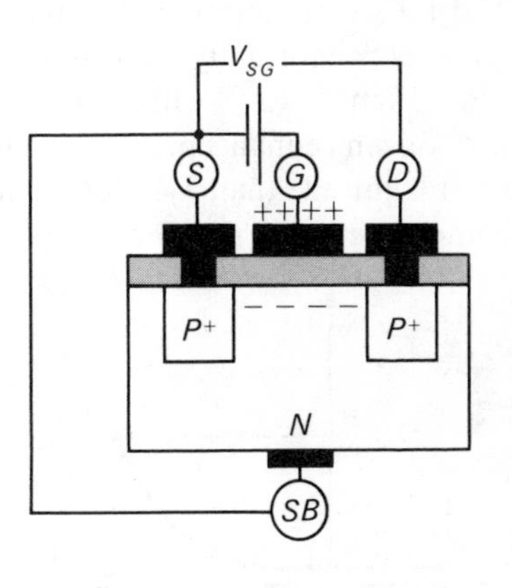

(*a*)

P-channel enhancement type being enhanced

(*b*)

FIGURE 12-21

gate to P material. At some point when V_{SG} is made sufficiently negative, the induced positive charges will overcome the negative charges, and a conducting path will be formed between source and drain. This value of V_{SG} which just causes the channel to commence conduction is called the *threshold voltage*, V_{TH}. Figure 12-22 illustrates the drain characteristic for the P-channel

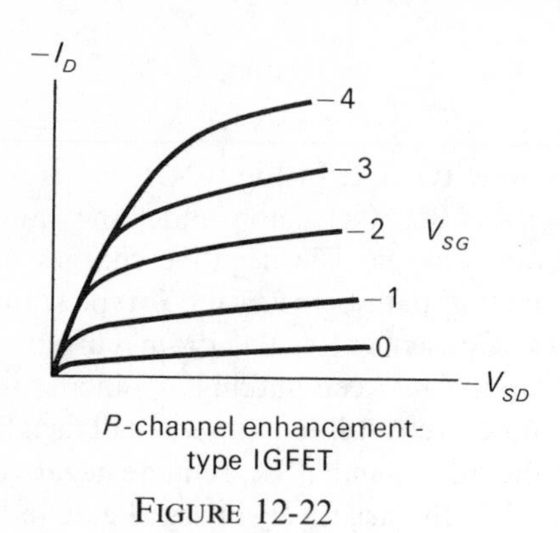

P-channel enhancement-type IGFET

FIGURE 12-22

enhancement-type IGFET. Note that in the enhancement-type IGFET we cannot swing V_{SG} in both directions since the channel is normally depleted and can only be enhanced. Figure 12-23 illustrates the transfer characteristic of an enhancement-type IGFET with V_{SD} greater in magnitude than V_P. Note that conduction does not commence until the threshold voltage, V_{TH},

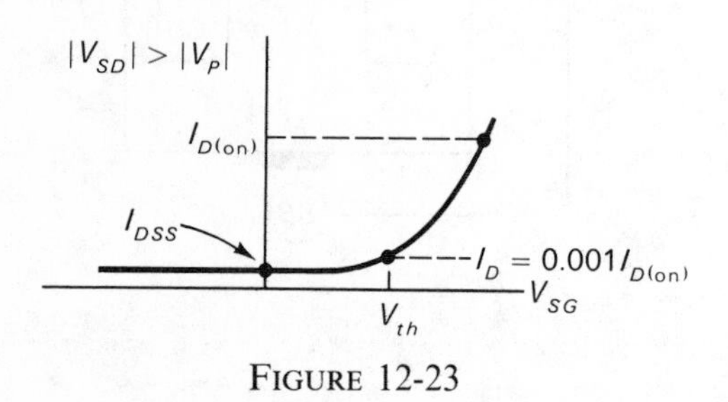

FIGURE 12-23

is reached, and that in the enhancement-type IGFET, I_{DSS} is very small as opposed to the depletion type. A drain current, $I_{D(on)}$, is sometimes defined for the enhancement type and V_{TH} is specified as that value of V_{SG} which permits a drain current equal to some small percentage of $I_{D(on)}$ to flow.

We have not, in some cases, bothered to indicate the specific polarities of V_{SG} or I_D on the characteristic curves, as only the general features are being emphasized. It should be understood that the P and N material in any of the IGFET types discussed can be interchanged so that both N- and P-channel devices may be fabricated. Naturally, the voltages will also have to be reversed.

In summary, then, depletion-mode IGFETs exhibit substantial channel conductance at zero gate bias. Application of a depleting gate bias reduces the charge carriers in the channel which raises the channel resistance. Enhancement mode IGFETs show very high channel resistance at zero gate bias, and an enhancing bias increases the charge carriers available in the channel so that channel resistance decreases.

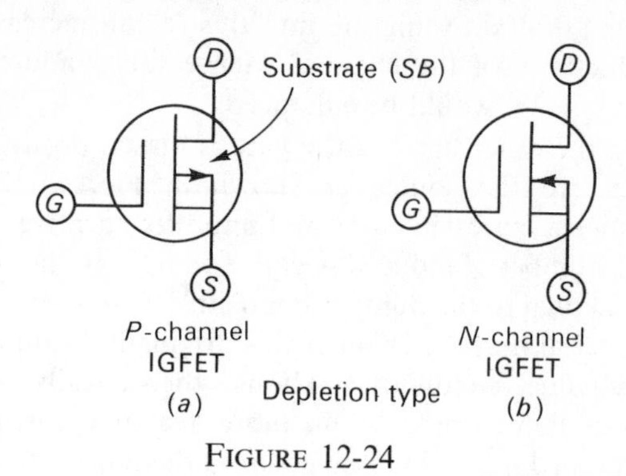

FIGURE 12-24

Figure 12-24 illustrates the schematic representation for the depletion-type IGFET, and Fig. 12-25 illustrates the schematic representation for the enhancement-type IGFET. Note that the channel in the depletion type is shown as a continuous line, whereas the channel in the enhancement type is shown as a broken line, which is

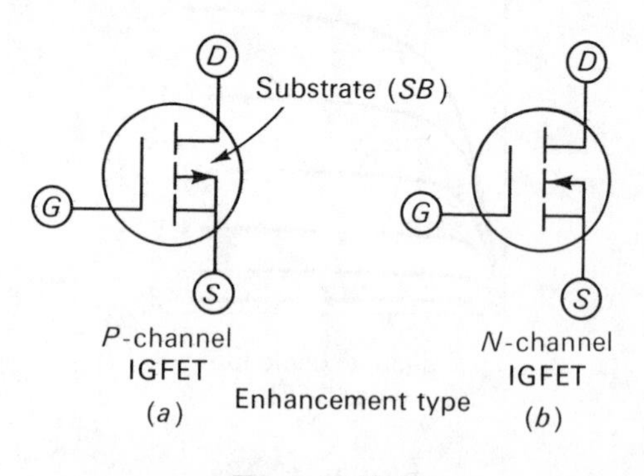

FIGURE 12-25

indicative of the mode of operation. The arrow is used to indicate the polarity of the substrate material. In these figures the substrate is shown tied to the source. This is not necessarily the case, and it is sometimes brought out as an external connection. An arrow indicated out of the device indicates an N-type substrate which implies a P-type channel, and an arrow indicated going into the device indicates a P-type substrate which implies an N channel. Sometimes the substrate is called the bulk or active bulk of the device.

It might be mentioned that in the depletion-type IGFET the gate can be positioned closer to the source than to the drain, which enables the interelectrode capacity between gate and drain to be minimized. This improves the high-frequency response of the device.

PROBLEMS WITHOUT ANSWERS

P 12-1 Describe the function of the gate in an FET.

P 12-2 Define V_P in terms of V_{SG}.

P 12-3 What is the physical significance of $|V_{SD}| = |V_P|$?

P 12-4 Why is the depletion region wedge-shaped in an FET that is carrying drain current?

P 12-5 Give I_{DSS} a physical interpretation.

P 12-6 What other drain-current specification might be used if I_{DSS} isn't given?

P 12-7 What type of IGFET exhibits a normally conducting channel?

P 12-8 Sketch the structure of a depletion-type IGFET.

P 12-9 Sketch the structure of an enhancement-type IGFET.

P 12-10 Is the JFET normally utilized in the enhancement or depletion mode?

P 12-11 Why does a JFET exhibit high input impedance?

P 12-12 Why does an IGFET exhibit high input impedance?

P 12-13 Sketch the schematic symbol of an *N*-channel JFET.

P 12-14 Sketch the schematic symbol of a *P*-channel JFET.

P 12-15 Sketch the schematic symbol of an *N*-channel depletion-type IGFET.

P 12-16 Sketch the schematic representation of a *P*-channel depletion-type IGFET.

P 12-17 Sketch the schematic representation of a *P*-channel enhancement-type IGFET.

P 12-18 Sketch the schematic representation of an *N*-channel enhancement-type IGFET.

P 12-19 What limits the magnitude of the gate-to-drain voltage in a JFET?

P 12-20 What limits V_{SG} or V_{GD} in an IGFET?

P 12-21 Discuss some relative advantages and disadvantages of JFETs versus IGFETs.

13 fet models

The analysis of electronic circuits involving active (amplifying) devices is usually enhanced if a device equivalent circuit or model is available. Generally this is because the model gives us a better feel for the device behavior than the equations (usually quite complex) which describe its physical properties. Also it is convenient to consider a large-signal or piecewise device model for large-signal or biasing considerations and a small-signal or incremental model for gain and other small-signal calculations. This was previously done with the transistor. The FET, however, poses some different considerations.

13-1 FET large-signal models

The drain characteristics of a JFET or IGFET will exhibit a family of curves with the general form shown in Fig. 13-1. Clearly these curves exhibit a high degree

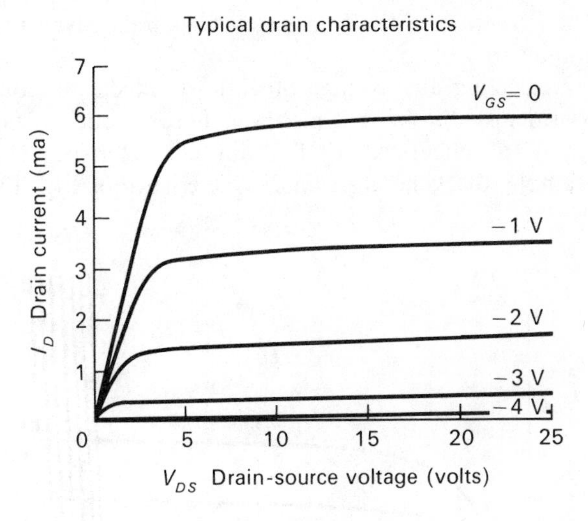

FIGURE 13-1

of nonlinearity between I_D and V_{SG}. For example, if V_{SD} is set at 15 volts and we consider how I_D varies with V_{SG}, it is evident that as V_{SG} decreases (becomes more negative) in this FET, the relative change in I_D becomes smaller and the curves crowd together.

Figure 13-2 is called a *transfer curve*, and it is a plot of I_D versus V_{SG} with, in this case, $V_{SD} = 15$ volts. The slope of this curve at any point is dI_D/dV_{SG}, or, approximately, a very small $\Delta I_D/\Delta V_{SG}$. This slope is a transfer function having the dimensions of a conductance and hence is called the *transconductance* and given the symbol g_m. Thus

$$g_m = \left.\frac{dI_D}{dV_{SG}}\right|_{dV_{SD}=0} \cong \left.\frac{\Delta I_D}{\Delta V_{SG}}\right|_{\Delta V_{SD}=0}$$

is a measure of the effectiveness of a change in gate

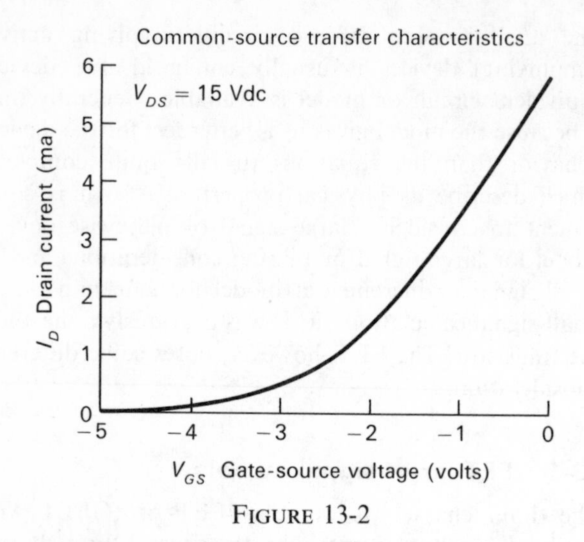

FIGURE 13-2

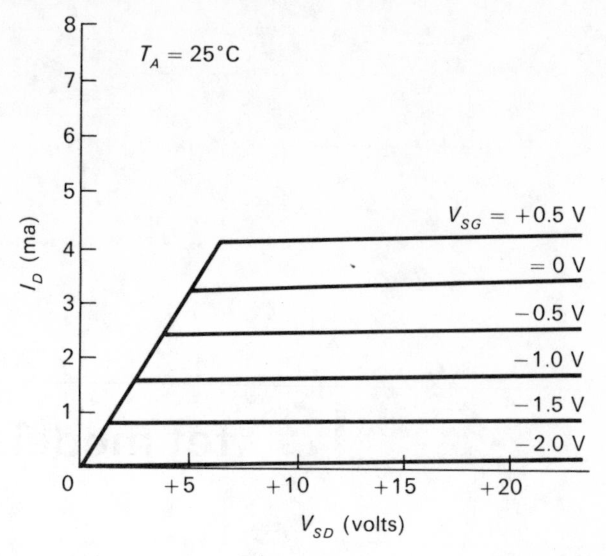

FIGURE 13-4

voltage *only* in causing a change in drain current.[1] The slope or g_m at $V_{SG} = 0$ is usually designated as g_{m0}. The curvature of the transfer curve is indicative of the nonlinearity of g_m.

Neglecting any voltage breakdown effects we could consider synthesizing a piecewise linear model to fit a family of linearized FET drain characteristics. For example, the general drain characteristics of Fig. 13-3

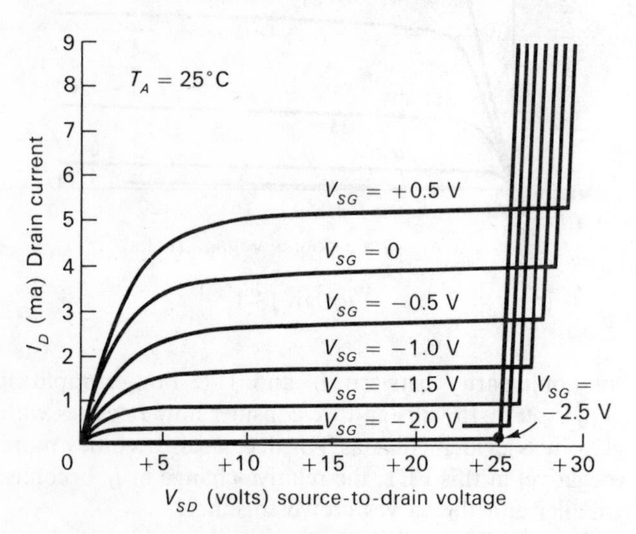

FIGURE 13-3

might be approximated by the straight line segments of Fig. 13-4. Whether it's a good or bad approximation is of no concern at the moment. All we want is a suitable piecewise model.

If we define the following points and slopes shown in Fig. 13-5: (1) V_{SD}' and I_{DSS}' are the breakpoint coordinates

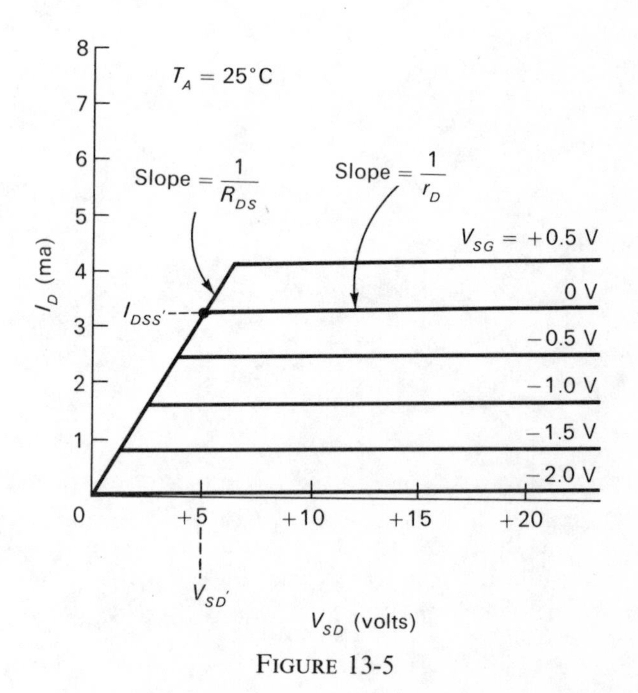

FIGURE 13-5

at $V_{SG} = 0$, (2) $r_D = \Delta V_{SD}/\Delta I_D$ as the reciprocal of the slope in the saturated region, (3) R_{DS} is the slope in the ohmic region, and (4) $g_m = \Delta I_D/\Delta V_{SG}$, we can construct the approximate *N*-channel piecewise model shown in Fig. 13-6. (A *P*-channel model would have the polarity of the diodes and the I_{DSS}' current source reversed.) The gate is assumed an open circuit for simplicity at this time. Inspection of this model indicates that when $V_{SG} \leq 0$ and $V_{SD} = 0$ (a short between *S* and *D*), diode D_p is ON due to the current sources driving it ON and D_o is OFF. With V_{SG} held at zero and V_{SD} increasing, the next break occurs at D_o turning ON. At this point $I_D = 0$ and $V_{SD} = 0$ corresponding to the origin of the

[1]The g_m is also given by the low-frequency value of y_{fs} if the *y* parameters of the FET are specified. At higher frequencies y_{fs} includes a reactive component.

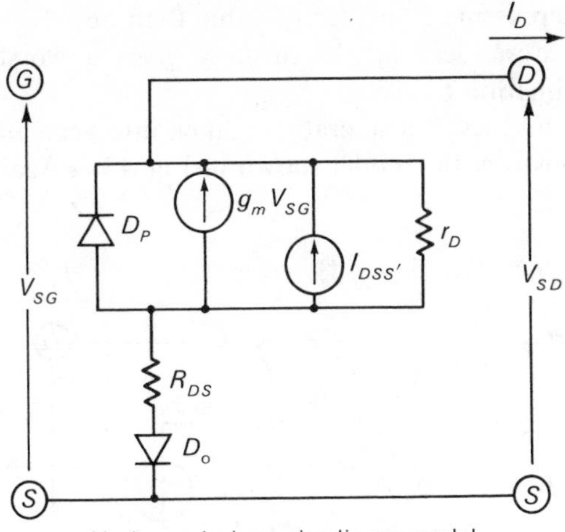

N-channel piecewise linear model. (Reverse diode and current sources polarities for *P* channel.)

FIGURE 13-6

drain characteristics. As V_{SD} continues to increase, a point is reached where D_p turns OFF. At this breakpoint

$$I_D = I_{DSS}' + g_m(0 \text{ volts}) = I_{DSS}'$$

and

$$V_{SD} = R_{DS}I_D = V_{SD}'$$

If V_{SG} is made negative

$$I_D = I_{DSS}' - g_m|V_{SG}|$$

which indicates a net decrease in I_D. Hence, the corresponding $V_{SD} = R_{DS}I_D$ is smaller and the breakpoint moves to the left, in agreement with the drain characteristics. Hence, the model seems to fit the curves. Negative values of I_D, that is, reverse current through D_0, are not permissible and actually mean the FET is cut off.

Going back to Fig. 13-5 and picking out the piecewise model parameters we have

$$I_{DSS}' = 3.2 \text{ ma}$$

$$V_{SD}' = 5 \text{ volts}$$

$$R_{DS} = \frac{5 \text{ volts}}{3.2 \text{ ma}} = 1.56 \text{ kilohms}$$

$$r_D \cong \frac{15 \text{ volts} - 5 \text{ volts}}{3.4 \text{ ma} - 3.2 \text{ ma}} = 50 \text{ kilohms}$$

$$g_m \cong \frac{3.2 \text{ ma}}{2 \text{ volts}} = 1.6 \text{ ma/volt} = 1600\ \mu\text{mho}$$

An alternative and better way to get g_m for this large-signal piecewise model is to sketch the transfer curve from Fig. 13-3 for some V_{SD} well into the saturated region ($|V_{SD}| \gg |V_p|$), say $V_{SD} = 15$ volts. This yields the following values:

V_{SG} (volts)	0	−0.5	−1.0	−1.5	−2
I_D (ma)	3.85	2.65	1.7	0.85	0.4

which are plotted for convenience on the drain characteristic as shown in Fig. 13-7. A separate plot of the transfer curve could also have been made without involving the drain characteristic on the same axis. The transfer curve is then approximated by a straight line, shown as the broken line of Fig. 13-7, to yield an average value of $g_m = 3.5$ ma/2 volts = 1.75 ma/volt = 1750 μmho. Of course the error becomes very large as I_D decreases since the actual transfer curve is leveling off, but normally we wouldn't operate a linear amplifier this close to cut-off anyway with large signals.

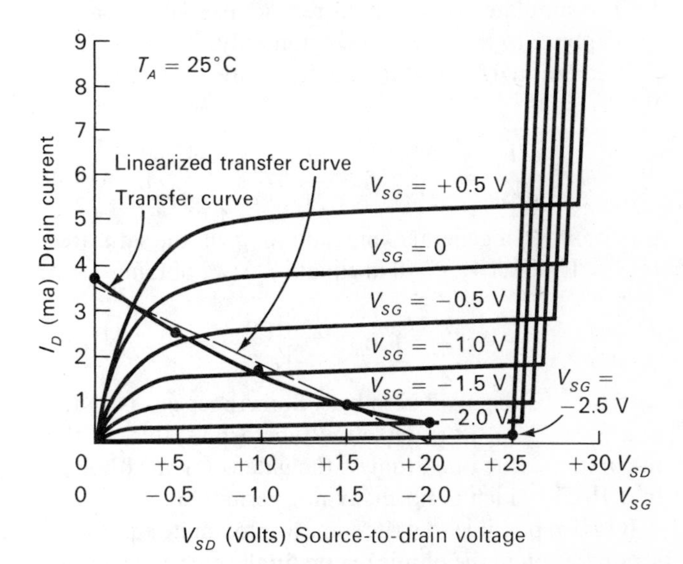

FIGURE 13-7

We might just check out the model by assuming some arbitrary V_{SG} and V_{SD} values in the saturated region to see if we can predict I_D. For example, let $V_{SG} = -1$ volt and $V_{SD} = 10$ volts. Since 10 volts $> V_{SD}' = 5$ volts the operating point is in the saturated region. From Fig. 13-6 with D_p OFF, D_0 ON, and noting that $r_D \gg R_{DS}$ we obtain

$$I_D \cong I_{DSS}' + g_m(V_{SG}) + \frac{V_{SD}}{r_D}$$

$$= 3.2 \text{ ma} + \frac{1.75 \text{ ma}}{\text{volts}}(-1 \text{ volts}) + \frac{10 \text{ volts}}{50 \text{ kilohms}}$$

$$= 1.65 \text{ ma}$$

which is in close agreement with the actual curves of Fig. 13-3.

At this point it is suggested that the reader work his way through problems PS 13-1, PS 13-2, and PS 13-3.

It turns out that there exists some fairly well-defined equations which relate the drain-current and channel resistance to the gate voltage V_{SG} and pinch-off voltage V_p. Thus

$$1)\qquad I_D = I_{DSS}\left(1 - \frac{V_{SG}}{V_P}\right)^2 \qquad (13\text{-}1)$$

The exponent (2) may actually lie between 1.5 and 2.5, but 2 in general is a good approximation for most FETs. Equation 1 is valid *only in the saturated region.*

Equation 1 may be solved for the V_{SG} required to yield a given I_D in the *saturated region.* Thus

$$2)\qquad V_{SG} = V_P\left(1 - \sqrt{\frac{I_D}{I_{DSS}}}\right) \qquad (13\text{-}2)$$

Equation 13-2 is particularly useful if we wish to use a FET to simulate a constant-current source. This equation also applies to the saturated region only.

If equation 13-1 is differentiated with respect to V_{SG}, there results

$$3)\qquad g_m = \frac{dI_D}{dV_{SG}} \cong \frac{\Delta I_D}{\Delta V_{SG}} = \frac{-2I_{DSS}}{V_P}\left(1 - \frac{V_{SG}}{V_P}\right) \qquad (13\text{-}3)$$

Equation 3 is a general expression for g_m in the saturated region. If we set $V_{SG} = 0$ in equation 3, we obtain

$$4)\qquad g_m = g_{m0} = \frac{-2I_{DSS}}{V_P} \qquad (13\text{-}4)$$

where g_{m0} is the g_m at zero bias. Sometimes g_{m0} is called out as $g_{m(\max)}$. For depletion-type FETs this is the maximum value of g_m (unless the gate is forward-biased in a JFET, which is normally not permitted).

It is also possible to write some approximate equations for operation in the ohmic region. In the piecewise model of Fig. 13-6 the values of R_{DS} was obtained from Fig. 13-5 as the inverse slope of the drain characteristic in the ohmic region. Note in Fig. 13-5 that all the V_{SG} segments are assumed to intersect this R_{DS} line. However, in truth, each V_{SG} curve actually curves into the origin as shown in Fig. 13-3. This can be allowed for by modifying the previous expression for $R_{DS} = V_{SD}'/I_{DSS}'$ to

$$5)\qquad R_{DS} = \frac{V_{SD}'}{I_{DSS}'}\,\frac{1}{1 - \sqrt{V_{SG}/V_P}} \qquad (13\text{-}5)$$

Note I_{DSS}' in the R_{DS} calculation is the I_D at V_{SD}', and $V_{SG} = 0$.

If the drain curves are not given and it is necessary to estimate V_{SD}', a fair approximation is

$$6)\qquad V_{SD}' \approx 2V_P \qquad (13\text{-}6)$$

The slight increase in I_D with V_{SD} in the saturated region is not allowed for in equation 13-1. If this is of concern, a shunt resistance $r_D = \Delta V_{SD}/\Delta I_D$ may be evaluated from the slope of the $V_{SG} = 0$ curve in the saturated region. This slope actually varies depending on the V_{SG} value under consideration, but since the $V_{SG} = 0$ curve represents a smaller r_D value than at a V_{SG} value which causes less I_D, this usually gives a worst-case approximation to r_D.

With all these considerations taken into account, one may construct the model shown in Fig. 13-8. Again the

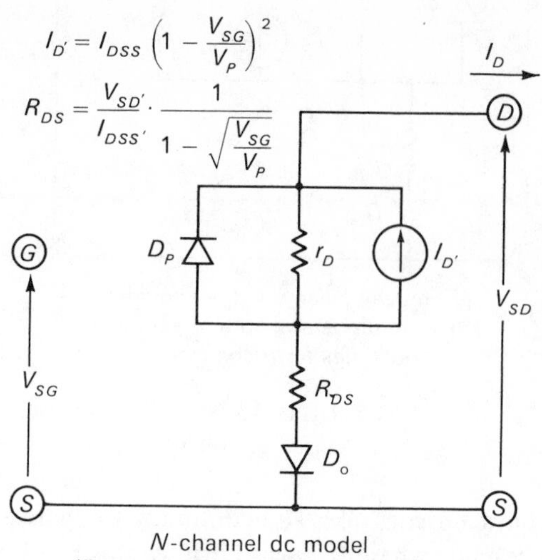

N-channel dc model
(Reverse diode and current source polarity for *P* channel.)

FIGURE 13-8

gate is assumed to be reverse-biased and exhibit negligible leakage current. We will call this the dc model since it is useful in computing quiescent operating conditions. However, it could also be used for large-signal conditions by simply calculating maximum and minimum I_D values corresponding to maximum and minimum V_{SG} values. The swing or change in I_D is then $I_{D(\max)} - I_{D(\min)}$.

Now it may turn out that V_P is not given. In that case it may be estimated from the drain characteristics by noting $V_{SG(\text{off})}$, which is the V_{SG} which just cuts off I_D. Alternatively, V_P may be estimated from the transfer curve (which may be given or constructed from the drain characteristics) by noting the intercept point on the V_{SG} axis of a line tangent to the transfer curve at $V_{SG} = 0$. This intercept point is $V_P/2$ as shown in Fig. 13-9. Use the value of I_{DSS} corresponding to the V_P in equation 13-1.

At this point you should carefully work through problem PS 13-4.

In some cases the leakage current associated with the reverse-biased gate in a JFET or the leakage current of the gate in an IGFET must be considered. IGFET leakage currents are exceedingly small, say 1 to 50 pa, and relatively insensitive to temperature, and what temperature sensitivity exists is relatively unpredictable. The reverse current of a silicon JFET, although much larger, is still typically less than 1 na at room temperature

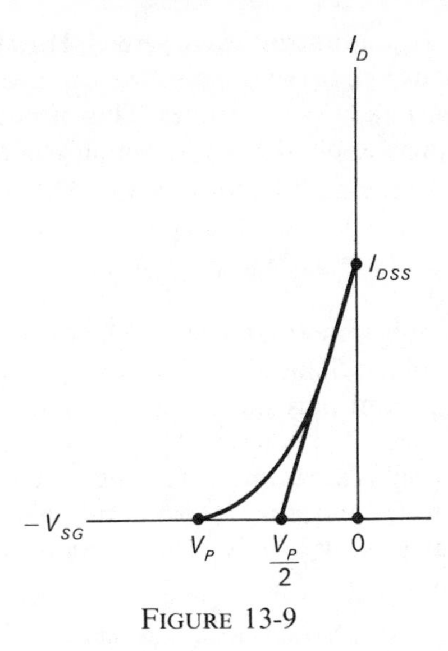

FIGURE 13-9

(27°C) and is perhaps up to 1000 times greater at 100°C depending on the FET type. Manufacturer's data should be consulted for specific information, or, if unavailable, use a rule of thumb similar to determining I_{CBO} in bipolar transistors.

It is usually unnecessary but if it proves desirable to include gate-leakage-current effects in either the piecewise or dc model, we might begin by recalling the general form of a silicon diode *V-I* curve as shown in Fig. 13-10.

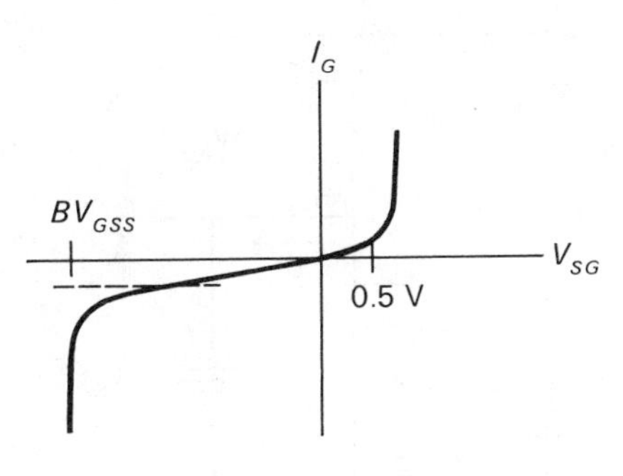

FIGURE 13-10

The forward-direction breakpoint is taken as 0.5 volt rather than the usual 0.6 volt since we want to be more conservative with regards to maintaining high input impedance if the gate diode becomes forward-biased. The leakage current of a JFET in the reverse direction is designated as I_{GSS} and it may be measured at some specific reverse voltage with the circuit of Fig. 13-11. If the reverse voltage is sufficiently increased, breakdown occurs at a voltage designated as BV_{GSS}. Circuits which measure other types of gate breakdown are shown in Fig. 13-12.

For our immediate purposes we can assume we stay out of the breakdown region so that the diode *V-I* curve of Fig. 13-10 can be piecewise linearized as shown in

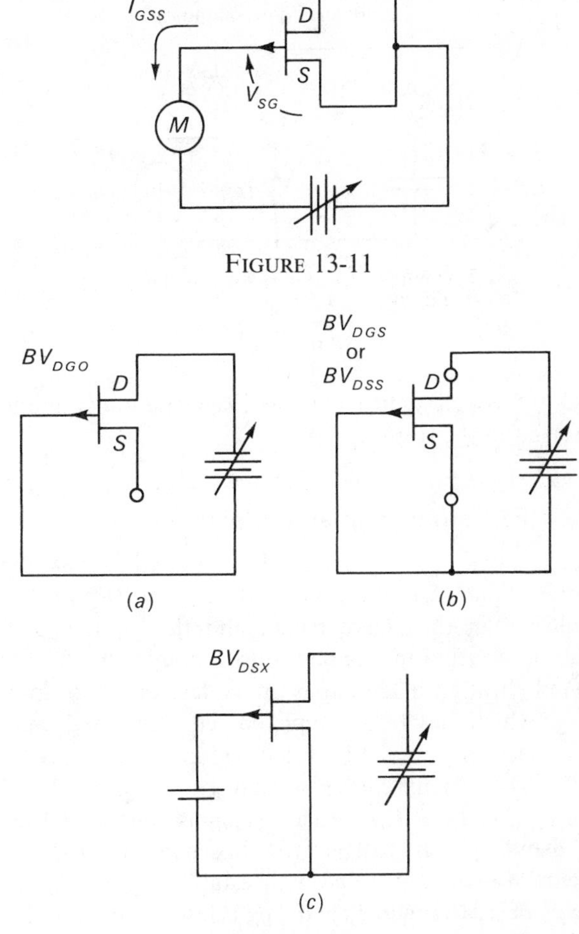

FIGURE 13-11

FIGURE 13-12

Fig. 13-13. A piecewise model to fit Fig. 13-13 is shown in Fig. 13-14. An *N*-channel JFET is being considered.

For values of V_{SG} large and negative, D_{G3} is OFF, D_{G2} is ON, and D_{G1} is OFF. At $V_{SG} = 0$, I_G is indeterminate, lying between zero and I_{GSS}; so assume I_{GSS} for any V_{SG} slightly less than zero, or zero for V_{SG} exactly equal to zero. If V_{SG} swings positive but less than 0.5 volt, D_{G2} turns OFF and I_{GSS} sinks itself in D_{G1} and $I_G = 0$. For

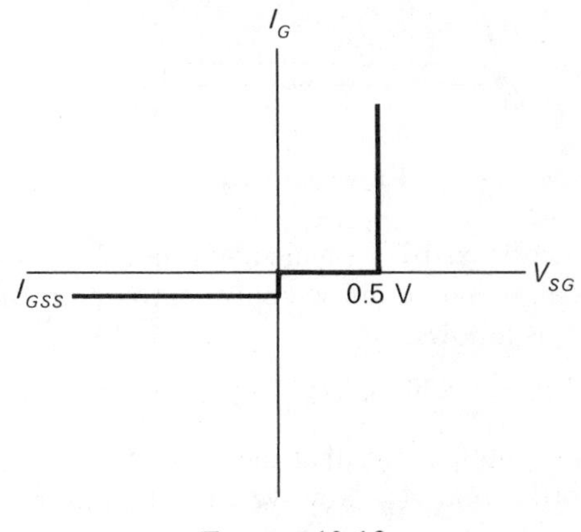

FIGURE 13-13

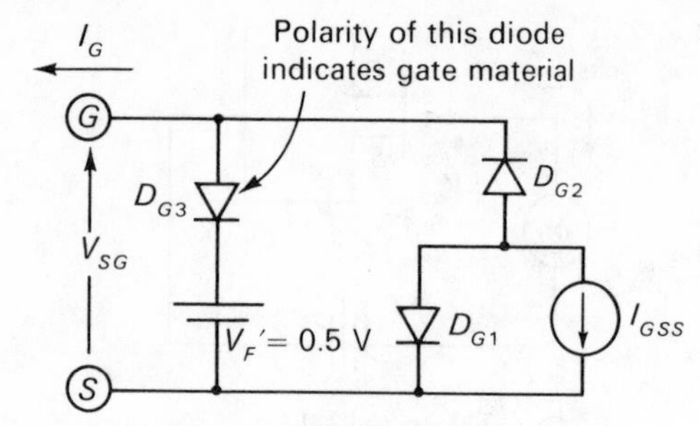

N-channel JFET input side model
(Reverse diodes and sources for *P* channel)

FIGURE 13-14

$V_{SG} > 0.5$ volt, D_{G3} turns ON and the gate looks like it is shorted to the source.

13-2 FET small-signal models

In most linear applications the FET is used as a *small-signal amplifier*. A quiescent operating point (Q point) is established in the active region and the input signal causes a variation in I_D about this Q point. This ΔI_D is directed through a load resistor to develop an output voltage which may be an amplified version of the input.

The active region of a FET is its saturated region with the Q point usually chosen so that $V_P < V_{SD} < BV_{DGX}$ and $I_{D(\min)} < I_D \leqq I_{DSS}$, where BV_{DGX} is defined in Fig. 13-12*a* and $I_{D(\min)}$ is an arbitrarily chosen minimum drain current. We are concerned with establishing an $I_{D(\min)}$ since g_m decreases with I_D and if g_m is low, the gain of the amplifier will be low. Thus the Q point might be located in the shaded area shown in Fig. 13-15.

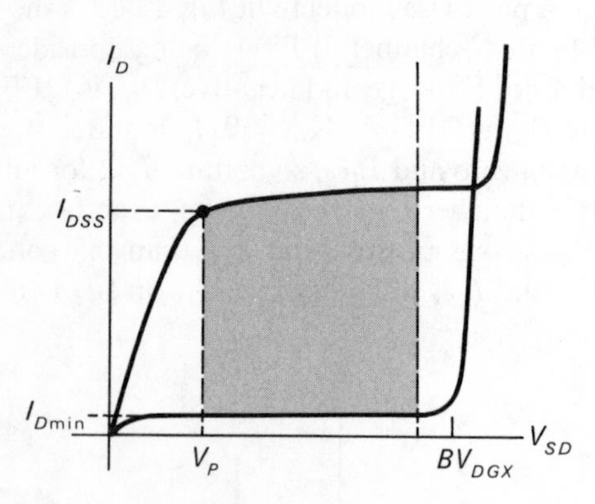

FIGURE 13-15

The generalized FET small-signal model (also called incremental, dynamic, or ac equivalent circuit) may be developed as follows:

1) $$I_D = f(V_{SG}, V_{SD})$$

Equation 1 simply states that the drain current, I_D, is a function of V_{SG} and V_{SD}. Now we know that in the saturated or amplifying region I_D is not a linear function of V_{SG} (or even V_{SD}, although less severely). However, for small variations about the Q point, we can assume the curves are essentially straight lines. Thus if linearity is assumed, we may apply the superposition theorem and write[1]

2) $$\Delta I_D = g_m \Delta V_{SG} + \frac{1}{r_d} \Delta V_{SD}$$

Equation 2 simply states that the total change in I_D is the superposition of the change in I_D due to V_{SG} only (which is the $g_m \Delta V_{SG}$ term) plus the change in I_D due to V_{SD} only (which is the $\Delta V_{SD}/r_d$ term). The transconductance, g_m, you will recall is a parameter relating a change in I_D to a change in V_{SG} only, and r_d relates the change in I_D due to a change in V_{SD} only. This is summarized in Fig. 13-16*a* to *c*.

[1] The reader familiar with the calculus may recognize equation 2 as

$$dI_D = \frac{\partial I_D}{\partial V_{SG}} dV_{SG} + \frac{\partial I_D}{\partial V_{SD}} dV_{SD}$$

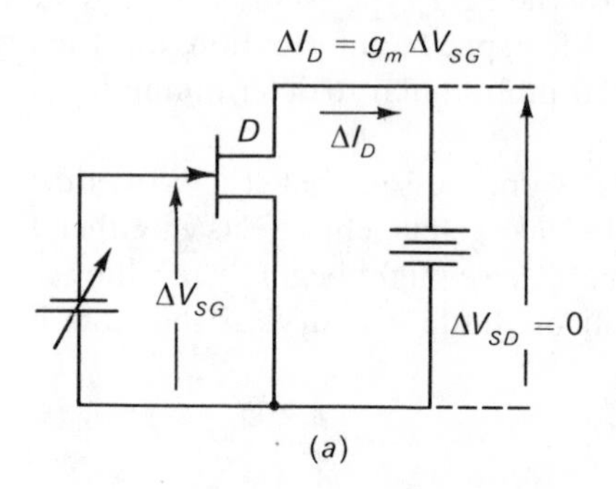

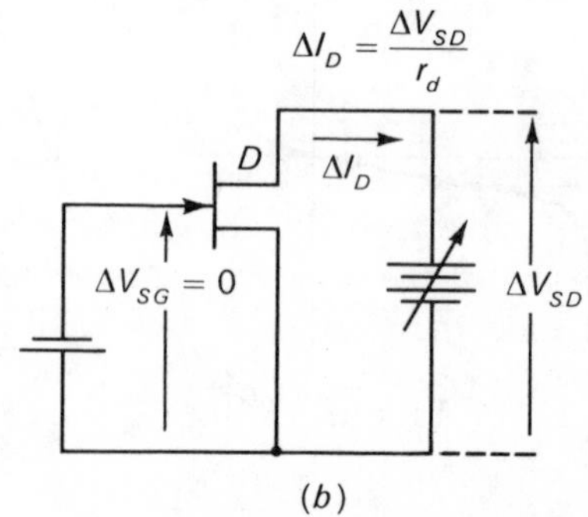

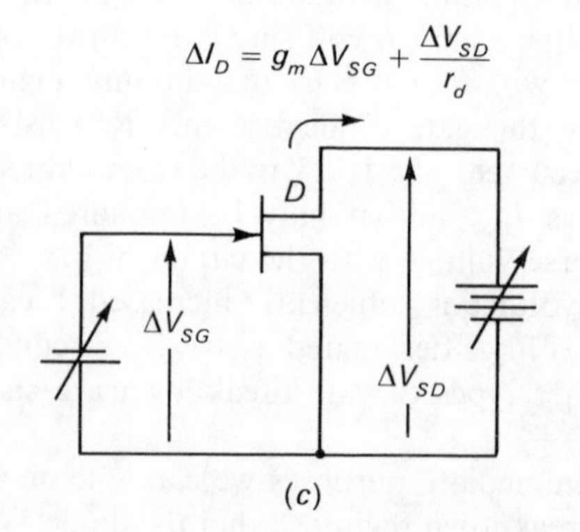

FIGURE 13-16

Now to avoid writing Δ all the time let us change our notation as follows

3) $\left.\begin{aligned} i_d &= \Delta I_D \\ v_{sg} &= \Delta V_{SG} \\ v_{sd} &= \Delta V_{SD} \end{aligned}\right\}$ Small-signal variations about the quiescent values

Substituting the above into equation 2 yields

4) $$i_d = g_m v_{sg} + \frac{1}{r_d} v_{sd} \qquad (13\text{-}7)$$

It should be understood that g_m and r_d in equation 13-7 are small-signal values taken at a specific Q point and therefore are not necessarily the same as g_m and r_D used in our piecewise model of Fig. 13-6.

An equivalent circuit which fits the nodal equation 13-7 is shown in Fig. 13-17.[1] That this model fits equation

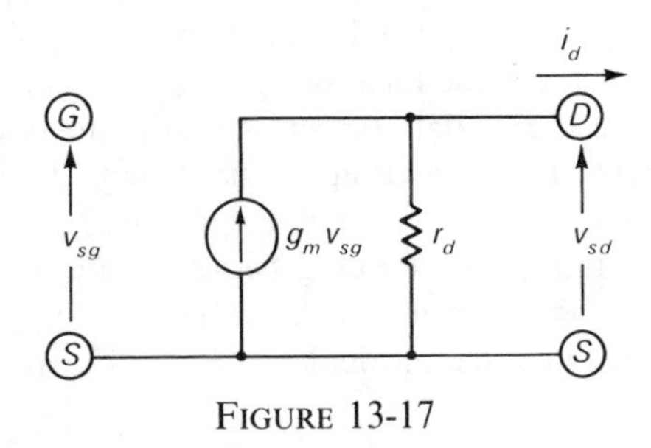

FIGURE 13-17

13-7 is easily verified by expressing Kirchhoff's current law for i_d. This incremental model fits all types (*P* or *N* channel) of JFETs and IGFETs in their active (saturated) region.

On the input side, the gate is assumed isolated from the source and drain by infinitely high resistances. This is usually a permissible approximation in most cases.

Figure 13-17 contains *Norton's equivalent circuit*. It is sometimes more convenient to work with the Thévenin's equivalent circuit of Fig. 13-18 which is derived by thevenizing Fig. 13-17 to obtain

1) $$v_{sd(oc)} = -r_d g_m v_{sg}$$

Let

2) $$\mu = r_d g_m \qquad (13\text{-}8)$$

Whenever v_{sg} is a positive quantity, the $g_m v_{sg}$ dependent current source in Fig. 13-17 will try to force more I_D. This tries to drive the drain negative (ac component only) relative to the source, and the polarity of the μv_{sg} dependent voltage source in Fig. 13-18 does likewise. If v_{sg} becomes negative we have a negative quantity for the

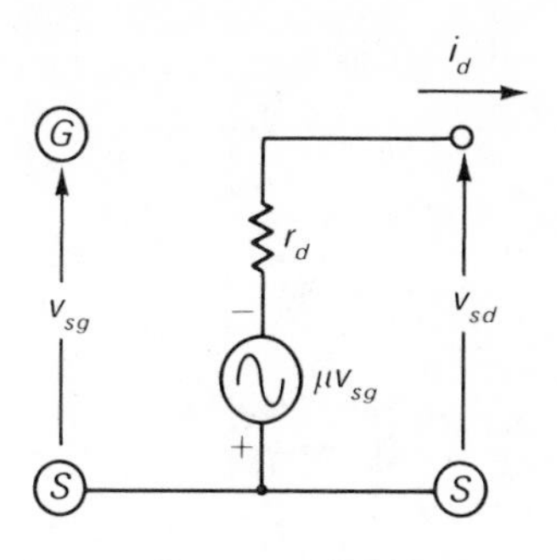

FIGURE 13-18

current source in Fig. 13-17 which is equivalent to reversing the current-source direction. Thus v_{sd} would tend to swing positive as v_{sg} goes negative. This 180° phase shift between v_{sd} and v_{sg} is also evident in Fig. 13-18, as a negative v_{sg} is equivalent to reversing the μv_{sg} source polarity, which means v_{sd} tends to swing positive.

An alternative way of denoting the phase reversal between v_{sd} and v_{sg} is shown in the model of Fig. 13-19. Remember, these are only signal variations and not actual bias polarities or dc quantities.

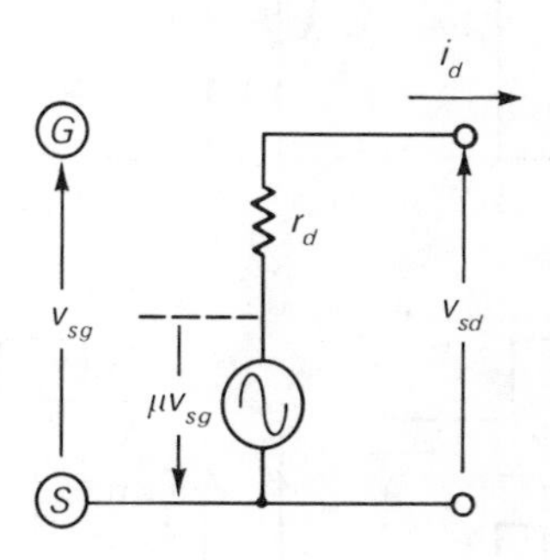

FIGURE 13-19

The FET may also be characterized in terms of a *y* parameter equivalent circuit as shown in Fig. 13-20 which models the equations

1) $$i_g = y_{is} v_{sg} + y_{rs} v_{sd} \qquad (13\text{-}9)$$

2) $$i_d = y_{fs} v_{sg} + y_{os} v_{sd} \qquad (13\text{-}10)$$

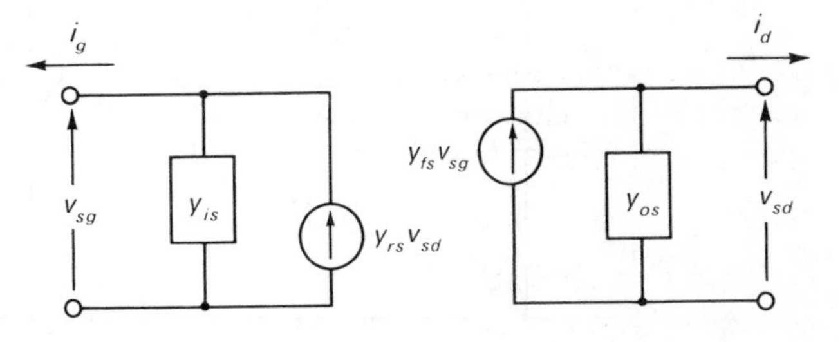

FIGURE 13-20

[1] For review on equivalent circuit development see Phillip Cutler, "Outline for DC Circuit Analysis with Illustrative Problems," chap. 15, McGraw-Hill Book Company, New York, 1968.

where

$$\begin{aligned} &3) \quad y_{is} = \left.\frac{i_g}{v_{sg}}\right|_{v_{sd}=0} \\ &4) \quad y_{rs} = \left.\frac{i_g}{v_{sd}}\right|_{v_{sg}=0} \\ &5) \quad y_{fs} = \left.\frac{i_d}{v_{sg}}\right|_{v_{sd}=0} \\ &6) \quad y_{os} = \left.\frac{i_d}{v_{sd}}\right|_{v_{sg}=0} \end{aligned} \tag{13-11}$$

Figure 13-21*a* through *d* illustrates the physical significance of the *y* parameters in equation 13-11. The biasing details are omitted for clarity. Just assume the FET is properly biased. Parameter y_{is}, for example, is a measure of how much ac gate current flows due to v_{sg}. At low frequencies where interelectrode capacities may be neglected $y_{is} \approx 0$. Remember the *y* is an admittance. At higher frequencies interelectrode capacities including *PN*-junction capacities come into play and y_{is} becomes a complex number due to the reactive effects. A study of y_{fs} indicates that at low frequencies $y_{fs} \approx g_m$ in our model. Similarly $y_{os} \approx r_d$.

A more detailed FET model which includes the effects of leakage and capacitance between gate and drain, and gate and source, is shown in Fig. 13-22. Normally g_{SG} and g_{DG} are negligibly small and only the interelectrode capacitances need be considered in the high-frequency range of the FET. Under these conditions, the simplified model of Fig. 13-23 may be used. The capacitances C_{SG}, C_{DG}, and C_{DS} are expressed in terms of the *y* parameters normally available in manufacturer's data. Also $g_m = g_{fs} = y_{fs}$ at low frequencies, and $1/r_d = g_{os} = y_{os}$ at low frequencies.

The $g_m v_{sg}$ current source and r_d of Fig. 13-23 may of course also be thevenized while maintaining C_{SG}, C_{DG}, and C_{SD} in the same location.

Manufacturer's data sheets usually include the *y* parameters as a function of frequency, and/or C_{iss} (also C_{is}), C_{rss} (also C_{rs}), and C_{oss} (also C_{os}). C_{iss} is the net capacitance seen looking between gate and source with the drain shorted (to the ac signal only) and the FET biased at same particular *Q* point. C_{rss} is identical

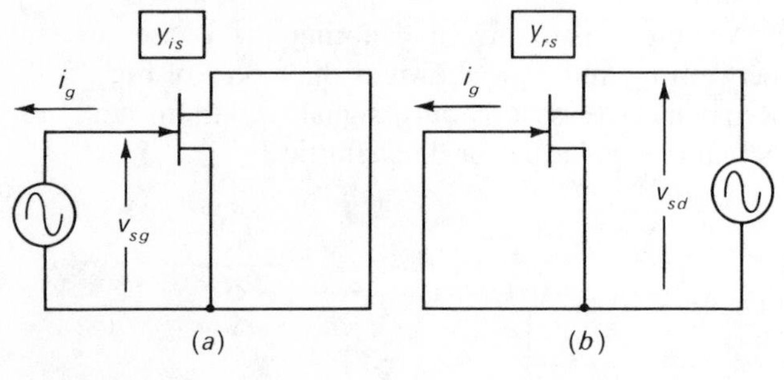

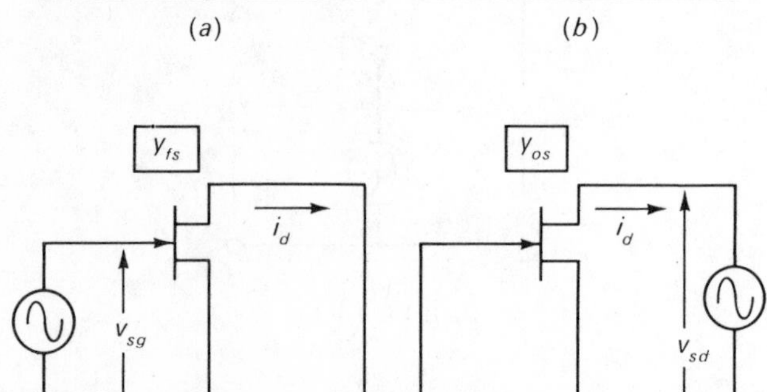

FIGURE 13-21

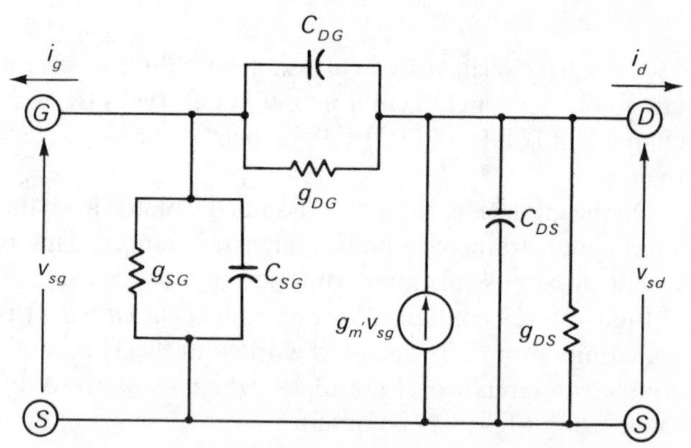

FIGURE 13-22

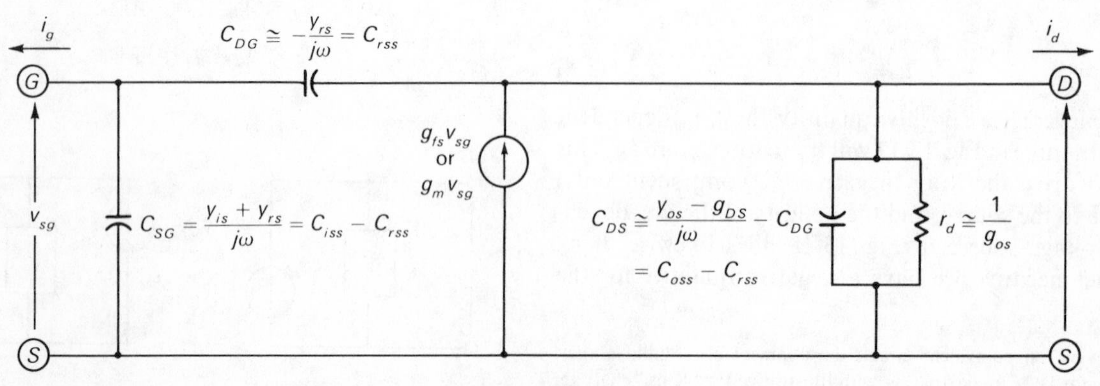

FIGURE 13-23

with C_{DG}, and C_{OSS} is the capacitance seen looking between source and drain (the output side) with the input (between gate and source) shorted. Other capacitance values are sometimes specified where one of the FET terminals is left open.

From a qualitative viewpoint, it would seem that the impedance seen looking between source and gate would decrease with increasing frequency due to C_{SG} and C_{DG}. The effect of C_{DG} may also be magnified due to Miller effect if v_{sd} increases due to voltage gain between v_{sg} and v_{sd}. Thus y_{is} increases with frequency. In a similar manner it is apparent that $|y_{rs}|$ also increases with frequency due to C_{DG}. Parameter y_{fs} is substantially constant and equal to g_m except at relatively higher frequencies where $|y_{fs}|$ may increase due to feed-through via C_{DG}. However, bulk and lead resistances also start coming into play to complicate the picture. Parameter y_{fs} should show an increase with increasing frequency due to C_{DG} and C_{DS}. These y parameters are usually given in the data sheets in the form of real (g) and reactive (b) components of admittance.

Figure 13-24 illustrates the manner in which the y parameters may be graphically displayed on a FET data sheet. Be sure to work through the Problems with

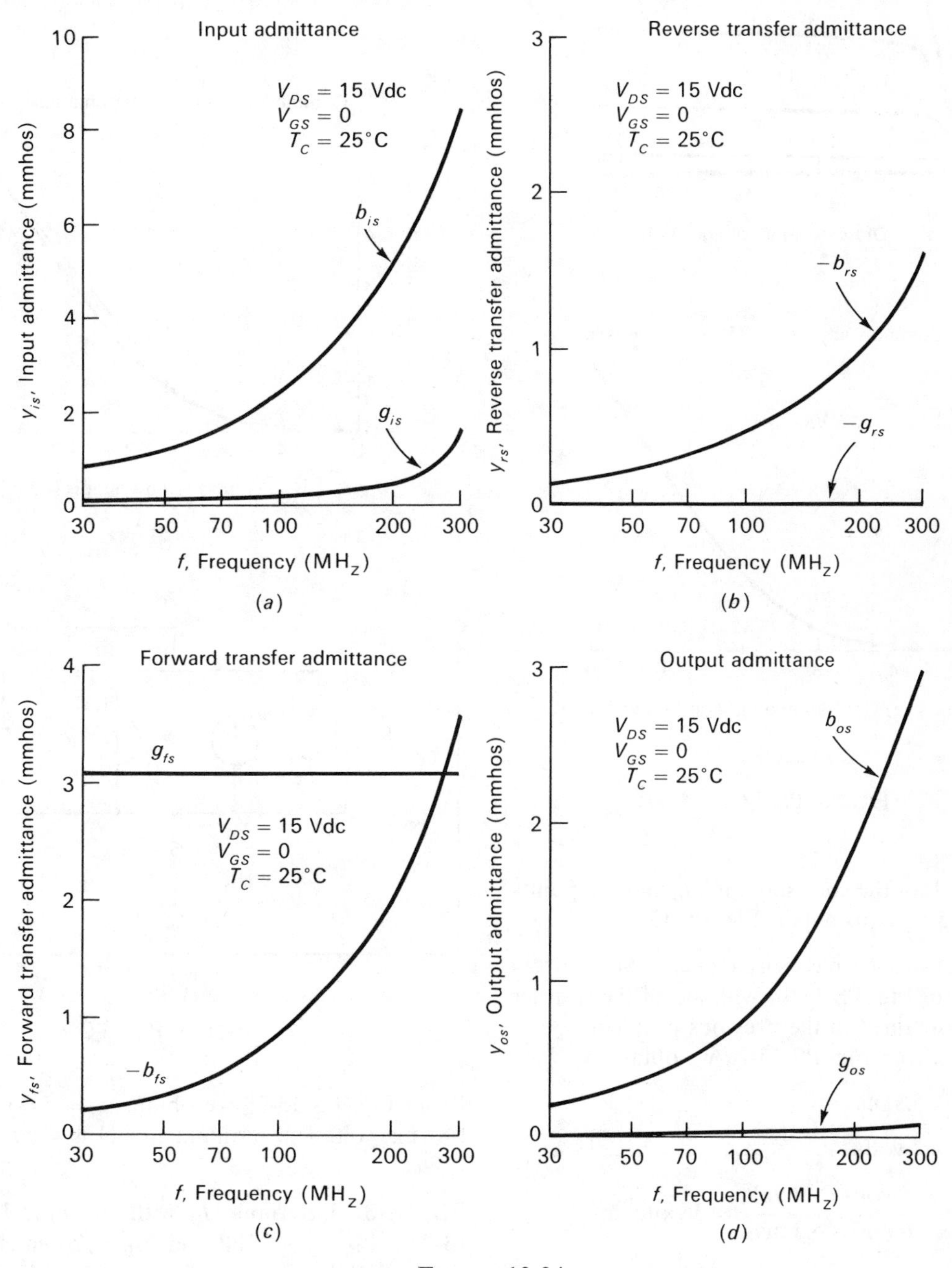

FIGURE 13-24

Solutions section in this chapter as additional small-signal theory is developed there.

PROBLEMS WITH SOLUTIONS

PS 13-1 Construct the transfer curve for V_{SD} = 15 volts from the drain curves of Fig. PS 13-1*a*.

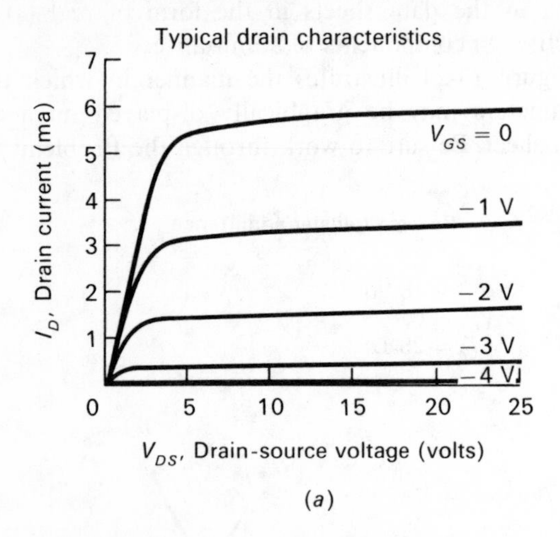

(*a*)

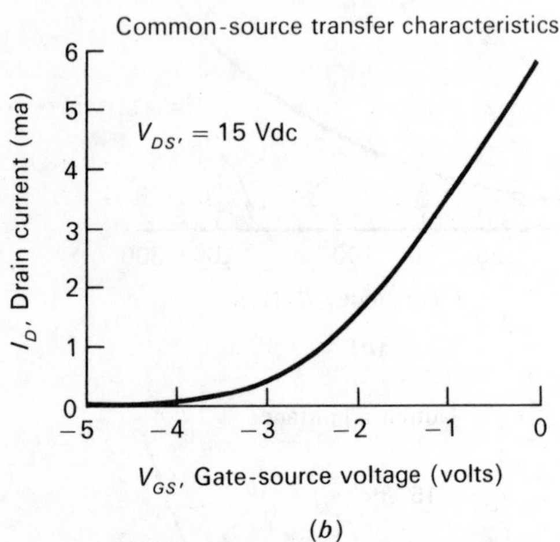

(*b*)

FIGURE PS 13-1

SOLUTION Plot the corresponding I_D and V_{SG} points along V_{SD} = 15 volts to obtain Fig. PS 13-1*b*.

PS 13-2 Construct a piecewise model to fit the drain characteristics of Fig. PS 13-1*a* with aid of the transfer characteristic obtained in the previous problem.

SOLUTION From Fig. PS 13-2*a* we obtain

$$I_{DSS}' = 5.5 \text{ ma}$$

$$V_{SD}' = 7 \text{ volts}$$

$$r_D \approx \frac{25 \text{ volts} - 7 \text{ volt}}{6 \text{ ma} - 5.5 \text{ ma}} = 36 \text{ kilohms}$$

$$R_{DS} = 7 \text{ volts}/5.5 \text{ ma} = 1.27 \text{ kilohms}$$

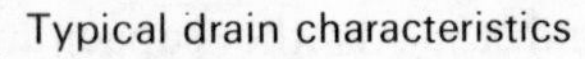

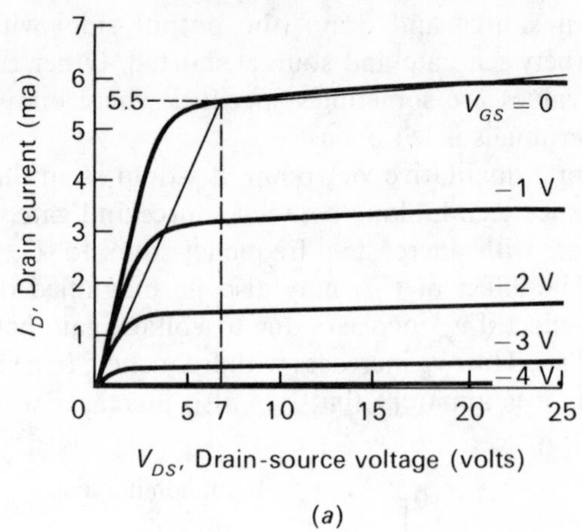

(*a*)

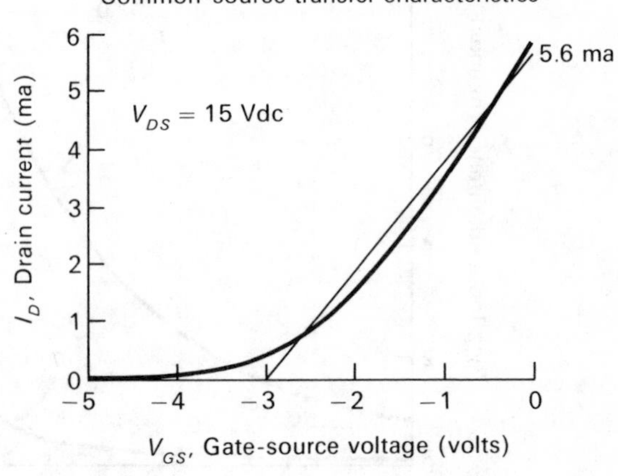

(*b*)

(*c*)

FIGURE PS 13-2

From Fig. PS 13-2*b* we obtain g_m = 5.6 ma/3 volts = 1.87 ma/volt. The resultant model is shown in Fig. PS 13-2*c*.

PS 13-3 Determine I_D with the model of Fig. PS 13-2*c* if $V_{SG} = -1$ volt and V_{SD} = 20 volts.

SOLUTION First check that D_p is OFF. The breakpoint for D_p occurs essentially at $V_{SD} = 1.27$ kilohms I_D where

$$I_D = 5.5 \text{ ma} + 1.87 \text{ ma/volts} (-1 \text{ volt}) = 3.63 \text{ ma}$$

or

$$V_{SD} = 1.27 \text{ kilohms} (3.63 \text{ ma}) = 4.6 \text{ volts}$$

Since 4.6 volts < 20 volts, D_P is OFF and operation is in the saturated region. Hence, duly noting that $r_D \gg R_{DS}$ as is almost always the case,

$$I_D \approx 5.5 \text{ ma} + 1.87 \text{ ma/volts} (-1 \text{ volt}) + \frac{20 \text{ volts}}{36 \text{ kilohms}}$$
$$= 4.19 \text{ ma}$$

The actual curves of Fig. PS 13-2*a* indicate I_D should be about 3.5 ma. The error is in part due to the r_D term (20 volts/36 kilohms) in the above equation whose graphic evaluation involved the most error. Well! You can't win them all, and it doesn't really matter very much since the actual devices vary considerably from the nominal curves published.

PS 13-4 Using the dc model of Fig. 13-8, determine I_D in the FET of Fig. PS 13-2*a* at $V_{SG} = -1$ volt, $V_{SD} =$ 15 volts. Compare the result with the solution obtained in PS 13-3 which utilized the piecewise model.

SOLUTION First check to see if the operating point is in the saturated region (D_p OFF, D_o ON). By inspection of Fig. PS 13-2*a* we see that any $V_{SD} > 7$ volts guarantees operation in the saturated region. Thus, neglecting r_D for the moment,

$$I_D = I_D' = I_{DSS}\left(1 - \frac{V_{SG}}{V_P}\right)^2$$

If we use $I_{DSS} = 5.5$ ma as in the piecewise analysis then $V_p = 7$ volts since I_{DSS} is defined as the I_D at $|V_{SD}| = |V_P|$. Thus

$$I_D = 5.5 \text{ ma}\left(1 - \frac{-1 \text{ volt}}{-7 \text{ volts}}\right)^2 = 4.04 \text{ ma}$$

This is a questionable calculation, however, since the I_D value in this region doesn't change much for a relatively large spread in V_{SD} which makes V_P hard to locate.

An alternative approach is to note that for $|V_{SG}| \geqq$ 4 volts, the drain current is essentially zero which would imply $|V_P| \approx 4$ volts since V_P is the V_{SG} required to cut off I_D. For a $|V_{SD}| = |V_p| = 4$ volts, the corresponding $I_{DSS} \approx 5.3$ ma. Thus

$$I_D = 5.3 \text{ ma}\left(1 - \frac{-1}{-4}\right)^2 = 2.98 \text{ ma}$$

A third alternative to consider in estimating V_P is to recall that a tangent to the transfer curve at $V_{SG} = 0$ intersects the V_{SG} axis at $V_p/2$. A straight edge placed tangent to the transfer curve at $V_{SG} = 0$ intersects the V_{SG} axis at -2 volts. Therefore, $V_p/2 = -2$ volts or $V_p = -4$ volts, which is to be expected since $V_{SG} =$ -4 volts was cut off on the drain characteristics. An $I_D = 2.98$ ma is in closer agreement than 4.04 ma. Thus a reasonably good way to estimate V_p is to note $V_{SG(\text{off})}$, the V_{SG} for I_D cut-off from the drain characteristics, or the $V_p/2$ intercept on the transfer curve.

We haven't yet allowed for r_D or R_{DS}. Generally r_D is so much larger than R_{DS} it may be neglected. For example, in this problem

$$R_{DS} = \frac{7 \text{ volts}}{5.5 \text{ ma}} \frac{1}{1 - \sqrt{-1 \text{ volt}/-4 \text{ volts}}}$$
$$= \frac{7 \text{ volts}}{5.5 \text{ ma}} \frac{1}{1 - \sqrt{.25}} = 1.7 \text{ kilohms}$$

which is much less than $r_D = 36$ kilohms. Thus, considering r_D,

$$I_D \approx 2.98 \text{ ma} + \frac{15 \text{ volts}}{36 \text{ kilohms}} = 3.4 \text{ ma}$$

which is very close to the I_D value at $V_{SD} = 15$ volts, $V_{SG} = -1$ volt read from Fig. PS 13-2*a*, or the transfer curve of Fig. PS 13-2*b* which was also plotted for $V_{SD} = 15$ volts. This result is also more accurate than the piecewise model considered in the previous problem.

PS 13-5 Using the incremental FET model of Fig. 13-22 prove $y_{is} = (g_{SG} + g_{DG}) + j\omega(C_{SG} + C_{DG})$.

SOLUTION Since y_{is} relates i_g to v_{sg} with $v_{sd} = 0$ we can simplify the model of Fig. 13-22 to Fig. PS 13-5.

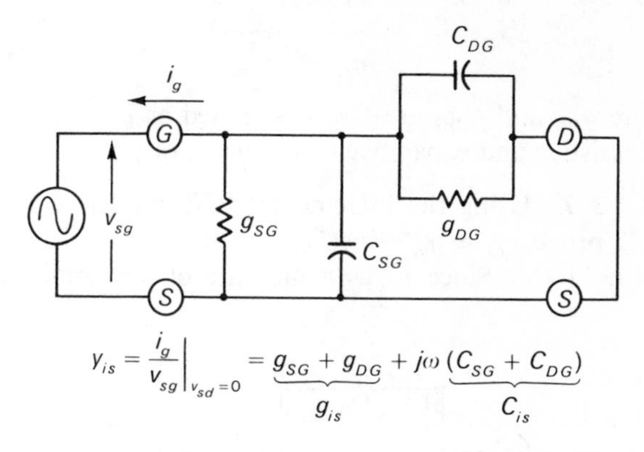

FIGURE PS 13-5

The short across the output sets $v_{sd} = 0$ and wipes out the effect of C_{DS}, g_{DS}, and the $g_m v_{sg}$ current source. Hence they may be omitted from Fig. PS 13-5 for clarity. Since all the remaining elements are in parallel the net admittance y_{is} is the vector sum of the conductances and susceptances. The net input conductance (resistive component of y_{is}) is called g_{is} and the capacitive component is C_{is}. Thus

$$y_{is} = \underbrace{(g_{SG} + g_{DG})}_{g_{is}} + j\omega\underbrace{(C_{SG} + C_{DG})}_{C_{is}} \quad (13\text{-}12)$$

PS 13-6 Using the incremental FET model of Fig. 13-22, prove $y_{rs} = -(g_{DG} + j\omega C_{DG})$.

SOLUTION Since y_{rs} relates i_g to v_{sd} with $v_{sg} = 0$ the circuit of Fig. 13-22 boils down to that of Fig. PS 13-6. The short across the input wipes out g_{SG} and C_{SG}, and the zero-impedance v_{sg} voltage source fixes the potential across C_{DG} and g_{DG} so that the $g_m v_{sg}$ current source and g_{SD} and C_{SD} are of no consequence. Since the reference direction of i_g is opposite to that in which a positive v_{sd} would actually cause i_g to flow, y_{rs} becomes a negative quantity given by

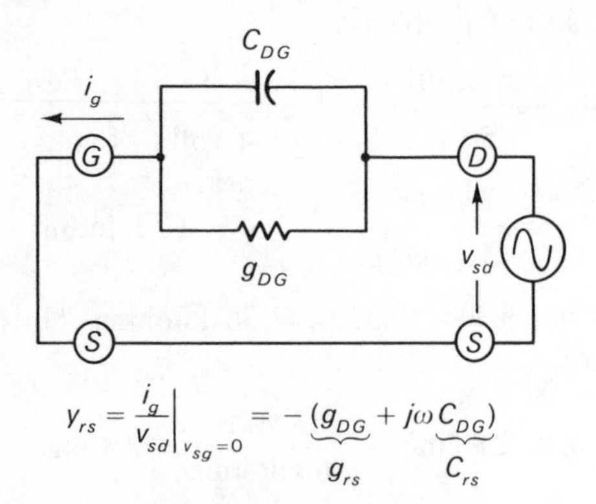

FIGURE PS 13-6

$$y_{rs} = -(\underbrace{g_{DG}}_{g_{rs}} + j\omega \underbrace{C_{DG}}_{C_{rs}}) \quad (13\text{-}13)$$

where g_{DG} and C_{DG} are also designated as g_{rs} and C_{rs}, the resistive and capacitive components of y_{rs}.

PS 13-7 Using the incremental FET model of Fig. 13-22 prove $y_{fs} \approx g_m - j\omega C_{DG}$.

SOLUTION Since y_{fs} is a measure of how much i_d results from a v_{sg} with the output terminals shorted ($v_{sd} = 0$) we may reduce the circuit of Fig. 13-22 to that of Fig. PS 13-7. The short across the output wipes out g_{SD} and C_{SD} but we have to keep the $g_m v_{sg}$ current source since it is a factor in determining i_d.

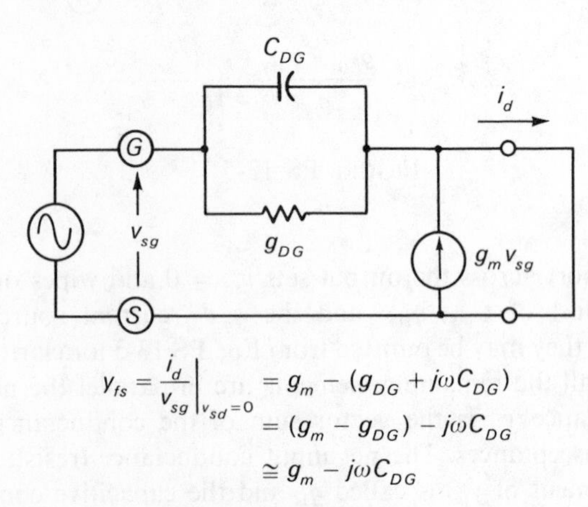

FIGURE PS 13-7

It may easily be shown by the superposition theorem that the i_d component caused by the $g_m v_{sg}$ current source only is $g_m v_{sg}$. The i_d component due to a positive v_{sg} only (the current source open-circuited) is directed opposite to the i_d reference direction, and the admittance is determined by C_{DG} and g_{DG}. Thus

$$i_d = g_m v_{sg} - (g_{DG} + j\omega C_{DG})v_{sg}$$

and

$$\therefore y_{fs} = \frac{i_d}{v_{sg}}\Big|_{v_{sd}=0} = g_m - (g_{DG} + j\omega C_{DG})$$

Collecting real and reactive components we have

$$y_{fs} = (g_m - g_{DG}) - j\omega C_{DG} \quad (13\text{-}14)$$

Since g_m is seldom less than 200 μmho and g_{DG} is typically a fraction of a micromho, which is the reciprocal of a resistance corresponding to several megohms, we can neglect the g_{DG} term to obtain

$$y_{fs} \approx g_m - j\omega C_{DG} = g_m - j\omega C_{rs} \quad (13\text{-}15)$$

In the low-frequency spectrum the reactive term is negligible but not so in the high-frequency spectrum. To pin down where the effect of C_{DG} becomes significant we can initially set the dividing line between the low- and high-frequency spectrum as the frequency at which the magnitude of the resistive and reactive term are identical. Thus

$$g_m = \omega' C_{rs}$$

or

$$\omega' = \frac{g_m}{C_{rs}} \quad (13\text{-}16)$$

Typically the larger g_m is the larger C_{rs} is, so the ratio is roughly constant. For a JFET with a $g_m = 3600$ μmho we might expect a $C_{rs} = 8$ pf. Thus

$$\omega' = \frac{3600 \times 10^{-6} \text{ mho}}{8 \times 10^{-12} \text{ farads}} = 450 \times 10^6 \text{ rad/sec}$$

or

$$f' = \frac{\omega'}{2\pi} = 71.6 \text{ megahertz}$$

Therefore, unless most any FET is to be operated in the high-frequency spectrum, say 10 megahertz or more, y_{fs} is essentially equal to g_m. For a rule of thumb we will arbitrarily set up the following relationship.

$$|y_{fs}| \cong g_m \quad \text{for } \omega \cong 0.1\omega' \quad (13\text{-}17)$$

If in doubt regarding whether or not g_m and y_{fs} can be used interchangeably, substitute in equation 13-16 to determine ω'. If the maximum frequency the FET is to be operated at is, say, less than $0.1\omega'$, it's safe to interchange y_{fs} and g_m. Of course, this is only true for y_{fs} specified at a test frequency much less than ω'.

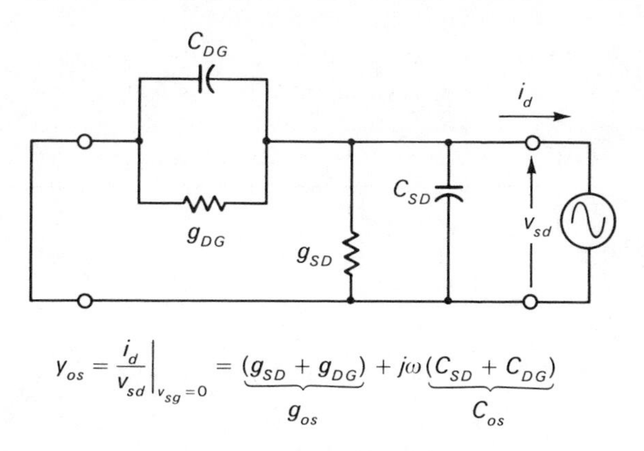

$$y_{os} = \left.\frac{i_d}{v_{sd}}\right|_{v_{sg}=0} = \underbrace{(g_{SD} + g_{DG})}_{g_{os}} + j\omega\underbrace{(C_{SD} + C_{DG})}_{C_{os}}$$

FIGURE PS 13-8

PS 13-8 Using the incremental FET model of Fig. 13-22 prove $y_{os} = (g_{SD} + g_{DG}) + j\omega(C_{SD} + C_{DG})$.
SOLUTION Since y_{os} is a measure of how much i_d will be drawn by a v_{sd} voltage source with $v_{sg} = 0$ we may reduce the circuit of Fig. 13-22 to that of Fig. PS 13-8. With $v_{sg} = 0$ (due to the shorted input) the $g_m v_{sg}$ current source is zero and may be replaced by an open circuit. The resultant i_d due to v_{sd} is determined by the admittance of the circuit which, by inspection, may be written as

$$y_{os} = \underbrace{(g_{SD} + g_{DG})}_{g_{os}} + j\omega\underbrace{(C_{SD} + C_{DG})}_{C_{os}} \quad (13\text{-}18)$$

Note y_{os} is sometimes expressed in terms of g_{os} and C_{os} with the equivalency indicated in the above equation.

PROBLEMS WITH ANSWERS

PA 13-1 A JFET exhibits a $g_m = 2100$ μmho at a given Q point. If $\Delta V_{SG} = 0.2$ volt while V_{SD} is held constant, what is ΔI_D approximately equal to?
ANSWER 0.42 ma

PA 13-2 A *P*-channel JFET exhibits $g_{m0} = 2500$ μmho, and $V_P = 6$ volts. Determine I_{DSS}.
ANSWER 7.5 ma

PA 13-3 If $V_P = -5$ volts and $I_{DSS} = 8$ ma, estimate I_D at $V_{SG} = -8$ volts in a JFET.
ANSWER $I_D = 0$

PA 13-4 Determine R_{DS} in the piecewise model if $I_{DSS}' = 2.8$ ma and $V_{SD}' = 3.6$ volts.
ANSWER 1.29 kilohms

PA 13-5 A change of 12 volts in V_{SD} causes a 0.4-ma change in I_D for an IGFET operated in its saturated region. What conclusion can you draw?
ANSWER $r_D = 30$ kilohms

PA 13-6 Summarize the steps you would follow in constructing a piecewise FET model from its drain (I_D versus V_{SD} for various V_{SG}) characteristics. Assume negligible gate leakage current.
ANSWER
1. Eyeball V_{SD}' and I_{DSS}'.
2. Compute R_{DS}.
3. Construct the transfer curve (saturated region).
4. Linearize the transfer curve to obtain an average g_m.
5. Determine r_D.
6. Construct the piecewise model observing proper diode and current-source polarities.

PA 13-7 A FET is characterized by $V_P = -4$ volts, $I_{DSS} = 8$ ma. Estimate I_D at $V_{SG} = -1.8$ volts.
ANSWER 2.42 ma

PA 13-8 What V_{SG} is required to bias a JFET at $I_{DSS} = 8$ ma. Estimate I_D at $V_{SG} = -1.8$ volts. and $V_p = 2.5$ volts?
ANSWER 0.732 volt

PA 13-9 The dc model data of some FET consists of $I_{DSS}' = 6$ ma, $V_{SD}' = 8$ volts, $V_p = -5$ volts, $r_D = 80$ kilohms. Estimate I_D at $V_{SD} = 25$ volts, $V_{SG} = -2$ volts.
ANSWER 2.47 ma

PA 13-10 Given the following parameters measured at $f = 1KH_z$, $V_{SD} = 15$ volts, and $V_{SG} = 0$:

$g_{is} = 0.02\ \mu$mho $\qquad g_{os} = 35\ \mu$mho
$C_{is} = 39$ pf $\qquad C_{os} = 21$ pf
$g_{rs} = 0.0008\ \mu$mho $\qquad y_{fs} = 3600\ \mu$mho
$C_{rs} = 8$ pf

Sketch the complete small-signal model.
ANSWER See Fig. PA 13-10

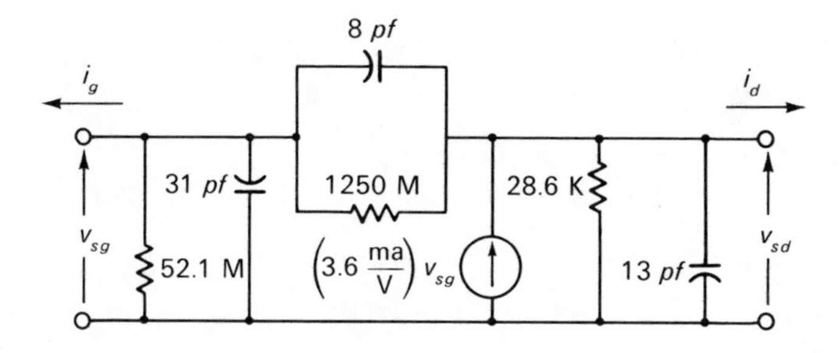

PROBLEMS WITHOUT ANSWERS

P 13-1 It takes a $\Delta V_{SG} = 0.1$ volt to cause a $\Delta I_D = 0.2$ ma with $\Delta V_{SD} = 0$. What is the g_m?

P 13-2 How would you determine V_p from the transfer characteristic?

P 13-3 If $V_p = -2$ volts, what type of JFET is involved?

P 13-4 If $V_p = 3$ volts and $I_{DSS} = 6$ ma, determine I_D at $V_{SG} = 0$ in the saturated region.

P 13-5 If $V_{TH} = +4$ volts, what type of FET is involved?

P 13-6 If $R_{DS} = 1$ kilohm in a piecewise JFET model and $I_{DSS}' = 4$ ma, determine V_{SD}' and estimate V_p.

P 13-7 Given the following piecewise model data:

$$\left.\frac{\Delta I_D}{\Delta V_{SG}}\right|_{\Delta V_{SD}=0} = 1.8 \text{ ma/volt}$$

$$\left.\frac{\Delta I_D}{\Delta V_{SD}}\right|_{\Delta V_{SG}=0} = 0.008 \text{ ma/volt}$$

$I_{DSS}' = 4$ ma $\qquad V_{SD}' = 5.5$ V

construct an N-channel piecewise model.

P 13-8 Determine I_D if $V_{SD} = 20$ volts, and $V_{SG} = -1$ volt for the FET in the previous problem.

P 13-9 If $V_p = 6$ volts and $R_{DS} = 800$ ohms at $V_{SG} = 4$ volts in the dc model, what is R_{DS} at $V_{SG} = 5$ volts?

P 13-10 Determine μ if $y_{fs} = 2000\ \mu$mho and $y_{os} = 10\ \mu$mho at 1 kilohertz at some Q point.

14 fet biasing

For the FET to function as a linear amplifier it must be biased into its active or amplifying region. The selected quiescent (Q) operating point must then be stabilized against temperature effects and device parameter variations, which may be quite severe due to manufacturing tolerances. This situation is somewhat similar to the tolerance spread in β or h_{FE} between junction transistors of the same type-number.

14-1 JFET biasing

The active or linear amplifying region of a JFET is represented by the shaded area shown in Fig. 14-1. The

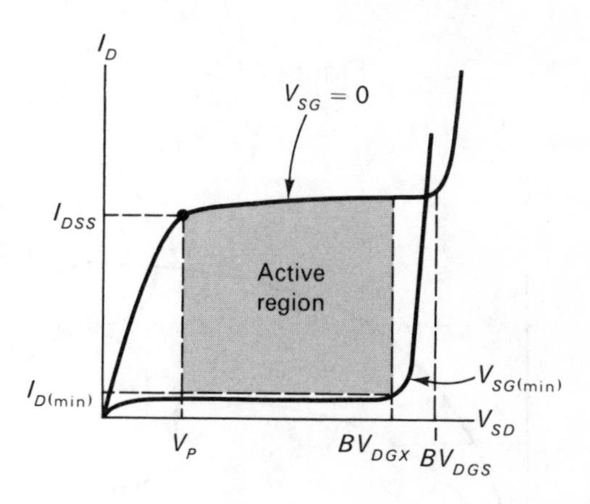

FIGURE 14-1

drain current is bounded by a maximum value equal to or slightly greater than I_{DSS} and an arbitrarily chosen minimum $I_{D(\min)}$. The choice of which $I_{D(\min)}$ depends upon the minimum useful g_m and distortion due to compression of the V_{SG} curves. The source-to-drain voltage, V_{SD}, is bounded by a maximum V_{SD} value equal to $BV_{DGX} = |BV_{DGS}| - |V_{SG(\min)}|$ and a maximum value of V_p. The magnitude of V_{SD} should preferably be more like $1.5|V_p|$ to $2|V_p|$ if distortion is critical.

A major problem in FET biasing is the large parameter spread in devices which even bear the same type-number. For example, Fig. 14-2 shows the $V_{SG} = 0$ drain characteristics of two similarly designated FETs, $Q1_A$ and $Q1_B$ respectively, at 25°C. Evidently $Q1_A$ is a much "hotter" FET in the sense that its I_{DSS} greatly exceeds that of $Q1_B$. In practice, the spread between $I_{DSS(\max)}$ and $I_{DSS(\min)}$ may range from three to one or even five to one.

Another problem is the sensitivity of the drain current with respect to temperature. Fig. 14-3 illustrates how the I_{DSS} point for a particular FET varies with temperature. Note that increasing temperature decreases I_D. Actually this negative temperature coefficient of I_D manifests itself over most, but not all, of the active region. We shall discuss this in more detail later; and,

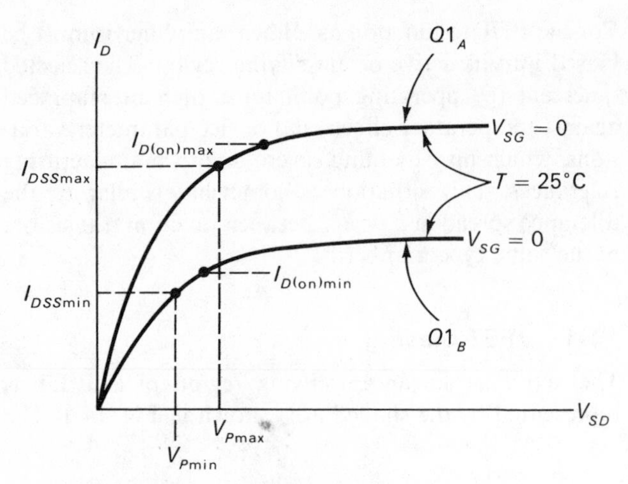

FIGURE 14-2

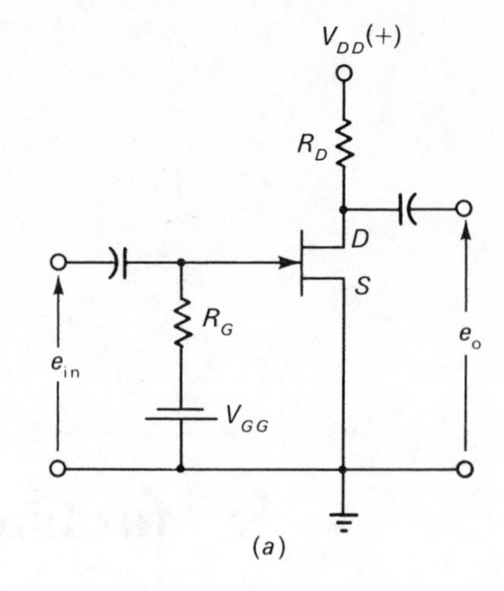

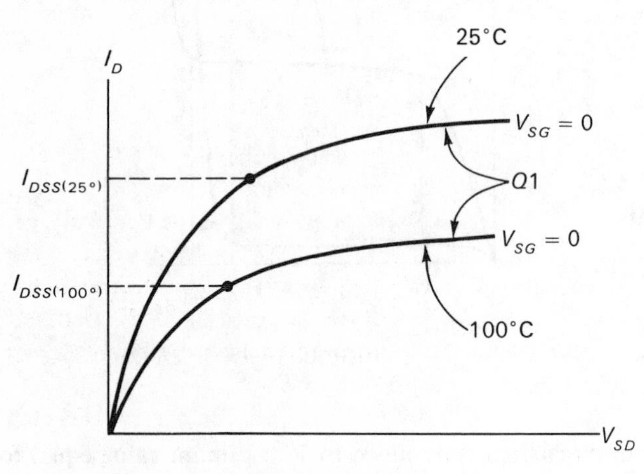

FIGURE 14-3

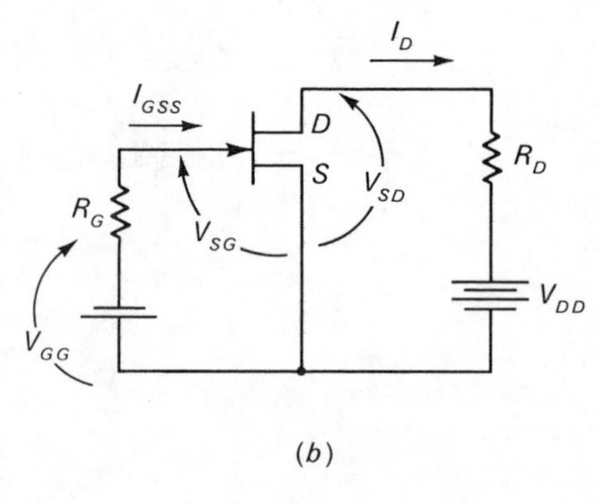

FIGURE 14-4

in the meantime, rejoice in the fact that this negative temperature coefficient of I_D means the FET is not plagued by the thermal runaway problem encountered with junction transistors, wherein V_{EB}', I_{CBO}, and β all tend to *increase* the collector current with an increase in temperature.

The gate leakage current, I_G, which may conservatively be taken as I_{GSS}, is also temperature-sensitive. If manufacturer's data is unavailable, I_{GSS} may be treated in a manner similar to I_{CBO} in a silicon junction transistor. Normally, I_{GSS} effects will be negligible.

14-2 Fixed bias

Figure 14-4*a* illustrates what is known as a *common-source amplifier*. It is shown to illustrate where the ac input signal is injected and where the output is extracted. What we are actually interested in at this time is the circuit of Fig. 14-4*b* which illustrates the essence of the biasing configuration. This particular biasing technique is called *fixed bias* because if we neglect I_{GSS}, the bias voltage $V_{SG} = V_{GG}$ where V_{GG} is a fixed gate supply voltage.

If R_G is very large and the temperature is high, I_{GSS} appreciably effects the V_{SG} bias since (Fig. 14-4*b*)

$$V_{SG} = -|V_{GG}| + R_G I_{GSS} \qquad (14\text{-}1)$$

Therefore, as I_{GSS} increases, $|V_{SG}|$ decreases which in turn causes I_D to increase and shifts the Q point upward. This increase in I_D, due to I_{GSS} and R_G, tends to offset the negative temperature coefficient of I_D. It is, in fact, possible to obtain an essentially zero change of I_D with temperature if R_G is properly selected and the FET is operated at a fairly constant temperature. Since a constant temperature environment is not the usual case, and since I_{GSS} effects are generally negligible, this matter will not be pursued further.

The quiescent operating point of a FET operating with constant- or fixed-bias voltage as shown in Fig. 14-4*b* may be determined by first noting that

$$V_{SD} = V_{DD} - R_D I_D \qquad (14\text{-}2)$$

This is the load line equation for this circuit and it may be conveniently plotted on the drain characteristics by plotting two points: one at $V_{SD} = V_{DD}$ and $I_D = 0$, the other at $V_{SD} = 0, I_D = V_{DD}/R_D$. The first case corresponds to the open-circuit voltage the FET looks into, and the second case corresponds to the maximum short-circuit current that could flow through it. The operating point will lie on the load line. The load line for two different values of R_D is plotted in Fig. 14-5. The slope of the load

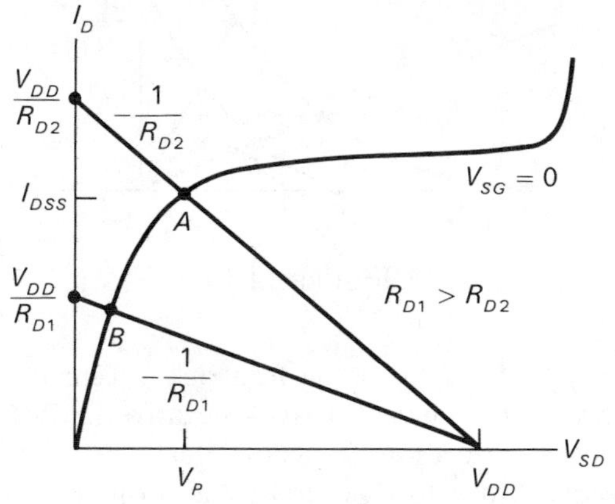

FIGURE 14-5

line is negative and equal to $-1/R_D$. Thus R_D is a smaller load (drain) resistor than R_{D1}. Note that if V_{SG} is ever permitted to equal zero, the R_{D1} load line moves out of the saturated (active) region since point B corresponds to a $|V_{SD}| < |V_P|$. Since the voltage gain, A_v, is proportional to R_D, it follows that the lower the V_P the larger we can make R_D and hence realize a higher A_v. This is one reason why a low V_P is desirable.

If the drain characteristics of the particular FET in question are available, say on a curve tracer, the Q point may be located by noting the intersection of the $V_{SG} = V_{GG}$ curve with the load line.

Work through problems PS 14-1 and PS 14-2 at this time.

As previously mentioned, FET production tolerances are such that a large spread in various parameters results. This is again illustrated in Fig. 14-6. Note that

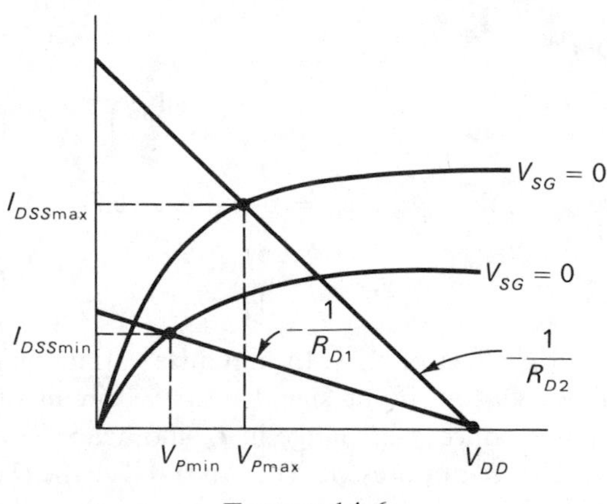

FIGURE 14-6

the correct load line corresponds to R_{D2} and not R_{D1} since R_{D2} lies in the active region for both the max and min FETs, whereas R_{D1} is only appropriate to the min FET.

A most useful way of analyzing bias circuits is with the aid of the transfer curves shown in Fig. 14-7 for a

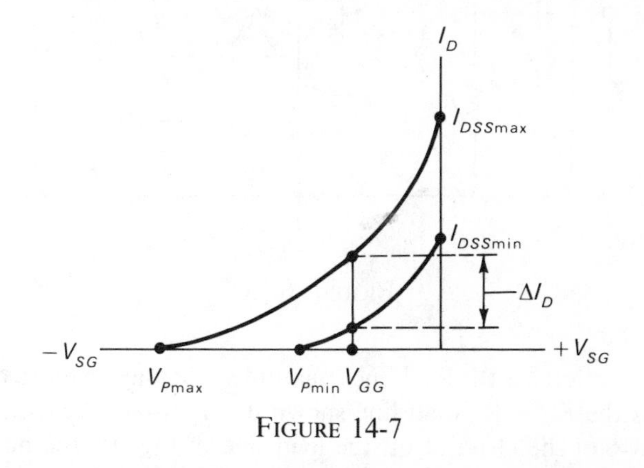

FIGURE 14-7

max and min FET.[1] Remember the transfer curve is plotted for some V_{SD} value well into the active region ($|V_{SD}| > |V_P|$). Since the drain current is relatively independent of V_{SD} in the saturated (active) region, the transfer curves are a good approximation to the variation of I_D with V_{SG} for any V_{SD} in saturation. Remember, saturation for a FET is opposite to what we consider as saturation for a bipolar transistor.

Since $V_{SG} = V_{GG}$ with fixed bias we can construct a vertical line at $V_{SG} = V_{GG}$ in Fig. 14-7 to determine the range of I_D values for the max and min cases. The actual quiescent value of drain current will lie somewhere within the ΔI_D range of I_D.

In general, fixed bias is satisfactory only if the temperature is held fairly constant, R_G is not too large, and careful attention is paid to the choice of circuit parameters.

14-3 Current-derived self-bias

The circuit of Fig. 14-8 illustrates a biasing arrangement commonly known as *self-bias*. More accurately it should be described as current-derived self-bias because the gate-bias voltage, V_{SG}, is actually developed by the flow of drain current through the source resistor R_S. Thus, neglecting any gate current,

1) $$V_{SG} = -R_S I_D \tag{14-3}$$

and

2) $$V_{SD} = V_{DD} - (R_S + R_D)I_D \tag{14-4}$$

[1]Although we shall not go into as detailed a worst-case analysis, the $I_{D(\max)}$ curve should be used at a minimum temperature and the $I_{D(\min)}$ curve at a maximum temperature.

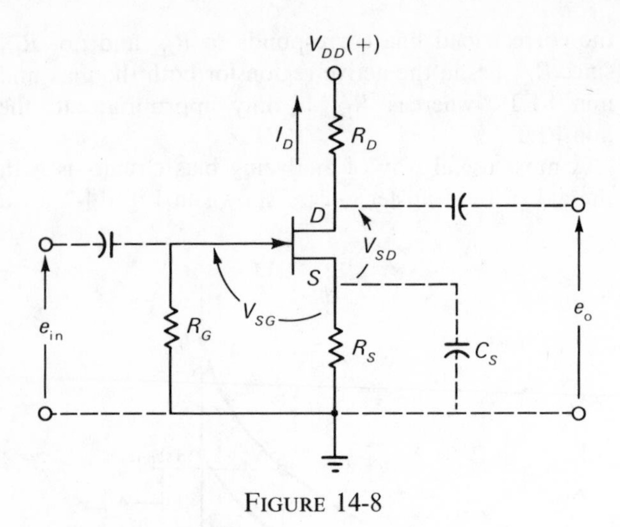

FIGURE 14-8

Equation 2 is the load line equation and it may be plotted as the $R_S + R_D$ load line shown in Fig. 14-9. The location of the Q point on the load line of Fig. 14-9 is not obvious as in the case of fixed bias. To locate the actual

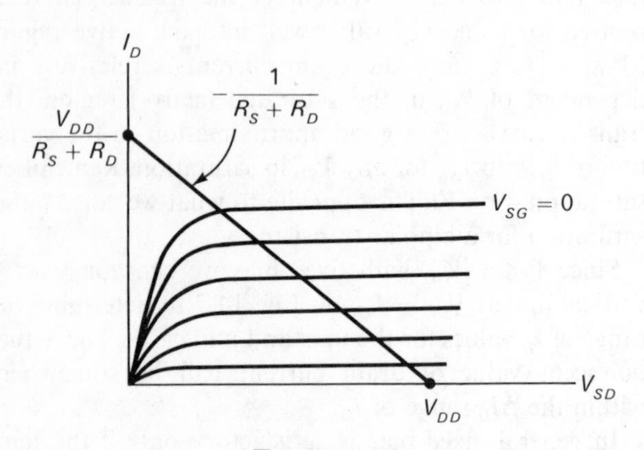

FIGURE 14-9

V_{SG} we use the transfer curve as shown in Fig. 14-10 and plot equation 1 preferably by assuming V_{SG} values and computing the corresponding I_D in

3) $$I_D = \frac{-V_{SG}}{R_S} \qquad (14\text{-}5)$$

Remember, V_{SG} will be a negative quantity so that I_D will turn out positive for the N-channel JFET of Fig. 14-8.

Equation 1 or 3 plots as a straight line, called the bias curve on the transfer characteristic shown in Fig. 14-10. The most important point to note in Fig. 14-10 is that the larger R_S becomes, the smaller the range of change in drain current due to device parameter variation. Thus with $R_{S2} > R_{S1}$, $\Delta I_{D2} < \Delta I_{D1}$.

A physical explanation of this is simply that since V_{SG} is developed by sampling I_D through R_S, any variation in I_D affects V_{SG}, and V_{SG} in turn controls I_D since

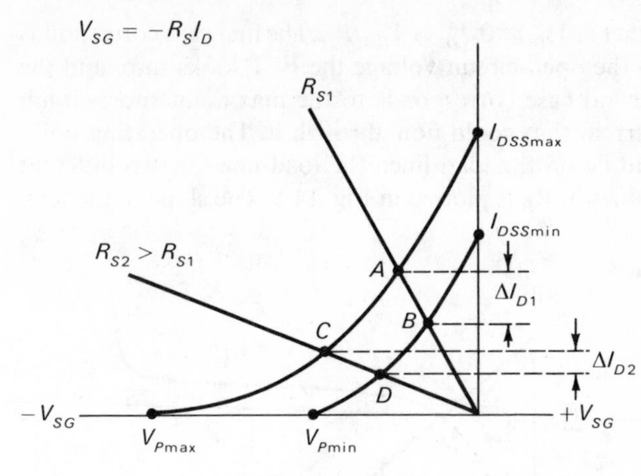

FIGURE 14-10

$\Delta I_D \approx g_m \Delta V_{SG}$. This is a feedback mechanism and specifically it is current-derived negative feedback. We shall learn more about feedback later. The point is that the feedback due to R_S exerts a stabilizing (negative or degenerative feedback) influence on I_D. For example, if the temperature increases or a low I_{DSS} FET is plugged in the drain current will, in general, tend to decrease. If I_D decreases, less bias voltage V_{SG} is developed across R_S, and this tends to cause I_D to increase, which in part compensates for the tendency of I_D to decrease. Therefore, with R_S present I_D will not decrease as much as it would if R_S were absent. Conversely, anything that tends to increase I_D causes more gate bias to be developed which tries to decrease I_D.

We can see from Fig. 14-10 that the larger R_S becomes, the smaller the resultant ΔI_D, which is desirable since this implies a more stable Q point. Unfortunately, the larger R_S becomes, the larger the bias voltage also becomes, and, since g_m decreases as the bias increases, we find that, for the large R_S values we would like to have, the g_m is too low.

The g_m, you will recall, is derived by differentiating

1) $$I_D = I_{DSS}\left(1 - \frac{V_{SG}}{V_P}\right)^2 \qquad (14\text{-}6)$$

to obtain

2) $$g_m = \frac{dI_D}{dV_{SG}} = g_{m0}\left(1 - \frac{V_{SG}}{V_P}\right) \qquad (14\text{-}7)$$

where g_{m0} is the g_m at $V_{SG} = 0$ and is given by

3) $$g_{m0} = \frac{-2I_{DSS}}{V_P} \qquad (14\text{-}8)$$

Since the effect of R_S is to minimize variations in I_D, it follows that an input signal into the circuit of Fig. 14-8 will produce less change in I_D and hence less gain than if R_S were not present. Thus to stabilize the Q point but still not degenerate the gain due to R_S, a bypass capacitor C_S may be shunted across R_S. If C_S looks like a short circuit at the operating signal frequency, the full amplifier gain will be realized simultaneously with improved Q-point predictability and stabilization.

Work through problems PS 14-3 and PS 14-4 at this time.

It was demonstrated that increasing R_S minimizes ΔI_D but, unfortunately, increases the bias and reduces I_D to a region of undesirably low g_m. Suppose our problem is such that we want to hold ΔI_D or the maximum and minimum values of I_D between points A and B on the transfer characteristic of Fig. 14-11. This can be achieved

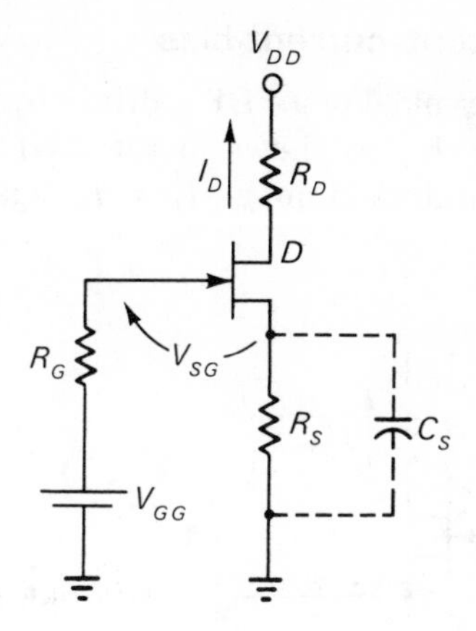

FIGURE 14-12

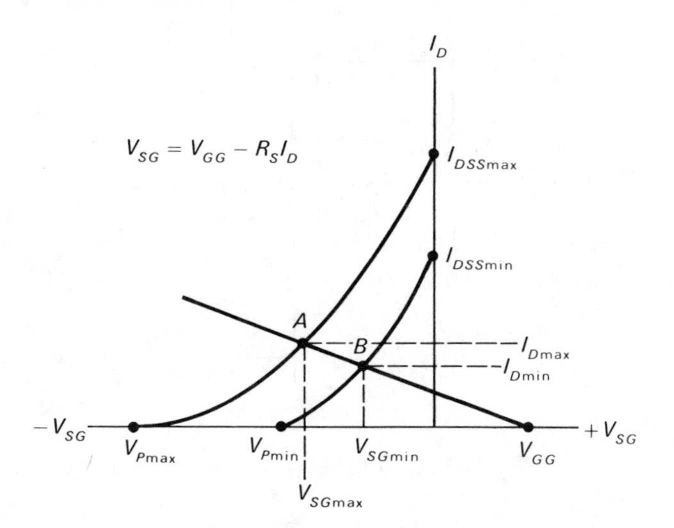

FIGURE 14-11

if we have a bias line which is tangent to points A and B as shown in Fig. 14-11. This fixes the slope and hence the value of R_S to

$$R_S = \frac{V_{SG(\max)} - V_{SG(\min)}}{I_{D(\max)} - I_{D(\min)}} \qquad (14\text{-}9)$$

Note, however, that to realize this value of R_S the bias line intersects the V_{SG} axis at a value V_{GG} which is of a polarity which tends to forward-bias the gate junction. Physically this implies a circuit as shown in Fig. 14-12 so that when $I_D = 0$, $V_{SG} = V_{GG}$.

The V_{GG} supply does not actually forward-bias the gate junction because the flow of drain current through R_S develops more reverse bias. Again R_S could be bypassed to maximize the ac signal gain.

A practical means of synthesizing the V_{GG} gate bias supply of Fig. 14-12 is shown in Fig. 14-13*a*. Resistors R_1 and R_2 form a voltage divider which establishes V_{GG} as shown in Fig. 14-13*b*. However, if R_1 and R_2 are very large in ohmic value, their respective resistances may appear somewhat unstable because of shunt leakage paths which vary with temperature, humidity, dust, and so forth. Consequently, the arrangement of

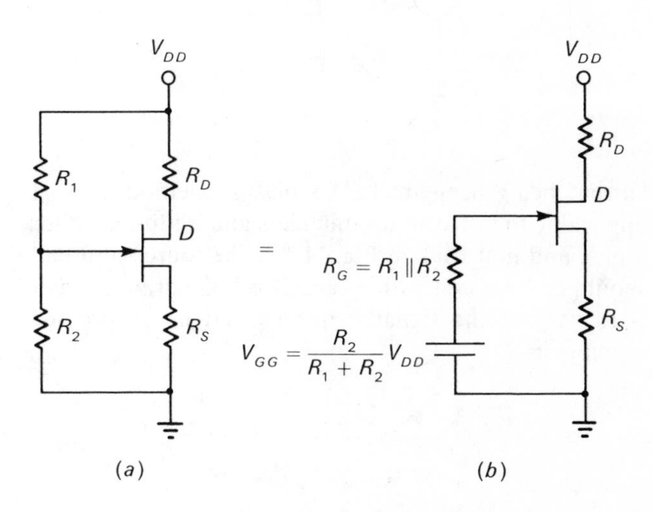

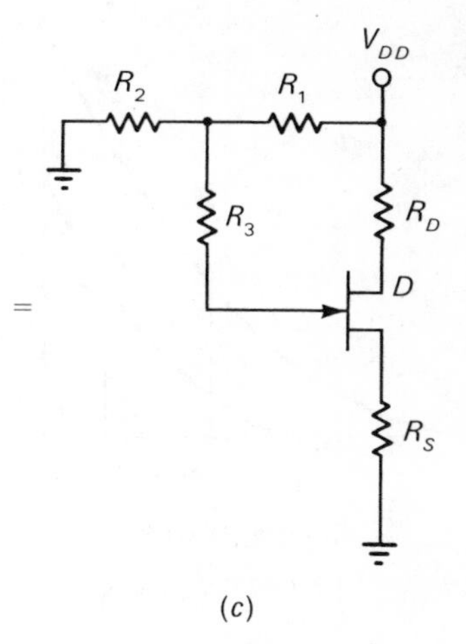

FIGURE 14-13

Fig. 14-13*c* may be more appropriate. Here the bias is set with R_1 and R_2 which may be held to a few hundred kilohms or less, while R_3 is a large resistor to raise the input impedance.

Now, carefully work through problems PS 14-5, PS 14-6, and PS 14-7.

14-4 Constant-current bias

If the source terminal of a FET is driven by a constant-current source, I_{SS}, as shown in Fig. 14-14, the drain current is also locked in at $I_D = I_{SS}$ (gate leakage current being negligible). This biasing method is most applicable to low-drift dc amplifiers and is illustrated for a max and min FET in Fig. 14-15. The source terminal should be bypassed with a capacitor if the drain current is to vary at the signal frequency and hence provide signal gain.

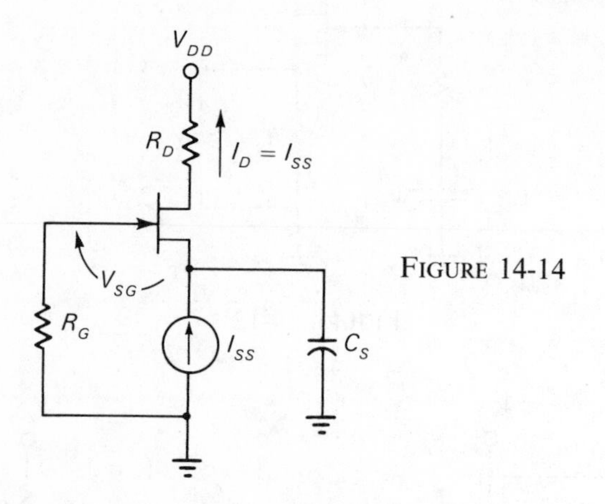

FIGURE 14-14

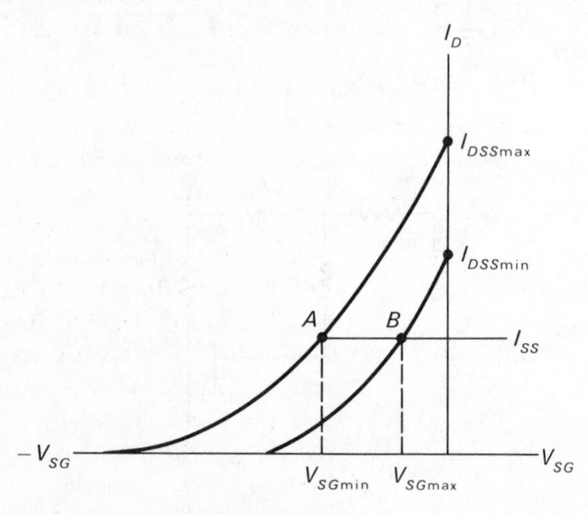

FIGURE 14-15

From Fig. 14-15 we see that the gate bias will fall between $V_{SG(\min)}$ and $V_{SG(\max)}$. To guarantee operation in the saturated (active) region we must be sure that $I_{SS} < I_{DSS}$ and $|V_{SD}| > |V_P|$ and, preferably, $|V_{SD}| > 2|V_P|$. Since this means

1) $$V_{SD} = V_{DD} - R_D I_{SS} \geqq 2|V_P| \qquad (14\text{-}10)$$

we can solve the above equation for V_{DD}, R_D, or I_{SS} depending upon what is given.

Fig. 14-16 illustrates how the constant-current source may be simulated by the high output impedance seen looking into the drain of a FET. Here $Q1$ is the constant-current source and $Q2$ is the amplifier. For this circuit to be effective, I_D must be $\leqq I_{DSS2}$. This guarantees $Q1$ will keep $Q2$ in the saturated region if sufficient V_{SD2} is maintained.

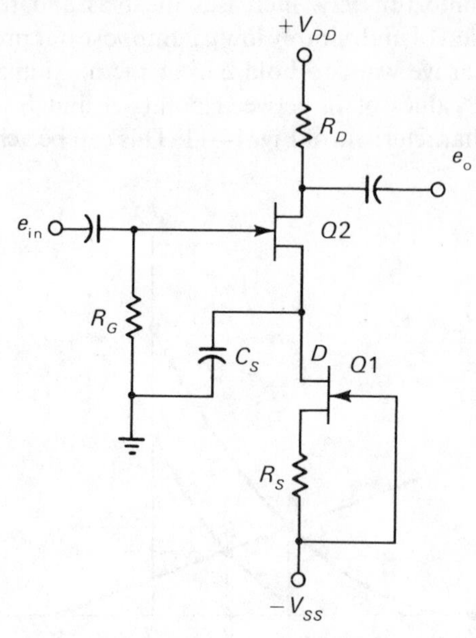

FIGURE 14-16

When analyzing or designing circuits such as in Fig. 14-16 or Fig. 14-17 it is useful to estimate the output current of the constant-current source. We already know how to do this for the transistor in Fig. 14-17. For the FET ($Q1$) in Fig. 14-16, the situation is more complicated. If we are designing this circuit and wish to estimate the

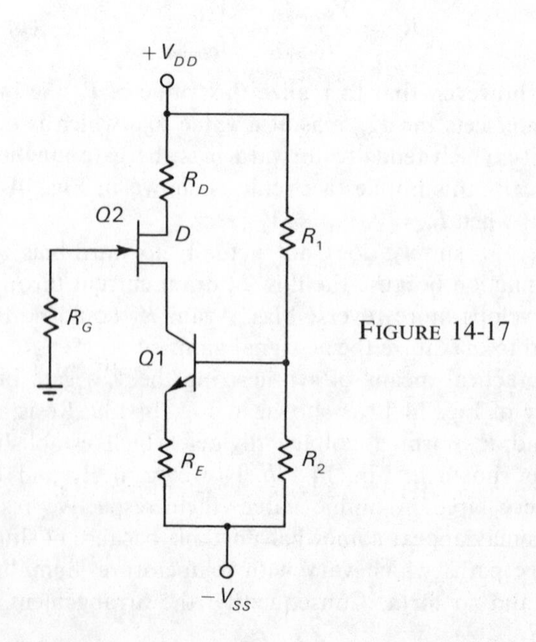

FIGURE 14-17

value of R_S to set I_D at a desired level in the saturated region, we would have to do the following: from the transfer or drain characteristics of $Q1$ determine the nominal value of V_{SG} required for the desired I_D. The value of R_S is then given by

$$1) \qquad R_S = \frac{V_{SG}}{I_D} \qquad (14\text{-}11)$$

If the transfer or drain characteristics are not given but V_P and I_{DSS} are known, we may solve the following equation

$$2) \qquad I_D = I_{DSS}\left(1 - \frac{V_{SG}}{V_P}\right)^2 \qquad (14\text{-}12)$$

for V_{SG} to obtain

$$3) \qquad V_{SG} = V_P\left(1 - \sqrt{\frac{I_D}{I_{DSS}}}\right) \qquad (14\text{-}13)$$

The above value of V_{SG} may then be substituted into equation 1 to determine I_D. If the max and min values of V_P and I_{DSS} are given, an average V_{SG} may be computed. Resistor R_S may then be adjusted for an exact I_D value if necessary.

At times the problem is one of analyzing a circuit such as shown in Fig. 14-16 for I_D. If the transfer characteristics are available (or are constructed from the drain characteristics), we plot an R_S bias curve to determine I_D. If the transfer or drain curves are not specified but I_{DSS} and V_P are given, the following analytic approach may be employed. If we substitute equation 14-11 into 14-12, we have

$$4) \qquad I_D = I_{DSS}\left(1 - \frac{R_S I_D}{V_P}\right)^2$$

Squaring, multiplying out, and collecting terms yields the following quadratic in I_D.

$$5) \quad 0 = I_{DSS}\frac{R_S{}^2}{V_P{}^2} I_D{}^2 - \left(\frac{2R_S I_{DSS}}{V_P} + 1\right) I_D + I_{DSS}$$

Solving 5 by the quadratic equation and considerable manipulation yields

$$6) \quad I_D = \frac{|V_P|}{2R_S{}^2 I_{DSS}}\left[2R_S I_{DSS} + |V_P| - \sqrt{|V_P|(4R_S I_{DSS} + |V_P|)}\right] \qquad (14\text{-}14)$$

Now work through problems PS 14-8, PS 14-9, and PS 14-10.

14-5 Zero-temperature-coefficient biasing

It turns out that if we plot I_D versus V_{SG} for different values of temperature in the saturated region ($|V_{SD}| > |V_P|$), we get a family of curves similar to Fig. 14-18. The

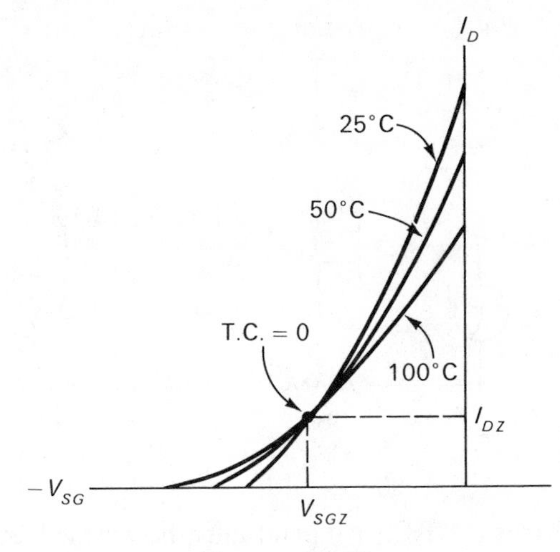

FIGURE 14-18

most significant point to note is that at the point where $V_{SG} = V_{SGZ}$, $I_D = I_{DZ}$ at any temperature. In other words, the temperature coefficient (T.C.) of drain current is zero. This means the Q point will not drift with temperature if the bias is held at V_{SGZ}. Actually the zero T.C. point in Fig. 14-18 isn't one single point that all the curves go through. In truth they are spread by a few millivolts of V_{SG} but for all practical purposes may be considered as intersecting at one point.

The reason we have a zero T.C. point is because there are two opposing mechanisms at work in the FET. First, the channel resistance increases with temperature due to reduced carrier mobility. This effect is similar to the increase of resistance with temperature in a metallic conductor and causes the drain current to exhibit a negative T.C. of about -0.7 percent/°C.

The second and opposing effect is the decrease of channel resistance due to the depletion regions shrinking as temperature increases. This effect is similar to the negative T.C. associated with a forward-biased diode and therefore the ΔI_D due to a ΔT from this cause is equivalent to a ΔV_{SG} of about 2.2 mv/°C. Since the channel is widening in this case, we have a positive T.C. for I_D.

For $|V_{SG}| > |V_{SGZ}|$ the positive T.C. due to depletion region shrinkage predominates, whereas for $|V_{SG}| < |V_{SGZ}|$ the negative T.C. of I_D predominates due to the increase of channel resistance. At $V_{SG} = V_{SGZ}$ these effects cancel and T.C. = 0. By adjusting V_{SG} to V_{SGZ} in a circuit such as shown in Fig. 14-19 it is possible to achieve an I_D drift of less than 10 μv/°C over a 100°C range.

Unfortunately, V_{SGZ} is fairly close to the pinch-off voltage, perhaps 0.7 volt above V_P. This means that g_m, and hence gain, will be low and that the linearity is poor. The lower the V_P of the FET, the further away V_{SGZ} is from cut-off, and hence low V_P units are desirable in this case. Since other biasing considerations (gain, linearity,

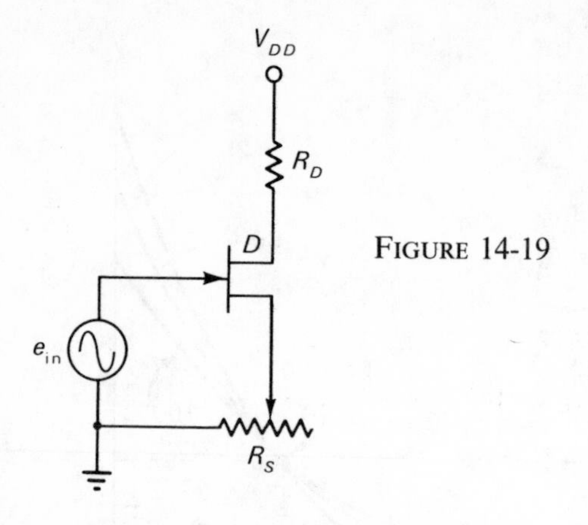

FIGURE 14-19

etc.) besides the T.C. = 0 point must be weighed, the Q point is usually not chosen at V_{SGZ} and I_{DZ} except in certain low-drift dc amplifier applications.

To quantitatively determine V_{SGZ} and I_{DZ}, we may consider the following relationships:[1]

1) $$I_D = f(T, V_{SG})$$

2) $$\frac{dI_D}{dT} = \left(\frac{\partial I_{DSS}}{\partial T}\right)\frac{dT}{dT} + \left(\frac{\partial I_D}{\partial V_{SG}}\right)\frac{dV_{SG}}{dT}$$

3) $$\frac{dI_D}{dT} = -0.007\frac{I_D}{°\text{C}} + g_m(2.2 \text{ mv}/°\text{C})$$

For T.C. = 0, $\frac{dI_D}{dT} = 0$. Therefore,

4) $$0 = -0.007\frac{I_D}{°\text{C}} + g_m(2.2 \text{ mv}/°\text{C})$$

or

5) $$0.007\, I_D = 0.0022 \text{ volt } g_m$$

6) $$I_D = \frac{0.0022 \text{ volt}}{0.007} g_m = (0.314 \text{ volt}) g_m$$

but

7) $$I_D = I_{DSS}\left(1 - \frac{V_{SG}}{V_P}\right)^2$$

from which we obtain

8) $$g_m = \frac{dI_D}{dV_{SG}} = \frac{-2I_{DSS}}{V_P}\left(1 - \frac{V_{SG}}{V_P}\right)$$

Substituting 8 into 6 and equating the result to 7 yields

9) $$(0.314 \text{ volt})\left[\frac{-2I_{DSS}}{V_P}\left(1 - \frac{V_{SG}}{V_P}\right)\right] = I_{DSS}\left(1 - \frac{V_{SG}}{V_P}\right)^2$$

[1]The reader unfamiliar with calculus need only consider the results developed in equations 14-15, 14-16, and 14-17.

Simplifying 9 yields

10) $$0.314\left(\frac{-2}{V_P}\right) = 1 - \frac{V_{SG}}{V_P}$$

or

11) $$-0.628 = V_P - V_{SG}$$

Since V_{SG} in 11 is actually V_{SGZ}, we have

12) $$V_{SGZ} = V_P \pm 0.628 \text{ volt} \begin{Bmatrix} + \text{ for } N \text{ channel} \\ - \text{ for } P \text{ channel} \end{Bmatrix} \qquad (14\text{-}15)$$

Substituting 12 into 7 yields

13) $$I_{DZ} = I_{DSS}\left(1 - \frac{V_P + 0.628 \text{ volt}}{V_P}\right)^2 = (0.628 \text{ volt})^2 \frac{I_{DSS}}{V_P^2} \qquad (14\text{-}16)$$

Substituting 12 into 8 yields the g_m at the T.C. = 0 point.

14) $$g_{mZ} = \frac{-2I_{DSS}}{V_P}\left(1 - \frac{V_P + 0.628 \text{ volt}}{V_P}\right)$$

or

15) $$g_{mZ} = \frac{2I_{DSS}}{V_P}\left(\frac{0.628 \text{ volt}}{V_P}\right) = g_{m0}\frac{(0.628 \text{ volt})}{V_P} \qquad (14\text{-}17)$$

Equation 15 demonstrates the desirability of a low V_P FET if appreciable g_m is to be realized at the T.C. = 0 point.

Now do problem PS 14-11.

14-6 Miscellaneous biasing methods

Figure 14-20 illustrates a biasing technique for stabilizing the Q point which employs both current- and voltage-derived self-bias (feedback) simultaneously. Resistor R_S senses changes in I_D and varies V_{SG} in a manner which tends to oppose the change in I_D. Resistor R_1, which would normally be returned to the V_{DD} supply to establish some effective V_{GG} supply, is instead returned to the drain. Thus R_1 and R_2 sample the voltage V_D. If I_D increases, V_D decreases due to the increased voltage drop across R_D. The decrease in V_D means less voltage across R_2, which increases the gate bias (gate becomes less positive in this case), which tends to decrease I_D in opposition to the initial increase. The net effect of both types of feedback is to stabilize the Q point.

The advantage of this combined current- and voltage-derived biasing method over that of straight current-derived biasing is that R_S can be smaller for the same

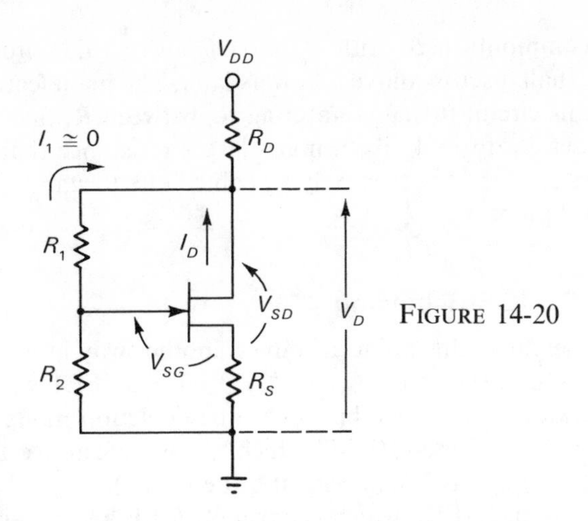

FIGURE 14-20

amount of Q-point stabilization. This may be important when it is necessary to maximize R_D with a low-voltage V_{DD} supply. In that case we could not tolerate a big voltage drop across R_S.

Unfortunately, returning R_1 to the drain instead of the V_{DD} supply tends to reduce the input impedance seen between gate and ground. However, the input impedance can still be sufficiently high for many applications so the advantages outweigh the disadvantages.

Since the negative feedback from the drain would also tend to reduce the signal gain, it is possible to break R_1 into two equal or unequal resistors and bypass their junction with C_B as shown in Fig. 14-21.

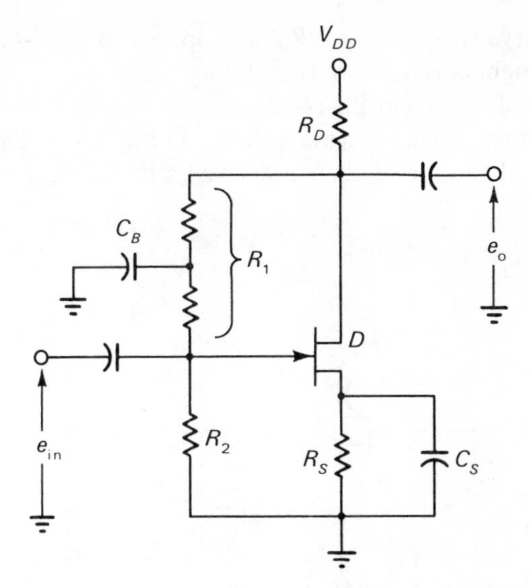

FIGURE 14-21

For an analysis of the circuit we can first reduce Fig. 14-20 to Fig. 14-22 by Thévenin's theorem. The current through R_1 is so small compared to I_D that it may be assumed negligible. Thus we may write for Fig. 14-22,

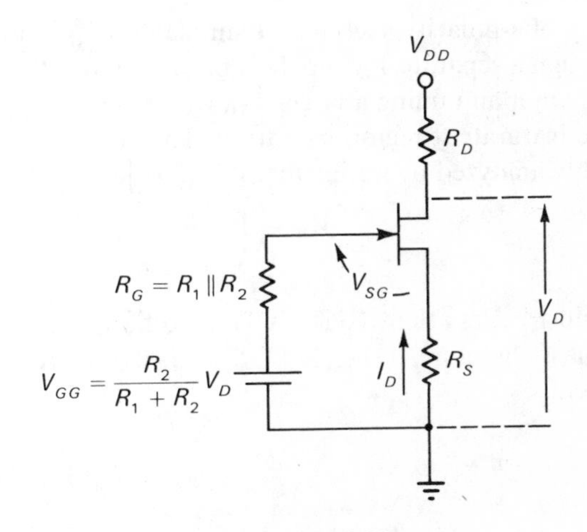

FIGURE 14-22

1) $$V_D \approx V_{DD} - R_D I_D$$

2) $$V_{SG} = -R_S I_D + \alpha V_D$$

where

3) $$\alpha = \frac{R_2}{R_1 + R_2} = \frac{1}{1 + (R_1/R_2)} \qquad (14\text{-}18)$$

Substituting 1 into 2 yields

4) $$V_{SG} = -R_S I_D + \alpha(V_{DD} - R_D I_D)$$

or

5) $$V_{SG} = -(R_S + R_D)I_D + \alpha V_{DD} \qquad (14\text{-}19)$$

Equation 5 may be used to plot a bias curve as shown in Fig. 14-23. Note from equation 5 that the slope of the

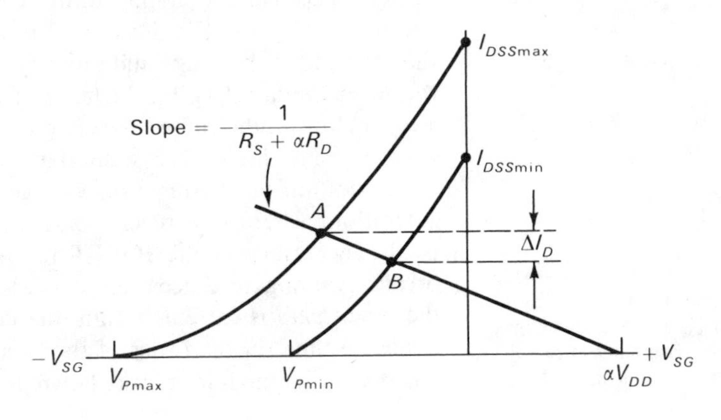

FIGURE 14-23

bias curve is $-1/(R_S + \alpha R_D)$ as opposed to just $-1/R_S$ for straight current-derived self-bias.

Now do problem PS 14-12.

Another biasing scheme is shown in Fig. 14-24. This is essentially a case of current-derived self-bias with a V_{SS}

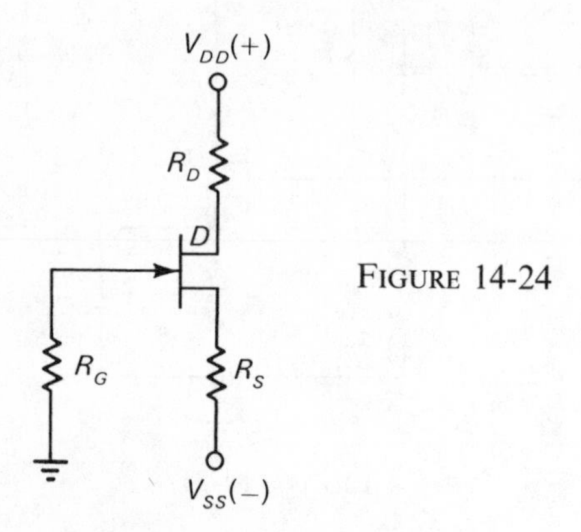

FIGURE 14-24

supply of a polarity such that it simulates a V_{GG} supply. Although a separate V_{SS} supply is needed, it aids the V_{DD} supply in maintaining a larger V_{SD} which is favorable to active (saturated) region operation. The circuit may be roughly analyzed by noting that if $|V_{SS}| \gg |V_P|$, and since

$$1) \qquad I_D = \frac{V_{SS} - V_{SG}}{R_S}$$

V_{SG} must be less than V_P if any I_D is to flow to develop self-bias. Therefore, if $|V_{SS}| \gg |V_P|$, say five or more times greater,

$$2) \qquad I_D \approx \frac{V_{SS}}{R_S}$$

Equation 2 implies that V_{SS} and R_S simulate a constant-current source driving the source terminal.[1]

Figure 14-25 illustrates another biasing scheme which

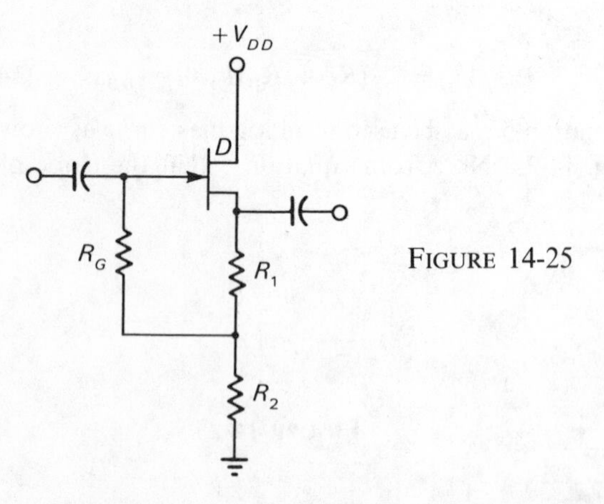

FIGURE 14-25

[1]A more meaningful way of concluding that $V_{SG} \approx 0$, or why the source tends to follow the gate, is to consider V_{SQ} as the error voltage in a negative feedback system, in which case, the error tends to be nulled out to an extent depending upon the gain of the amplifier.

is commonly used with a *source-follower* FET circuit. We shall discuss source followers later. The main feature of this circuit is that connecting R_G between R_1 and R_2 causes R_G to look like a much larger resistance than it physically is. However, only R_1 contributes to stabilizing the Q point.

14-7 IGFET biasing

We might at this point introduce another way in which FETs are designated. A type "*A*" FET is a JFET which operates with reverse bias or in the depletion mode. A type "*B*" FET is an IGFET which can operate above and below zero bias (enhancement or depletion). A type "*C*" FET is an IGFET which is normally OFF and requires enhancement or forward bias to become active. Remember, in an IGFET we don't have to worry about a gate diode turning ON when forward-biased because of the insulated gate structure. Figure 14-26 illustrates the

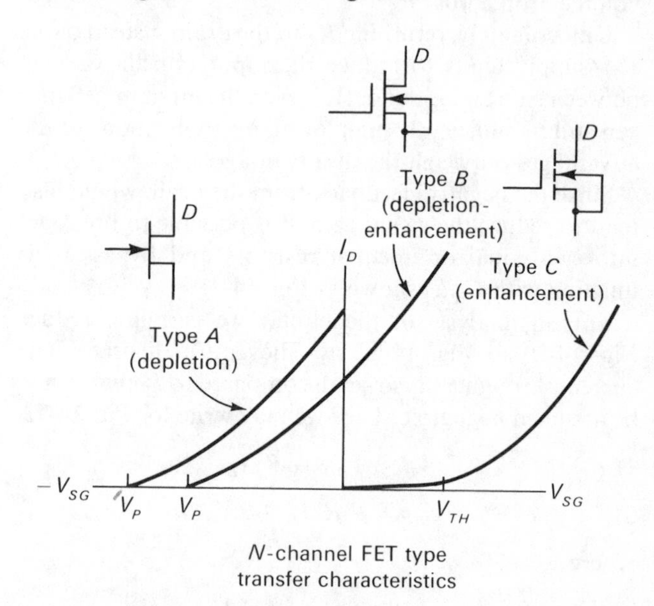

N-channel FET type
transfer characteristics

FIGURE 14-26

transfer curves for an *N*-channel type *A*, *B*, or *C* FET. The V_{SG} values would be reversed 180° for a *P*-channel device. The corresponding drain characteristics are shown in Fig. 14-27*a*, *b*, and *c*, respectively. Notice that except for the range and polarity of V_{SG} values and the lesser value of r_D (or higher g_{os}) indicated by the steeper slope of the curves, the characteristics of the type *B* and *C* FETs are otherwise similar to a JFET (type *A*).

In addition to *PN*-junction voltage-breakdown considerations wherever a reverse-biased junction exists, it is also necessary in the IGFET to consider dielectric breakdown of gate. Since the silicon dioxide layer under the metal gate is extremely thin, the field intensity can easily become on the order of thousands of volts/in. If the insulating medium breaks down, it is punctured and

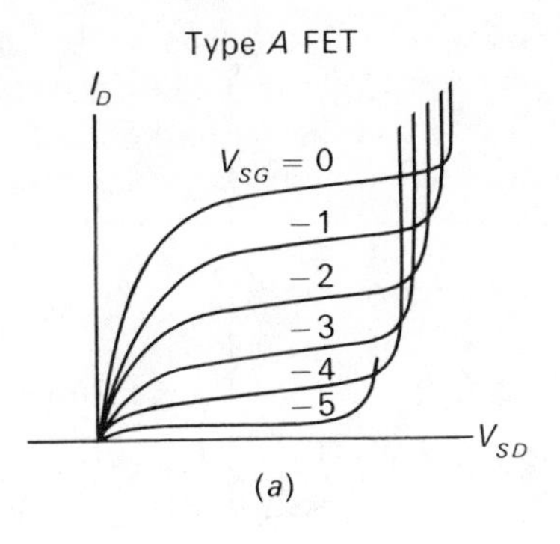

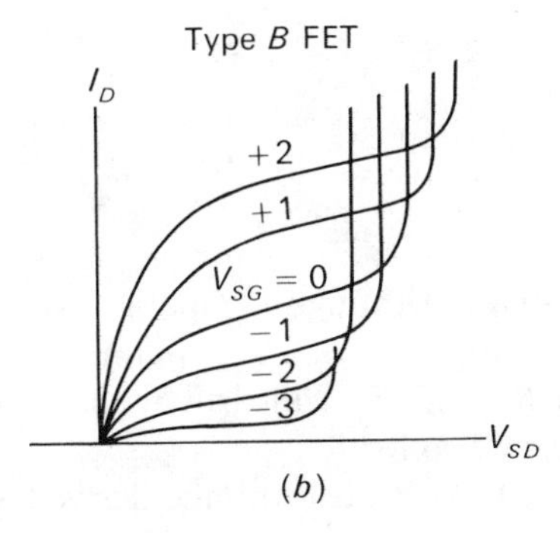

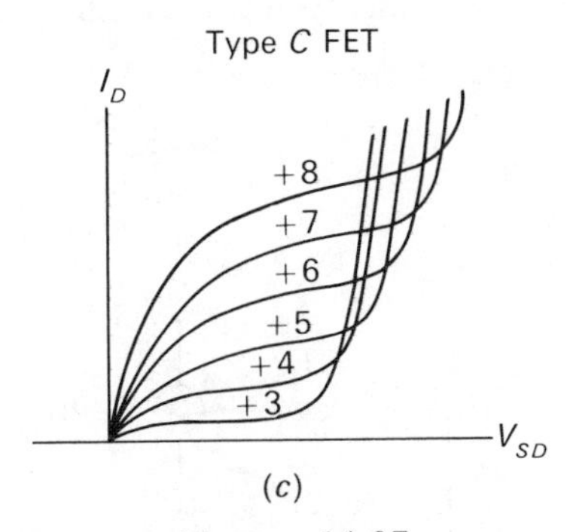

FIGURE 14-27

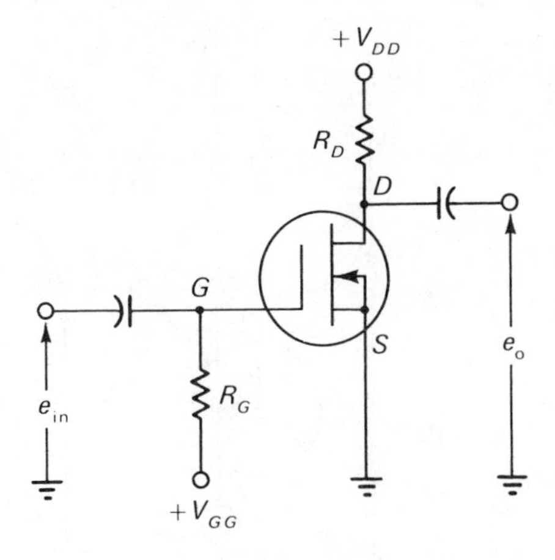

FIGURE 14-28

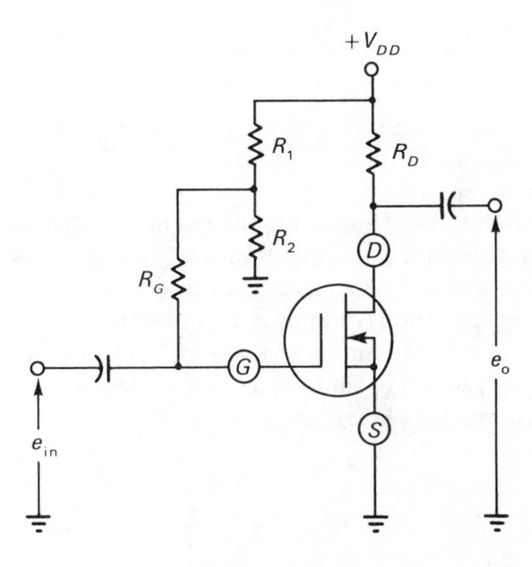

FIGURE 14-29

the device is ruined. Even static charges developed by sliding an IGFET out of plastic foam packing may, on a dry day, cause gate breakdown. For this reason the FET leads should be kept shorted until the device is actually mounted. Some manufacturers build in zener diodes across the gate which are designated to break down below the voltage at which the gate punctures. This, however, reduces the input impedance of the device to that of the zener diode reverse impedance which may be several megohms but is still much less than the impedance of the insulated gate itself.

Figure 14-28 illustrates fixed bias applied to an *N*-channel type *B* FET. Varying the polarity of V_{GG} sets the operating point in the depletion or enhancement portion of the transfer curve. Figure 14-29 is another example of fixed bias. By returning R_2 to a negative supply, the *Q* point could be shifted to the depletion portion of the transfer curve.

Figure 14-30 illustrates a case of current-derived self-

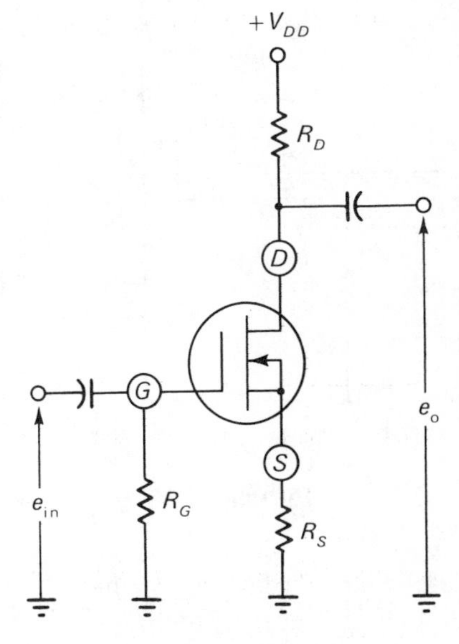

FIGURE 14-30

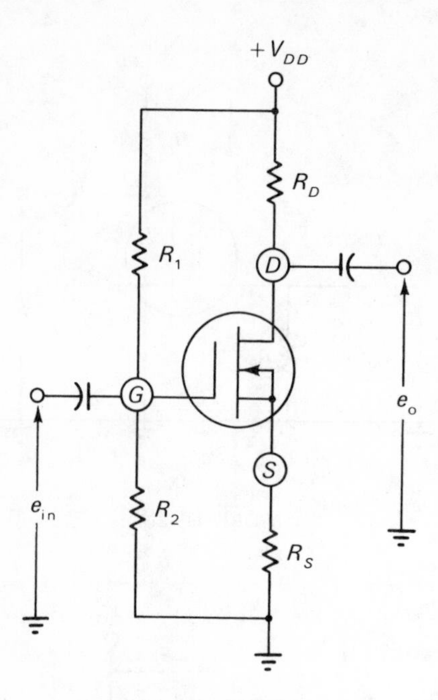

FIGURE 14-31

bias. Figure 14-31 illustrates the use of a voltage divider in the gate circuit to synthesize a V_{GG} supply. This may be necessary if R_O is so large that the Q point tends to be near cut-off. The V_{SS} supply in Fig. 14-32 serves the same purpose, or from another viewpoint, if $|V_{SS}| \gg |V_P|$, we may consider V_{SS} and R_S as simulating a constant-current source which fixes $I_D \approx V_{SS}/R_D$.

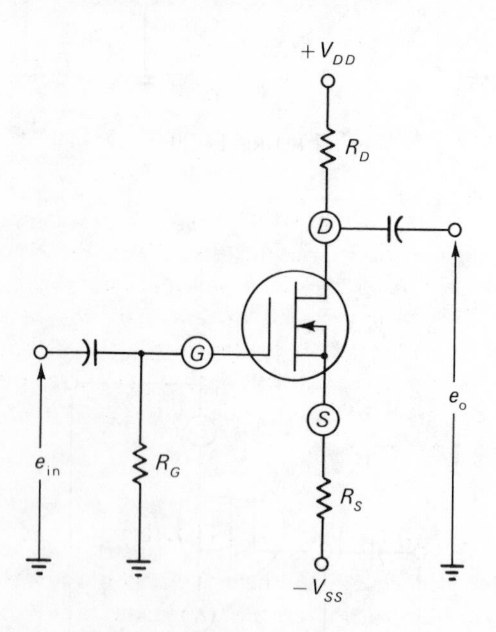

FIGURE 14-32

If the gate is just left floating, we have the circuit of Fig. 14-33 in effect, where R_1 and R_2 are leakage resistances. As you can see, this is essentially a case of voltage-

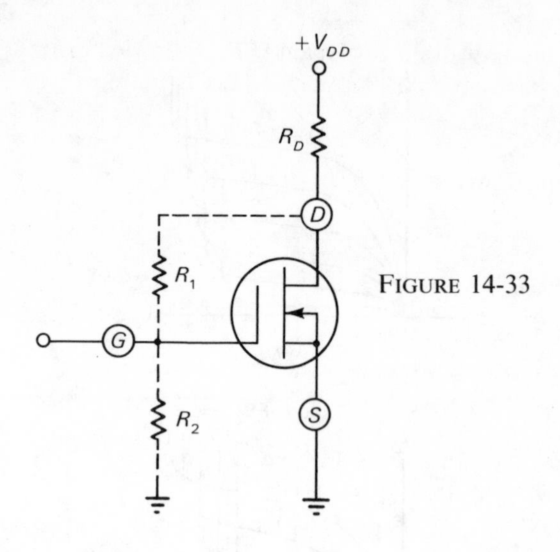

FIGURE 14-33

derived self-bias. Although a floating gate yields the highest input impedance, the Q point is relatively unstable since R_1 and R_2 will drift with humidity, dust, temperature, and so forth.

Figure 14-34 illustrates voltage-derived self-bias as applied to an N-channel type C FET, although a type B

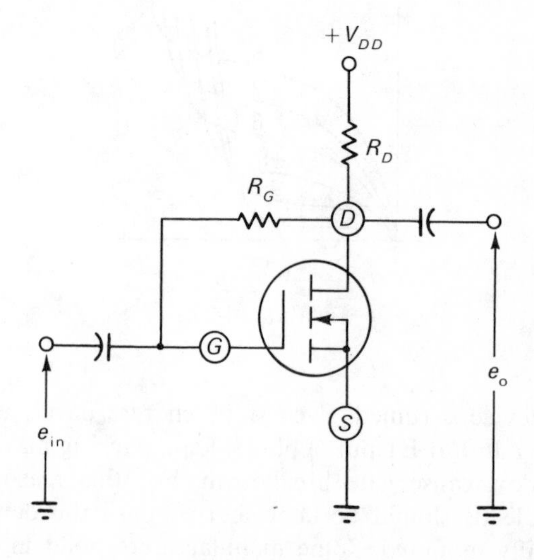

FIGURE 14-34

FET could also be biased in this manner. In designing this circuit, the value of R_G is immaterial (since $I_G \approx 0$) as long as R_G is much less than the drain-to-gate leakage resistance. The first step is to choose V_{SG} and I_D values in the saturated region. Note that V_{SG} will have to exceed the threshold voltage, V_{TH}, if any I_D is to flow. Since $I_G = 0$, it follows that $V_{SD} = V_{SG}$ and therefore

1)
$$R_D = \frac{V_{DD} - V_{SD}}{I_D} \tag{14-20}$$

If the problem is one of analyzing this circuit with R_D,

V_{DD}, and the drain curves given, the following procedure may be used. First plot the load line equation

1) $$V_{SD} = V_{DD} - R_D I_D$$

on the drain curves. Then since

2) $$V_{SG} = V_{SD}$$

plot the locus of points for equation 2. The intersection of equation 2 and the load line on the drain characteristics locates the Q point.

Now work through problem PS 14-13.

14-8 The ohmic region (voltage-controlled resistor)

Certain applications make it desirable to operate the FET in its ohmic region where $|V_{SD}| \ll |V_P|$ since it is in this region that the FET tends to simulate a voltage-controlled resistor. You will recall that for V_{SD} much less than $|V_P|$ and I_D small, the drain end of the channel is not appreciably constricted. Hence we see a reasonably linear resistance between source and drain whose ohmic value is readily controlled by V_{SG}. Figure 14-35 illustrates

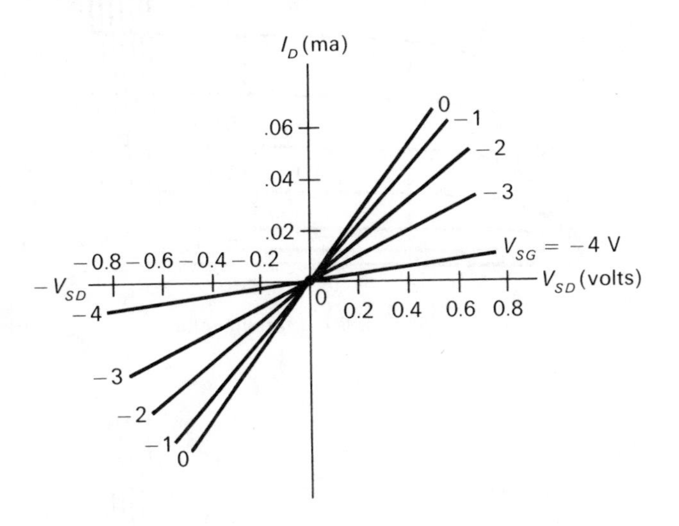

FIGURE 14-35

the drain characteristics in the vicinity of the origin where the channel resistance is essentially linear as evidenced by the almost straight-line volt-ampere curves. Note also the symmetry of the curves which is analogous to a linear resistor. Clearly, as the magnitude of the gate bias increases (more negative for the P-channel FET illustrated here) the slope (conductance) decreases, which implies an increase in resistance.

The drain-to-source channel resistance in the region about the origin is, to a good approximation, given by

1) $$r_{DS} \approx \frac{1}{g_m} \qquad (14\text{-}21)$$

where, as was previously shown,

2) $$g_m = g_{m0}\left(1 - \frac{V_{SG}}{V_P}\right) \qquad (14\text{-}22)$$

and

3) $$g_{m0} = \frac{-2I_{DSS}}{V_P} \qquad (14\text{-}23)$$

4) $$r_{DS} = \frac{-V_P}{2I_{DSS}[1 - (V_{SG}/V_P)]}$$

$$= \frac{r_{DS(\text{on})}}{1 - (V_{SG}/V_P)} \qquad (14\text{-}24)$$

where

5) $$r_{DS(\text{on})} = -\frac{1}{g_{m0}} \qquad (14\text{-}25)$$

A graphic insight into the above equations may be obtained from Fig. 14-36, wherein Fig. 14-35 represents

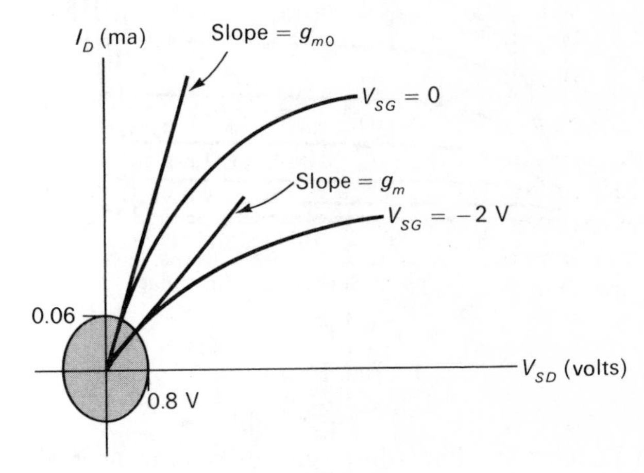

FIGURE 14-36

the shaded portion about the origin. Since the V-I curves are sensibly linear in this region and pass through the origin, the static (dc) and dynamic (ac) resistance are identical. The slope of any given V_{SG} curve about the origin is approximately equal to g_m at that particular V_{SG}. For $V_{SG} = 0$, the g_m is defined as g_{m0}, or g_{fs0}, or $g_{m(\max)}$.

PROBLEMS WITH SOLUTIONS

PS 14-1 The FET used in the circuit of Fig. PS 14-1*a* has the drain characteristics shown in Fig. PS 14-1*b*. Graphically determine the Q point. Assume I_{GSS} is negligible.

SOLUTION Since I_{GSS} is negligible

1) $$V_{SG} = \frac{30 \text{ kilohms}}{30 \text{ kilohms} + 60 \text{ kilohms}}(3 \text{ volts}) = 1 \text{ volt}$$

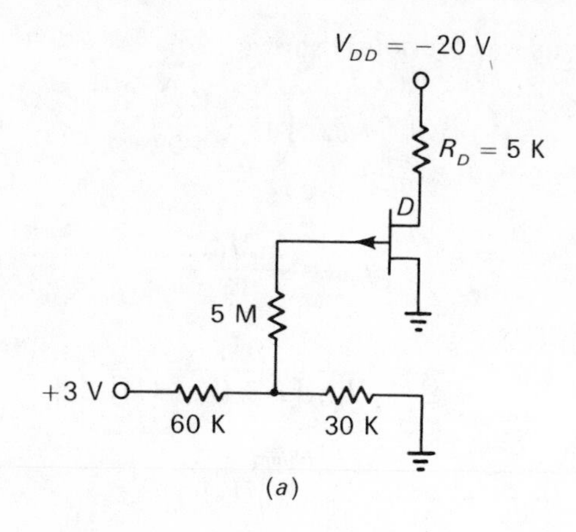

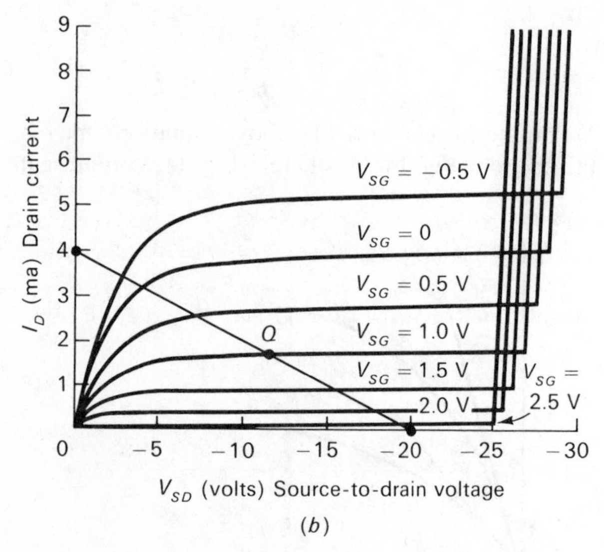

FIGURE PS 14-1

A 5-kilohm load line is drawn on the drain characteristics as shown in Fig. PS 14-1*b*, and the intersection of the $V_{SG} = 1$ volt curve and the load line locates the Q point.

PS 14-2 Assume $V_{SD}' = -7.5$ volts for the FET of Fig. PS 14-1*b*. What is the largest value of R_D that should be used if V_{SG} may vary from 0 to 2 volts in the circuit of Fig. PS 14-1*a*.

SOLUTION The $V_{SD}' = -7.5$ volts specification represents the lowest permissible V_{SD} value. The corresponding I_{DSS}' for $V_{SG} = 0$ at $V_{SD} = -7.5$ volts is about 3.7 ma. Therefore,

$$R_{D(\max)} = \frac{|V_{DD}| - |V_{SD}'|}{I_{DSS}'} = \frac{20 \text{ volts} - 7.5 \text{ volts}}{3.7 \text{ ma}}$$

$$= 3.4 \text{ kilohms}$$

PS 14-3 Graphically determine the Q point for the FET in the circuit of Fig. PS 14-3*a*. Assume the drain characteristics of Fig. PS 14-3*b*.

SOLUTION Plot the load line. Since the open-circuit $V_{SD} = V_{DD} = -20$ volts, and the short-circuit $I_D =$ 20 volts/(4.5 kilohms + 0.55 kilohm) = 3.96 ma ≈ 4 ma, we have the load line shown in Fig. PS 14-3*c*.

Next construct a transfer curve, say for $V_{SD} = -10$

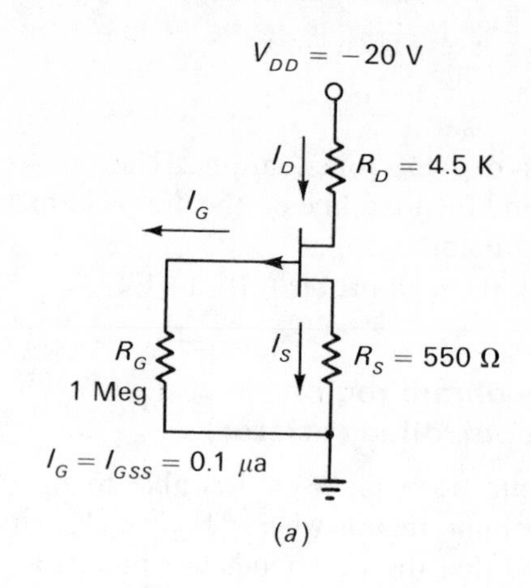

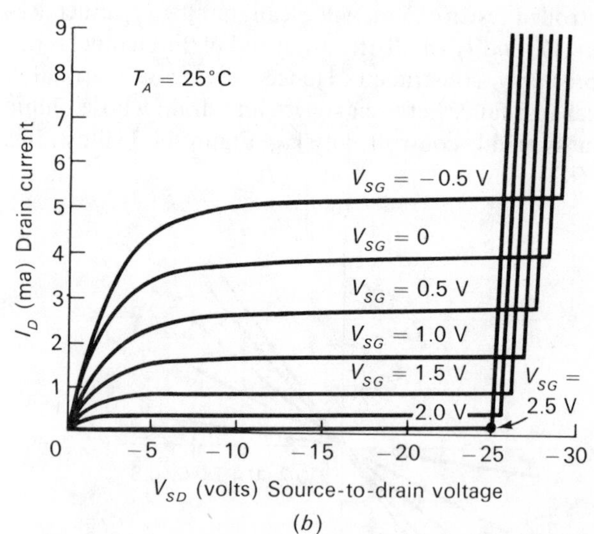

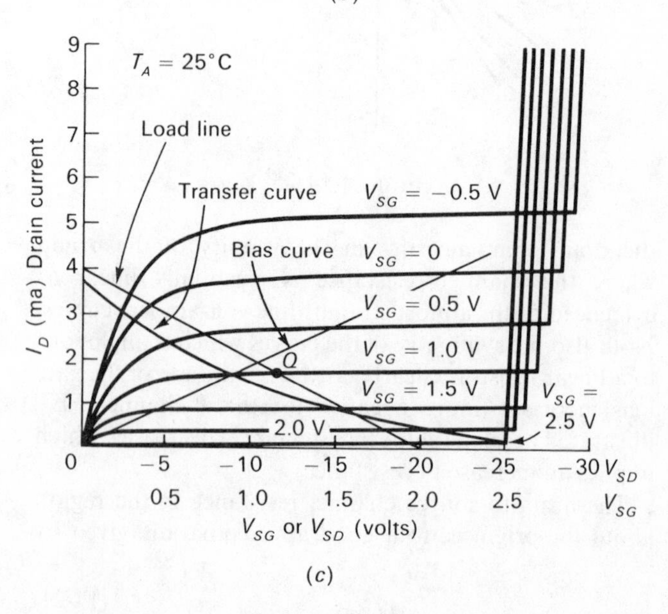

FIGURE PS 14-3

volts, and, instead of drawing a separate figure for the transfer curve, draw it right on the drain characteristics. Then change the V_{SD} axis to a V_{SG} axis as shown in Fig. PS 14-3*c*.

Now we construct a bias curve using the I_D, V_{SG} axis and the equation $I_D = V_{SG}/R_S$. Assuming V_{SG} values yields the table

V_{SG} (volts)	0	0.5	1.0	1.5
I_D (ma)	0	0.909	1.82	2.73

Connecting these points yields the bias curve (actually a line). The intersection of the *bias* curve and the *transfer* curve yields the quiescent V_{SG} and I_D values. Thus $V_{SGQ} \approx 0.9$ volt and $I_{DQ} \approx 1.7$ ma. If this I_{DQ} value is projected or otherwise located on the *load* line, the Q point on the drain characteristics is established. The quiescent source-to-drain voltage $V_{SDQ} = -11.5$ volts is determined by projecting down from the Q point onto the V_{SD} axis.

PS 14-4 Construct a dc model to fit the FET curves of Fig. PS 14-3*b*, and with this model verify the Q point graphically determined in the previous problem.

SOLUTION The dc model for a *P*-channel FET in the circuit of Fig. PS 14-3*b* is shown in Fig. PS 14-4. The

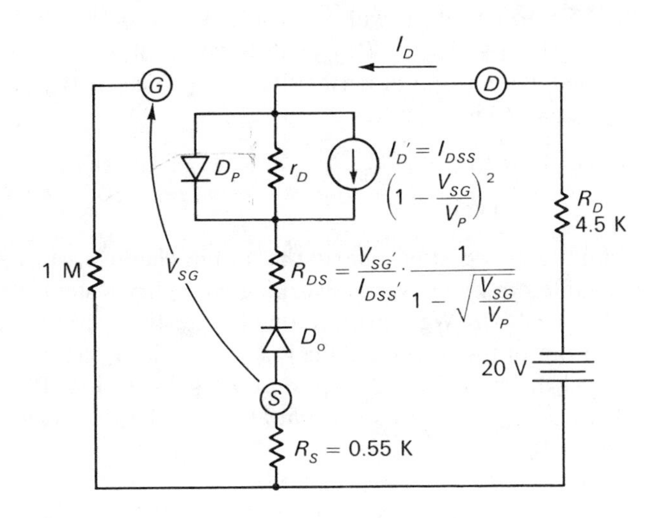

FIGURE PS 14-4

model parameters estimated from the drain curves of Fig. PS 14-3*b* are

$$V_{SD}' = -7.5 \text{ volts} \qquad V_P \approx \frac{V_{SD}'}{2} = 3.75 \text{ volts}$$

$$I_{DSS}' = 3.75 \text{ ma} \qquad I_{DSS} \approx 3 \text{ ma}$$

$$r_D = 36 \text{ kilohms}$$

Now let us assume that D_P is OFF and D_o is ON so that the Q point is in the active region. This assumption can be verified later. The following equations may be written by superposition.

1) $$I_D = \frac{r_D I_D'}{r_D + R_{DS} + R_S + R_D} + \frac{V_{DD}}{r_D + R_{DS} + R_S + R_D}$$

and

2) $$V_{SG} = R_S I_D$$

where

3) $$R_{DS} = \frac{V_{SD}'}{I_{DSS}'} \frac{1}{1 - \sqrt{V_{SG}/V_P}}$$

4) $$I_D' = I_{DSS}\left(1 - \frac{V_{SG}}{V_P}\right)^2$$

If equations 2, 3, and 4 are substituted into 1 and we try to solve for I_D, we would face a tedious and discouraging problem as the reader may prove to himself. Instead, let us make some reasonable simplifications. Since $r_D \gg R_{DS}$ in the active region, we can neglect R_{DS} in equation 1. Thus, substituting 2 and 4 into 1, we obtain

5) $$I_D = \frac{r_D I_{DSS}\,[1 - (R_S I_D/V_P)]^2 + V_{DD}}{r_D + R_S + R_D}$$

Squaring, multiplying out, and collecting terms yields

6) $$0 = \left[\frac{r_D I_{DSS}(R_S{}^2)}{V_P{}^2}\right] I_D{}^2 - \left(\frac{2 r_D I_{DSS} R_S}{V_P} + r_D + R_S + R_D\right) I_D + (V_{DD} + r_D I_{DSS})$$

Substituting values yields

7) $$0 = 2.32 \text{ kilohms}^2/\text{volt } I_D{}^2 - (31.67 \text{ kilohms} + 41.05 \text{ kilohms})\, I_D + (20 \text{ volts} + 108 \text{ volts})$$

8) $$0 = 2.32 \text{ kilohms}^2/\text{volt } I_D{}^2 - 72.7 \text{ kilohms } I_D + 128 \text{ volts}$$

9) $$I_D = \frac{72.7 \text{ kilohms} \pm \sqrt{(72.7 \text{ kilohms})^2 - 4(2.32 \text{ kilhoms}^2/\text{volt})(128 \text{ volts})}}{2(2.32 \text{ kilohms}^2/\text{volt})}$$

10) $$I_D = \frac{72.7 \text{ kilohms} \pm \sqrt{5285.3 \text{ kilohms}^2 - 1187.8 \text{ kilohms}^2}}{4.64 \text{ kilohms}^2/\text{volt}}$$

11) $$I_D = \frac{72.7 \text{ kilohms} \pm \sqrt{40.97 \times 10^2 \text{ kilohms}^2}}{4.64 \text{ kilohms}^2/\text{volt}} = \frac{72.7 \text{ kilohms} \pm 64.01 \text{ kilohms}}{4.64 \text{ kilohms}^2/\text{volt}}$$

12) $I_D = 29.5$ ma or 1.87 ma

Since the first root in the above is greater than I_{DSS}, it is rejected and we have for our solution

13) $I_D = 1.87$ ma

which is in close agreement to the 1.7-ma value graphically determined. Had we included R_{DS}, the solutions would have been closer still.

To check that D_O is ON and D_P is OFF, we note that I_D can only flow if D_O is ON. To check D_P we calculate V_{SD} from

14) $V_{SD} = R_S I_D - V_{DD} + R_D I_D$

$= 0.55$ kilohm (1.87 ma) $-$ 20 volts +

+ 4.5 kilohms (1.87 ma) $= -10.6$ volts

which reverse-biases D_P, as assumed.

The quiescent gate bias is

15) $V_{SG} = R_S I_D$

$= 0.55$ kilohm (1.87 ma) $= 1.03$ volts

This problem painfully illustrates that the graphic solution is generally easier than that using the dc model. A piecewise model could also have been used, and with less algebra but also less accuracy. The models are nevertheless useful in visualizing the device behavior or for computer-aided design and analysis.

PS 14-5 Determine the source bias resistor R_S and gate supply voltage if I_D is to be held between points A and B with current-derived self-bias.

SOLUTION The value of R_S may be determined from the slope of a line tangent to points A and B in Fig. PS 14-5.

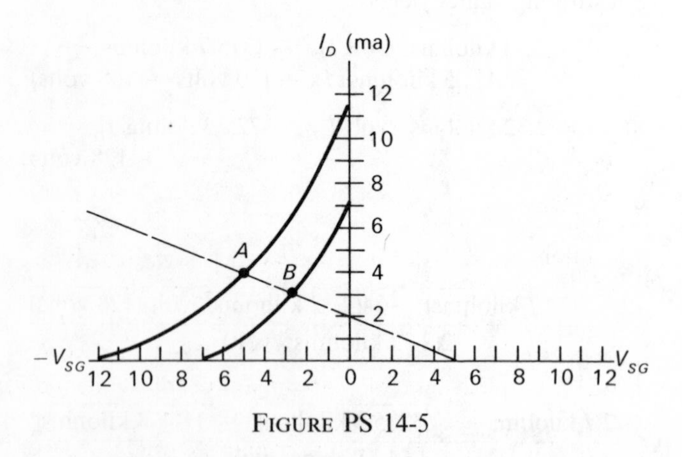

FIGURE PS 14-5

$$R_S = \frac{V_{SG(A)} - V_{SG(B)}}{I_{D(A)} - I_{D(B)}} = \frac{5 \text{ volts} - 2.6 \text{ volts}}{4 \text{ ma} - 3 \text{ ma}}$$

$= 2.4$ kilohms

The intercept of this bias line and the V_{SG} axis indicates $V_{GG} = 5$ volts.

PS 14-6 Suppose the positive side of the V_{SG} axis was not shown in Fig. PS 14-5. How could you analytically determine V_{GG}?

SOLUTION By similar triangles, we could write the following proportions

1) $$\frac{0 - I_{D(A)}}{V_{GG} - V_{SG(A)}} = \frac{0 - I_{D(B)}}{V_{GG} - V_{SG(B)}}$$

Substituting $I_{D(A)} = 4$ ma, $I_{D(B)} = 3$ ma, $V_{SG(A)} = -5$ volts, $V_{SG(B)} = -2.6$ volts, and solving for V_{GG} yields $V_{GG} = 4.6$ volts which is in close agreement with the previous solution.

PS 14-7 Assume the FET type of Fig. PS 14-5 has a minimum BV_{DGS} rating of 82 volts. With no further information, select a suitable value of R_D, V_{DD}, and gate divider resistors to guarantee an input impedance greater than 5 megohms and active-region operation. The drain current is to be held between points A and B with the gate bias circuitry evolved in the previous problem.

SOLUTION We previously established $R_S = 2.4$ kilohms, $V_{GG} = 5$ volts. From Fig. PS 14-5 we see that $V_{P(\max)} \approx -12$ volts. Therefore, if an input signal causes the FET to cut off, it will occur at a $V_{SG} = -12$ volts. Since $|V_{SG}| + |V_{SD}| \leq BV_{DGS}$, if breakdown is to be avoided, the maximum permissible V_{SD} (which is BV_{DGX} for $V_{SG} = V_P$) is given by

$$V_{SD(\max)} = BV_{DGS} - |V_{P(\max)}| = 82 \text{ volts} - 12 \text{ volts}$$

$= 70$ volts

Although we don't have the drain characteristics available, we know we are working with curves similar to Fig. PS 14-7a. We also know that to keep the operating point out of the ohmic region $V_{SD(\min)}$ should be greater than $|V_P|$, say at least $2|V_P|$ or 24 volts. From Fig. PS 14-5, $I_{DSS(\max)} = 11.5$ ma so that at $V_{SD} = 2|V_P|$ we can assume $I_{DSS(\max)} \approx 12$ ma.

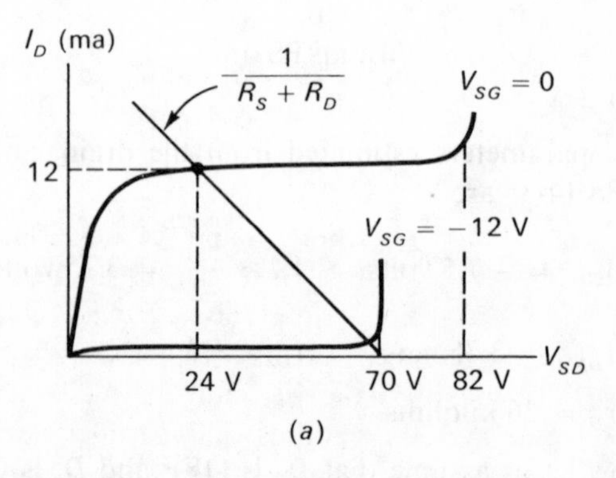

FIGURE PS 14-7

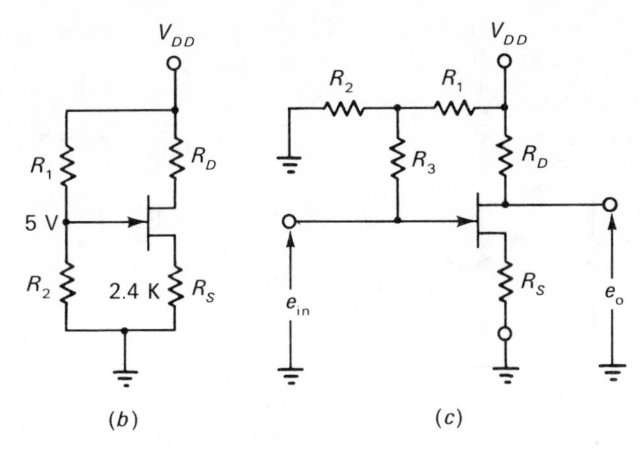

FIGURE PS 14-7 (*Continued*)

Now our circuit will look somewhat like Fig. PS 14-7*b*. The $R_S + R_D$ load line should lie as shown in Fig. PS 14-7*a* if R_D is to be maximized while still retaining active-region operation from $V_{SG} = 0$ to $V_{SG} = V_P$. Thus, from the slope of the load line, we may write

$$-\frac{1}{R_S + R_D} = \frac{0 \text{ ma} - 12 \text{ ma}}{70 \text{ volts} - 24 \text{ volts}}$$

Since $R_S = 2.4$ kilohms, we may solve the above to obtain $R_D = 1.4$ kilohm. With a nominal I_D of 3.5 ma (the midpoint between $I_{D(A)}$ and $I_{D(B)}$) we have

$$V_{SD} = 70 \text{ volts} - 3.5 \text{ ma } (2.4 \text{ kilohms} + 1.4 \text{ kilohms})$$
$$= 56.7 \text{ volts}$$

The above value of R_D was chosen to guarantee active-region operation even if V_{SG} should be driven to zero by some large input signal coupled into the gate. However, if the input signal is small so that the operating point does not shift appreciably, we could use a much larger value of R_D and thereby increase the stage voltage gain. Thus we could assume a $V_{SD(\min)} = 2|V_P| = 24$ volts and with a maximum $I_D \approx 4$ ma, we have

$$R_D = \frac{V_{DD} - R_S I_D - V_{SD(\min)}}{I_D}$$
$$= \frac{70 \text{ volts} - 2.4 \text{ kilohms (4 ma)} - 24 \text{ volts}}{4 \text{ ma}}$$
$$= 9.1 \text{ kilohms}$$

Although design equations could be developed for R_1 and R_2 in Fig. 14-7*b* it is easier to assume some values and check the design. If we choose $R_2 = 10$ megohms, the bleeder current (neglecting I_{GSS}) is 5 volts/10 megohms = 0.5 μa and R_1 = (70 volts − 5 volts)/0.5 μa = 130 megohms. The parallel combination of R_1 and R_2 is then surely greater than 5 megohms as required.

It turns out, however, that resistors of such large ohmic value tend to be unstable with time and temperature. Furthermore, leakage paths across these resistors may completely upset the V_{BB} bias voltage. Hence, a better arrangement is to use the circuit of Fig. PS 14-7*c* where R_1 and R_2 are less than 1 megohm and R_3 can be chosen greater than 5 megohms. Here R_3 has virtually no effect on V_{GG} unless its so large that I_{GSS} is significant.

PS 14-8 Given that $I_{DSS1} = 8$ ma, $I_{DSS2} = 6$ ma, $V_{P1} = -7$ volts, and $V_{P2} = -4$ volts, design the circuit of Fig. PS 14-8 for maximum small-signal gain and a quiescent drain current of 2 ma.

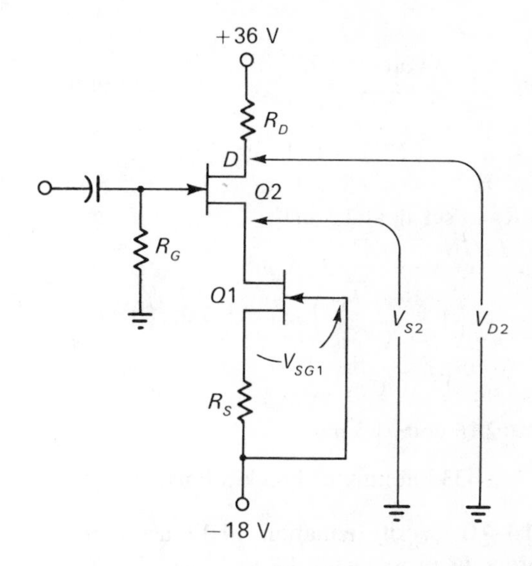

FIGURE PS 14-8

SOLUTION

$$V_{SG1} = V_{P1}\left(1 - \sqrt{\frac{I_D}{I_{DSS1}}}\right) = -7 \text{ volts}\left(1 - \sqrt{\frac{2 \text{ ma}}{8 \text{ ma}}}\right)$$
$$= -3.5 \text{ volts}$$

$$R_S = \frac{3.5 \text{ volts}}{2 \text{ ma}} = 1.75 \text{ kilohms}$$

We want both FETs to sit in the saturated region, which implies that $|V_{SD}| > 2|V_P|$ as a worst-case condition.

As a rule of thumb, the voltage at the source of $Q2$ will approximately equal that at the gate. Later when you have learned something about source followers and feedback, you will see why this is so. Since the gate of $Q1$ sits at zero volts relative to ground, we might expect $V_{S2} \approx 0$ volts. Actually, we can calculate V_{SGZ} from

$$V_{SG2} = V_{P2}\left(1 - \sqrt{\frac{I_D}{I_{DSS2}}}\right) = -4 \text{ volts}\left(1 - \sqrt{\frac{2 \text{ ma}}{6 \text{ ma}}}\right)$$
$$= -1.69 \text{ volts}$$

Therefore with the gate at zero volts relative to ground

$$V_{S2} = +1.69 \text{ volts}$$

Hence

$$V_{SD1} = -3.5 \text{ volts} + 18 \text{ volts} + 1.69 \text{ volts} = 16.2 \text{ volts}$$

Since $2\,|V_{P1}| = 14$ volts and $V_{SD1} = 16.2$ volts, we see that $Q1$ is in its saturated or constant-current source region.

Now for $Q2$ to be in its saturated region we require $V_{SD2} \geqq 2|V_{P2}| \geqq 2|4 \text{ volts}| = 8$ volts. Hence, let $V_{SD2} = 8$ volts. Therefore

$$V_{D2} = V_{S2} + V_{SD2} = 1.69 \text{ volts} + 8 \text{ volts} = 9.69 \text{ volts}$$

and

$$R_D = \frac{36 \text{ volts} - 9.69 \text{ volts}}{2 \text{ ma}} = 13.2 \text{ kilohms}$$

PS 14-9 An FET is to be used as a constant-current source. If $V_P = -4.5$ volts and $I_{DSS} = 5.5$ ma, determine R_S to set I_D at 1.5 ma.
SOLUTION

$$V_{SG} = V_P\left(1 - \sqrt{\frac{I_D}{I_{DSS}}}\right) = -4.5 \text{ volts}\left(1 - \sqrt{\frac{1.5 \text{ ma}}{5.5 \text{ ma}}}\right)$$

$$= -2.15 \text{ volts}$$

$$R_S = 2.15 \text{ volts}/1.5 \text{ ma}$$

$$= 1.433 \text{ kilohms} \approx 1.43 \text{ kilohms}$$

PS 14-10 Verify equation 14-14 using the data of the previous problem.
SOLUTION

$$I_D = \frac{4.5 \text{ volts}}{2(1.43 \text{ kilohms})^2\, 5.5 \text{ ma}}$$

$$\times \{2(1.43 \text{ kilohms})\, 5.5 \text{ ma} + 4.5 \text{ volts}$$

$$- \sqrt{4.5 \text{ volts}\,[4(1.43 \text{ kilohms})\, 5.5 \text{ ma} + 4.5 \text{ volts}]}\}$$

$$= 1.5 \text{ ma}$$

which checks (surprisingly)!

PS 14-11 Given the following FET parameters: $I_{DSS} = 8$ ma, $V_P = -5$ volts, determine R_S, I_{DZ}, and g_{mZ} for T.C. = 0 operation in the circuit of Fig. PS 14-11.
SOLUTION By equations 14-15, 14-16, and 14-18

1) $V_{SGZ} = V_P + 0.628 \text{ volt} = -5 \text{ volts} + 0.628 \text{ volt}$

$$= -4.37 \text{ volts}$$

2) $$I_{DZ} = (0.628 \text{ volt})^2 \frac{I_{DSS}}{V_P^{\,2}}$$

$$= (0.628 \text{ volt})^2 \frac{8 \text{ ma}}{(-5 \text{ volts})^2} = 0.126 \text{ ma}$$

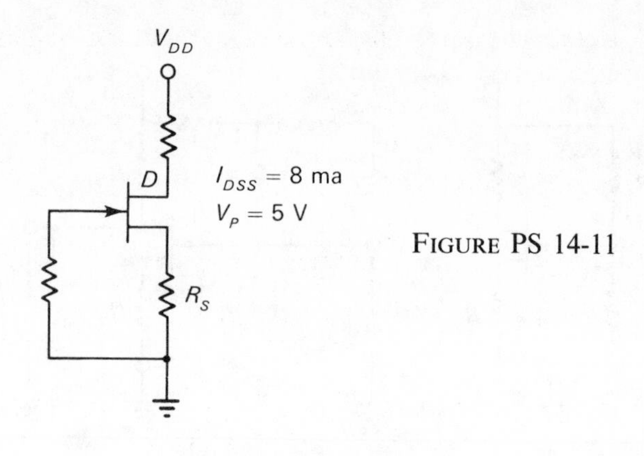

FIGURE PS 14-11

3) $$R_S = \frac{V_{SGZ}}{I_{DZ}} = \frac{4.37 \text{ volts}}{0.126 \text{ ma}} = 34.7 \text{ kilohms}$$

4) $$g_{mZ} = \frac{2I_{DSS}}{V_P}\frac{(0.628 \text{ volt})}{V_P} = 0.40 \frac{\text{ma}}{\text{volt}} = 40\ \mu\text{mho}$$

PS 14-12 A *P*-channel JFET has the following specifications: $I_{DSS(\text{max})} = 3$ ma, $V_{P(\text{max})} = 4.5$ volts, $I_{DSS(\text{min})} = 1.1$ ma, and $V_{P(\text{min})} = 2.5$ volts. It is desired to use this FET as a small-signal amplifier at a Q point of $I_{DQ} = 1 \text{ ma} \pm 0.25$ ma in the circuit of Fig. PS 14-12*a*.
SOLUTION To determine the bias curve we need the

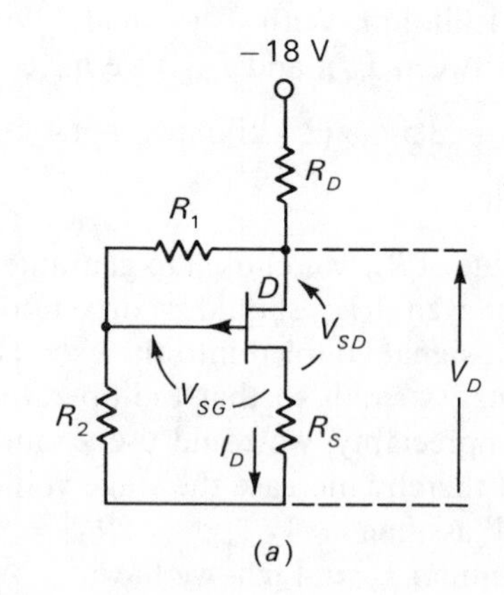

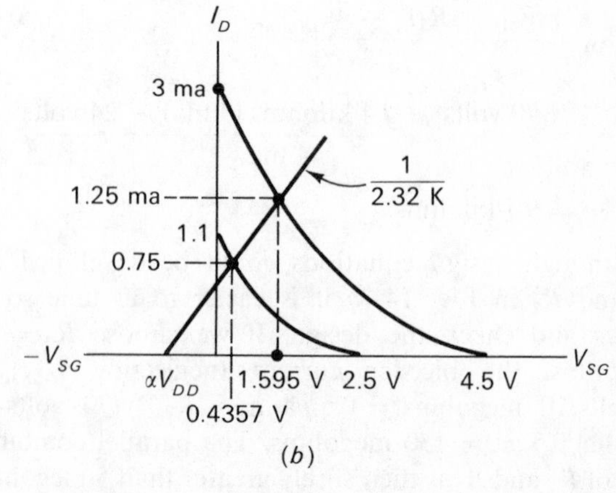

FIGURE PS 14-12

V_{SG} values corresponding to the I_D max and min values. By equation 14-13, we obtain

1) $$V_{SG(\max)} = V_{P(\max)}\left(1 - \sqrt{\frac{I_{D(\max)}}{I_{DSS(\max)}}}\right)$$

$$= 4.5 \text{ volts}\left(1 - \sqrt{\frac{1.25 \text{ ma}}{3 \text{ ma}}}\right)$$

$$= 1.595 \text{ volts}$$

2) $$V_{SG(\min)} = 2.5 \text{ volts}\left(1 - \sqrt{\frac{0.75 \text{ ma}}{1.1 \text{ ma}}}\right) = 0.4357 \text{ volt}$$

3) $$\Delta V_{SG} = V_{SG(\max)} - V_{SG(\min)} = 1.16 \text{ volts}$$

From the slope of the bias curve, equation 14-18 is then

4) $$R_S + \alpha R_D = \frac{\Delta V_{SG}}{\Delta I_D} = \frac{1.16 \text{ volts}}{0.5 \text{ ma}} = 2.32 \text{ kilohms}$$

Fig. PS 14-12*b* illustrates the nature of the bias curve.

The minimum V_{SD} value should be approximately twice V_P in magnitude to guarantee operation in the linear region. Thus using $V_{P(\max)}$ and $I_{D(\max)}$ for a worst case,

5) $$V_{SD(\min)} = -9 \text{ volts}$$

$$= -18 \text{ volts} + (R_S + R_D)\ 1.25 \text{ ma}$$

yields

6) $$R_S + R_D = \frac{-9 \text{ volts} + 18 \text{ volts}}{1.25 \text{ ma}} = 7.2 \text{ kilohms}$$

From Fig. 14-23, redrawn as Fig. PS 14-12*b* (since this is a *P*-channel FET), we may write by similar triangles or whatever,

7) $$\frac{1.25 \text{ ma}}{1.595 \text{ volts} - \alpha V_{DD}} = \frac{0.75 \text{ ma}}{0.4375 \text{ volt} - \alpha V_{DD}}$$

Solving for αV_{DD} yields

8) $$\alpha V_{DD} = -1.299 \text{ volts}$$

Therefore

9) $$\alpha = \frac{-1.299 \text{ volts}}{-18 \text{ volts}} = 0.07215$$

Substituting 9 into 4 yields

10) $$R_S + 0.07215 R_D = 2.32 \text{ kilohms}$$

And from 6

11) $$R_S + R_D = 7.2 \text{ kilohms}$$

Equations 10 and 11 are simultaneous in R_S and R_D and may be solved by substitution or other means (subtract 11 from 10 to eliminate R_S) to yield

12) $R_D = 5.26$ kilohms and $R_S = 1.94$ kilohms

The values of R_1 and R_2 may be determined by first noting that

13) $$\alpha = 0.07215 = \frac{1}{1 + (R_1/R_2)}$$

Solving for R_1/R_2 yields

14) $$\frac{R_1}{R_2} = \frac{1 - \alpha}{\alpha} = 12.86$$

Then if we arbitrarily choose $R_2 = 5$ megohms, we obtain

15) $R_1 = 12.86(5 \text{ megohms}) = 64.3$ megohms

If R_1 is divided into, say, 30-megohm and 34.3-megohm resistors with the tie-point bypassed as shown in Fig. 14-21, the input impedance in the mid-frequency range as seen between gate and ground is 5 megohms $\|$ 30 megohms $\approx$ 4.3 megohms, which is sufficiently high in most cases.

PS 14-13 Determine the Q point in the circuit of

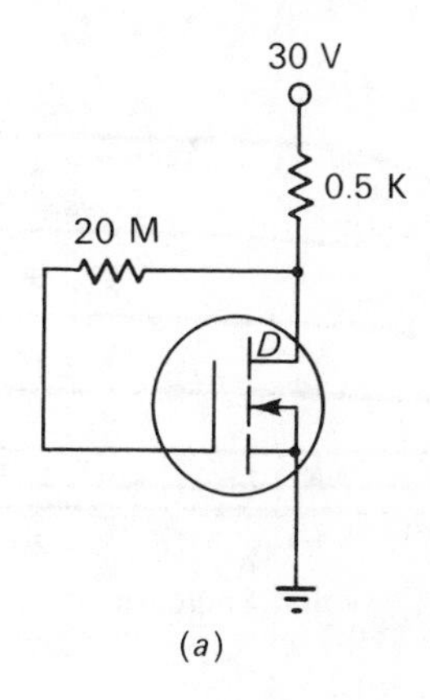

(*a*)

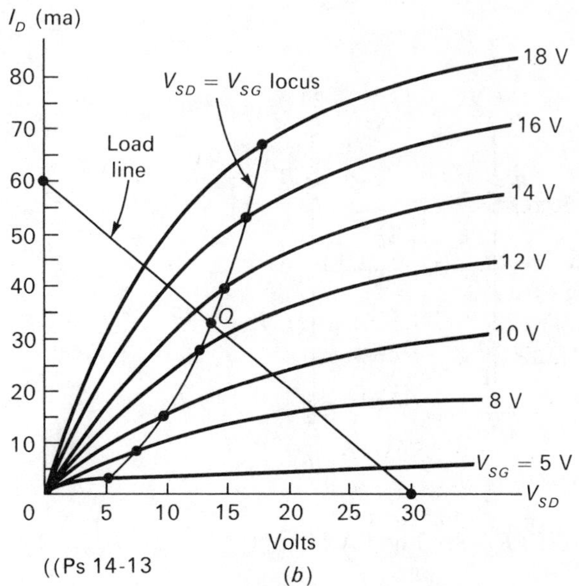

(*b*)

FIGURE PS 14-13

Fig. PS 14-13*a*. Assume the IGFET drain characteristics of Fig. PS 14-13*b* are applicable.

SOLUTION Construct a load line on the drain family of curves. When $I_D = 0$, $V_{SD} = 30$ volts, and when $V_{SD} = 0$, $I_D = 30$ volts/0.5 kilohm $= 60$ ma. Next determine the locus of points for $V_{SG} = V_{SD}$ by assuming the indicated V_{SG} values and locating the corresponding V_{SD} values. The intersection of the $V_{SG} = V_{SD}$ locus and the load line is the Q point, about 33.8 ma of I_D and 13.2 volts of V_{SD} and V_{SG}.

PROBLEMS WITH ANSWERS

PA 14-1 The *P*-channel FET having the drain characteristics shown in Fig. PA 14-1*a* is to be operated at the indicated Q point with fixed bias; +6- and −20-volt supplies are available. The input impedance should be about 5 megohms.

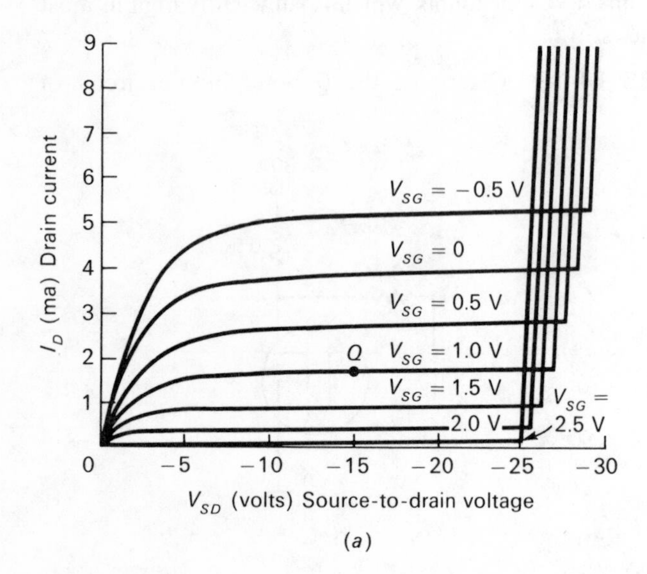

(*a*)

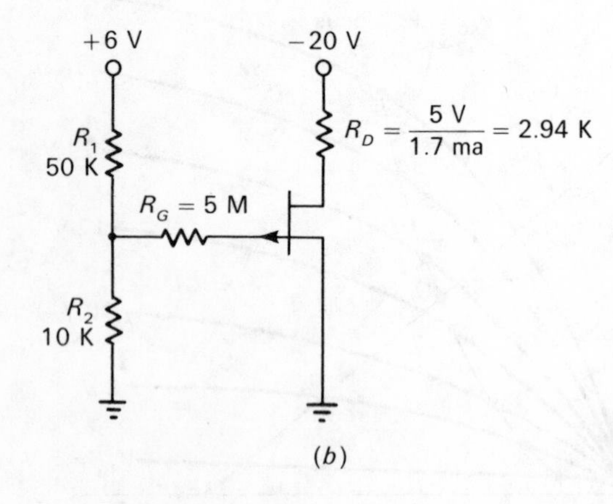

(*b*)

ANSWER See Fig. PA 14-1*b*

PA 14-2 Figure PA 14-2*a* represents a piecewise approximation to some FET curves. Develop the piecewise model. Assume $I_{GSS} = 0$.

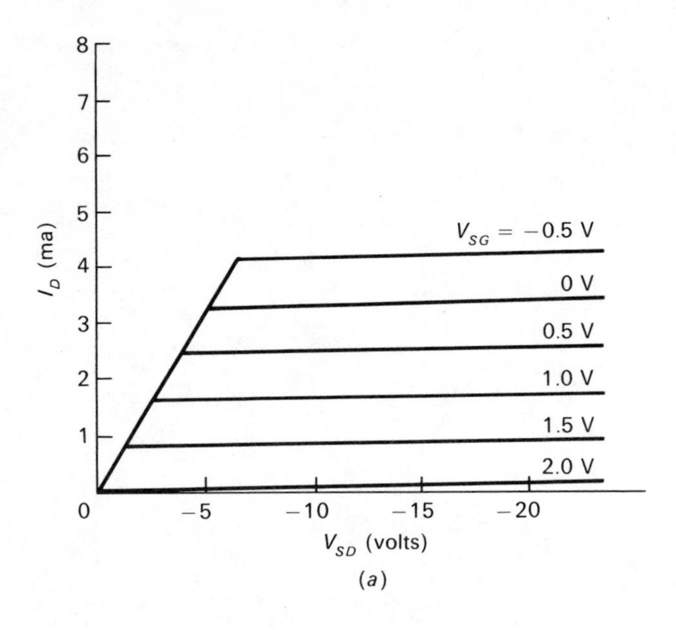

(*a*)

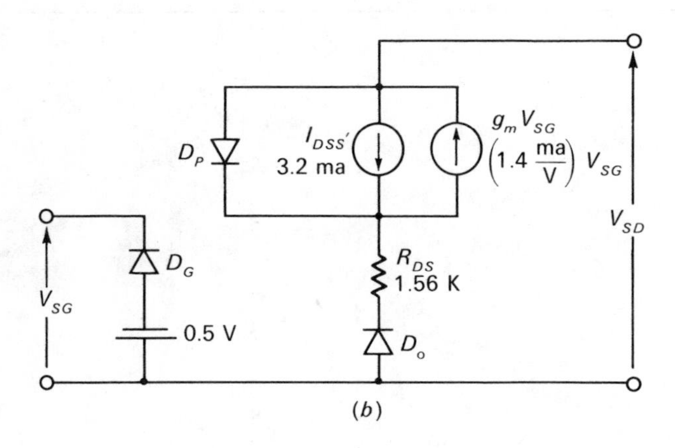

(*b*)

ANSWER See Fig. PA 14-2*b*

PA 14-3 Using the piecewise model of Fig. PA 14-2 determine the Q point in the circuit of Fig. PA 14-3.

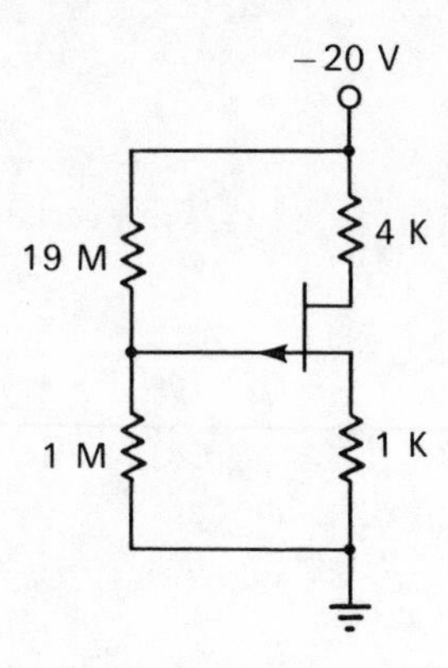

ANSWER $I_D = 1.92$ ma
$V_{SD} = -10.4$ volts

PA 14-4 Using the piecewise model of Fig. PA 14-2, determine the Q point in the circuit of Fig. PA 14-4.

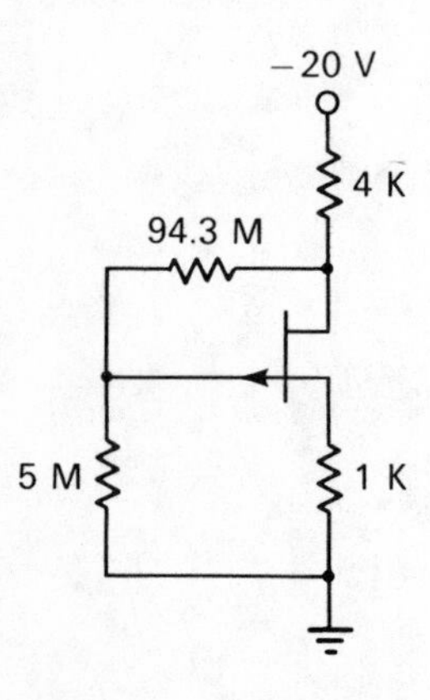

ANSWER $I_D = 1.72$ ma
$V_{SD} = -11.4$ volts

PA 14-5 Using the drain curves of Fig. PA 14-5*b*, determine the circuit parameters in Fig. PA 14-5*a* for a Q point at $I_D = 1.7$ ma, $V_{SG} = 1$ volt.

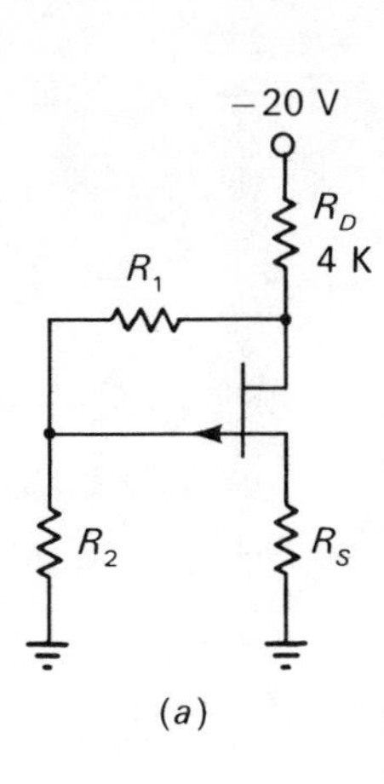

(*a*)

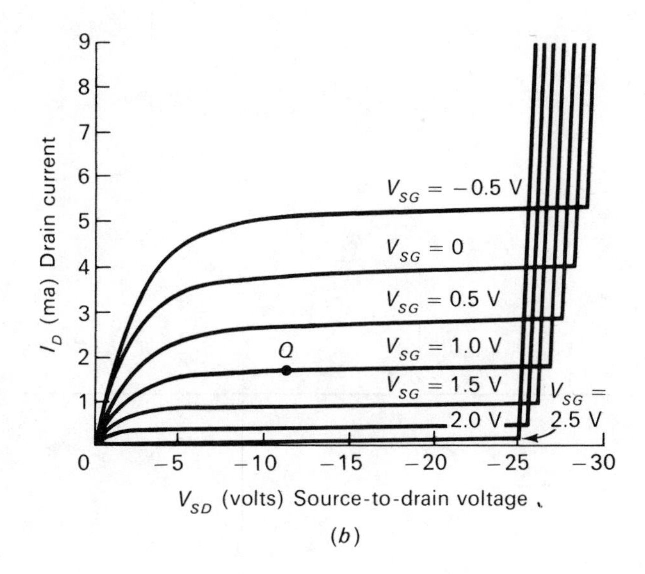

(*b*)

ANSWER $R_S = 1$ kilohm

$R_1/R_2 = 17.86$

PA 14-6 Design the circuit of Fig. PA 14-6 so that the FET ($Q1$) looks like a 2-ma constant-current source to the silicon transistor $Q2$ which is to be biased at $V_{EC} = +8$ volts. Assume $V_P = -3.5$ volts, $I_{DSS} = 4.8$ ma, $\beta = 50$. Assume a bleeder current of 0.1 ma through R_1.

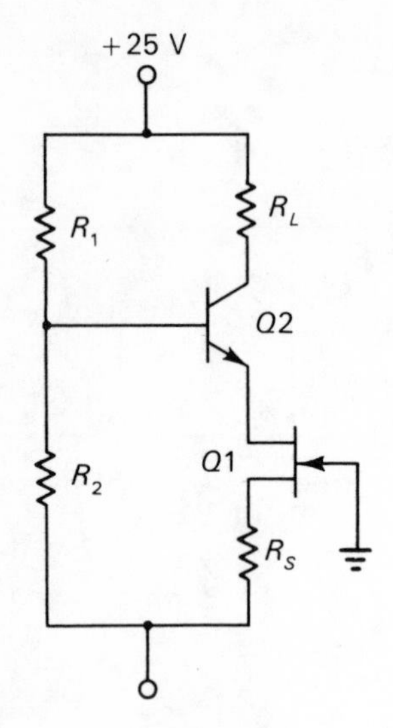

ANSWER $R_S = 10$ kilohms
$R_L = 5.1$ kilohms
$R_1 = 174$ kilohms
$R_2 = 454$ kilohms

PA 14-7 An N-channel JFET has the following parameters: $I_{DSS} = 5$ ma, $V_P = -6$ volts. Determine the gate bias, V_{SGZ}, and drain current, I_{DZ}, required for a zero temperature coefficient of drain current I_D.

ANSWER $V_{SGZ} = -5.37$ volts
$I_{DZ} = 0.0548$ ma

PROBLEMS WITHOUT ANSWERS

P 14-1 Determine the quiescent operating point of the FET in Fig. P 14-1*a*. Assume the curves of Fig. P 14-1*b* apply.

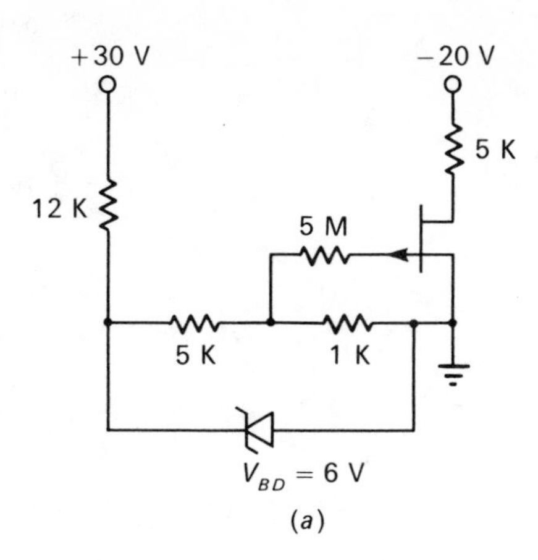

(*a*)

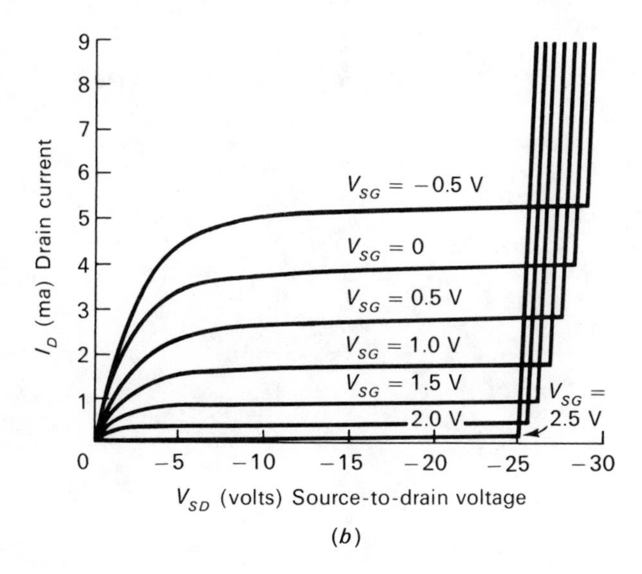

(*b*)

P 14-2 Design the circuit of Fig. P 14-2*a* for a Q point at $I_D = 5$ ma $\pm$ 1 ma, $V_{SD} \geqq 10$ volts. Assume the transfer curves of Fig. P 14-2*b* apply.

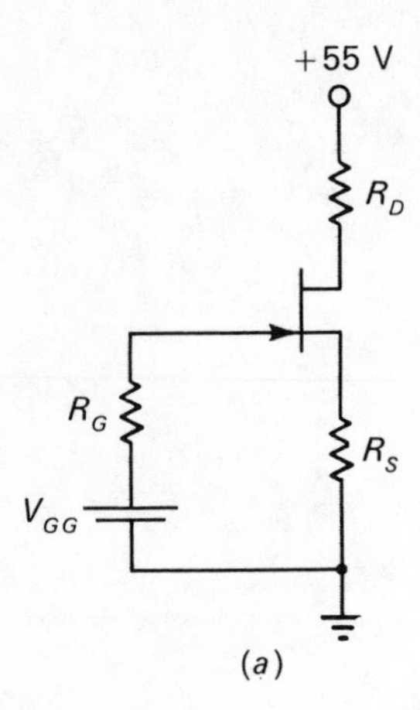

(*a*)

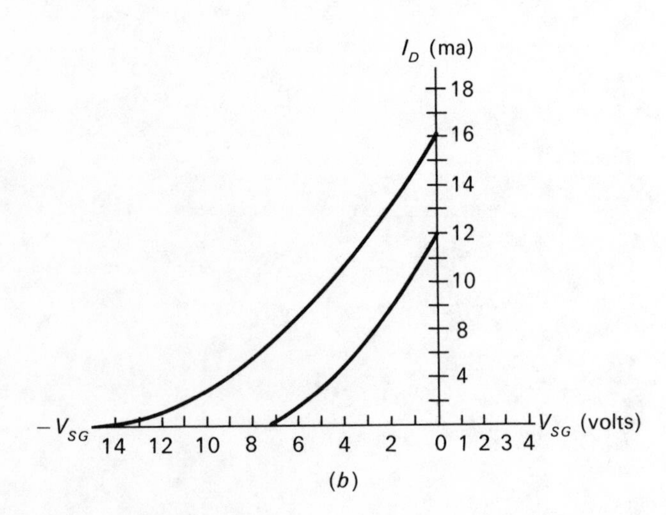

(*b*)

P 14-3 FETs are sometimes encapsulated in a single two-terminal package called a constant-current diode or current limiter and in this sense may be considered the dual of a zener diode (voltage limiter). Generally the gate is internally tied to the source as shown in Fig. P 14-3. Suppose it is desired to have the 8-kilohm load see approximately a 2-ma constant-current source and that $V_P = -5$ volts. What other FET parameter must be specified, and estimate the E_{XX} supply voltage.

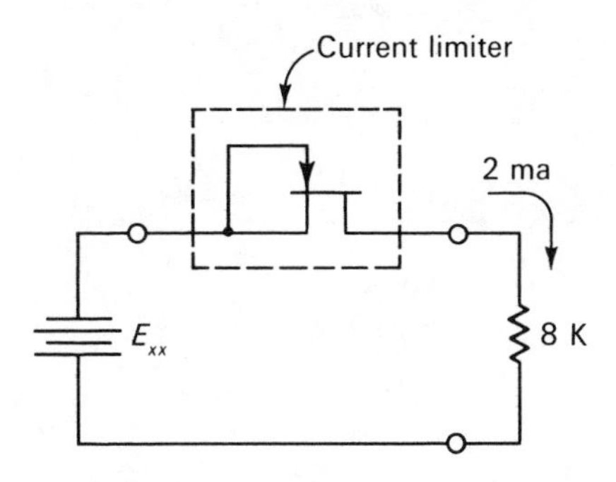

P 14-4 Estimate I_D in Fig. P 14-4.

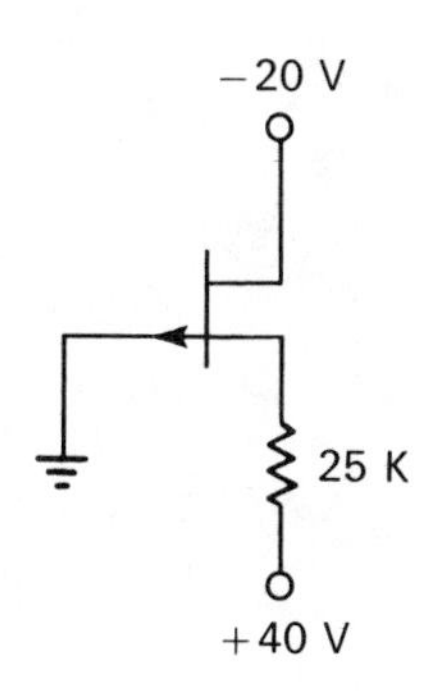

P 14-5 If $I_{DSS} = 3$ ma, estimate the maximum value of V_{SS} in Fig. P 14-5.

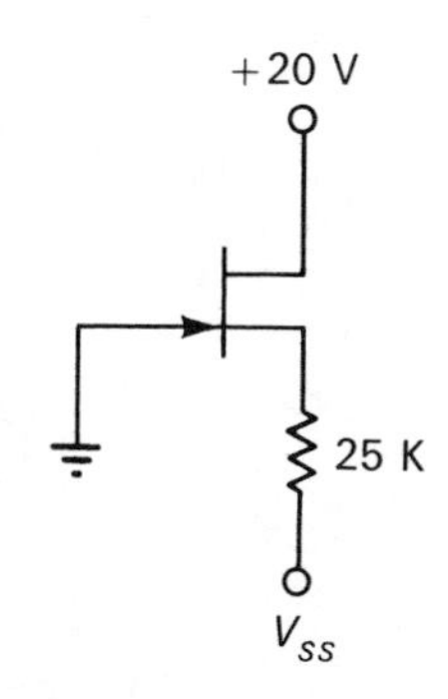

P 14-6 Determine the Q point of the FET in the circuit of Fig. P 14-6a. Assume the curves of Fig. P 14-6b apply.

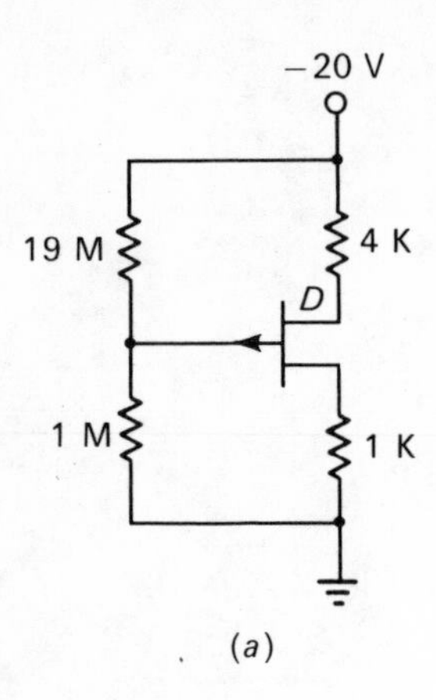

(a)

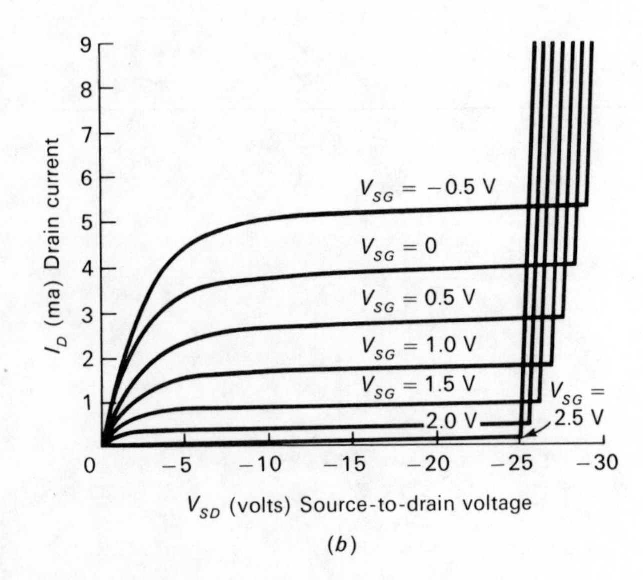

(b)

P 14-7 Design the MOSFET circuit shown in Fig. P 14-7*a* so that the Q point is located as shown on the drain curves of Fig. P 14-7*b*.

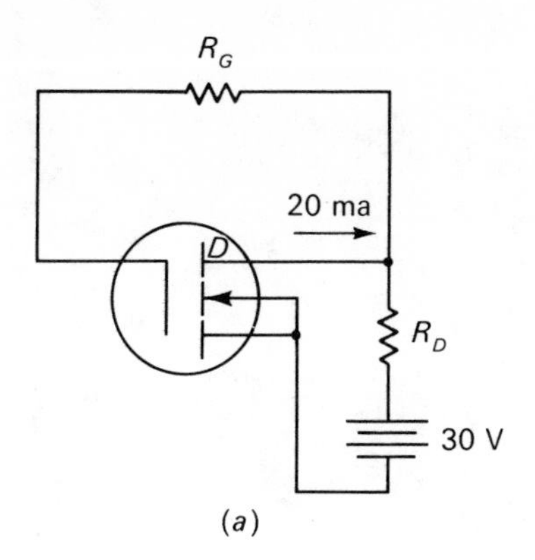

(*a*)

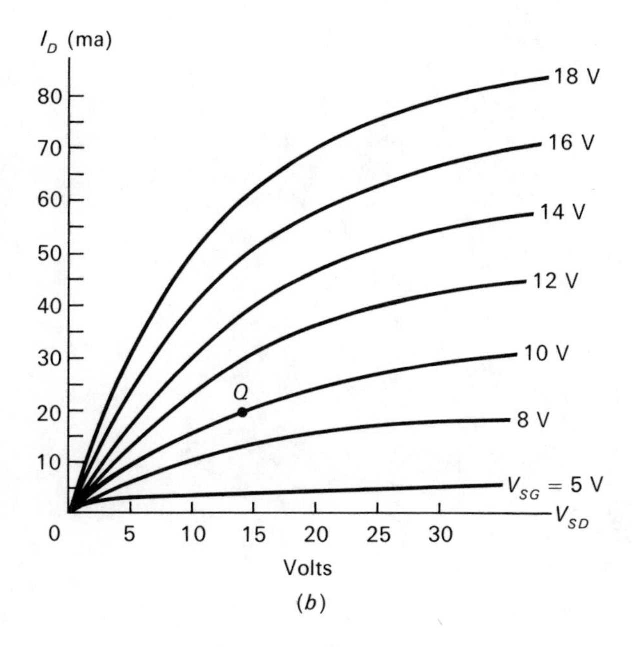

(*b*)

15 basic fet amplifiers

In this chapter we will consider some basic FET amplifier configurations which in many ways are analogous to basic vacuum tube amplifiers. The emphasis is somewhat heavier on the JFET than the IGFET because the JFET, in general, makes a better linear amplifier, whereas the IGFET tends to be superior in switching applications. Nevertheless the device models and equivalent circuits of this chapter are applicable to both JFETs and IGFETs when operated in the active region. The only difference will be in the actual parameter values. Thus, germanium JFETs might be most applicable in a low-temperature environment because of very low internal noise, and the IGFET in a high-frequency application because of its lower input capacity, and for most cases the silicon JFET is usually a wise choice.

15-1 Incremental model review

To develop the gain and impedance relationships of the basic amplifiers we must use the incremental model of the device. Figure 15-1 is a review of the practical FET

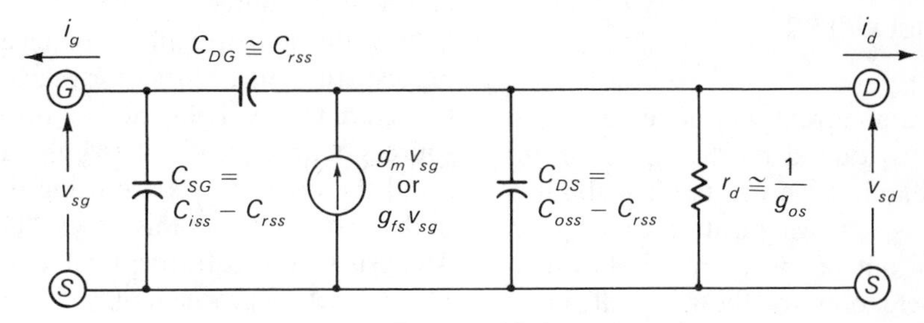

FIGURE 15-1

model. The internal resistances between source and gate and drain and gate are so high as to be negligible for most practical purposes. The device parameters such as C_{SG}, for example, are given in terms of the externally measured parameters available on a data sheet such as C_{iss} and C_{rss}. The device parameters were also related to the y parameters in Chapter 13.

If the FET is used in that part of the frequency spectrum where the internal capacitances may be neglected, we have Fig. 15-2a as an equivalent circuit. Since the gate diode is reverse-biased in the active region, we can assume the gate current is negligible. Hence the gate terminal appears completely isolated as shown. However, any voltage v_{sg} causes a current $g_m v_{sg}$ to appear in the dependent current source with the reference direction shown. The reference direction of this current source depends only upon the reference direction of v_{sg} and has nothing to do with the fact that a P- or N-channel device may be used. The current and voltages we are discussing are ac or incremental quantities which ride superimposed upon the average or quiescent level. A little thought will convince you that if v_{sg} swings positive in Fig. 15-2a and we are talking about an N-channel

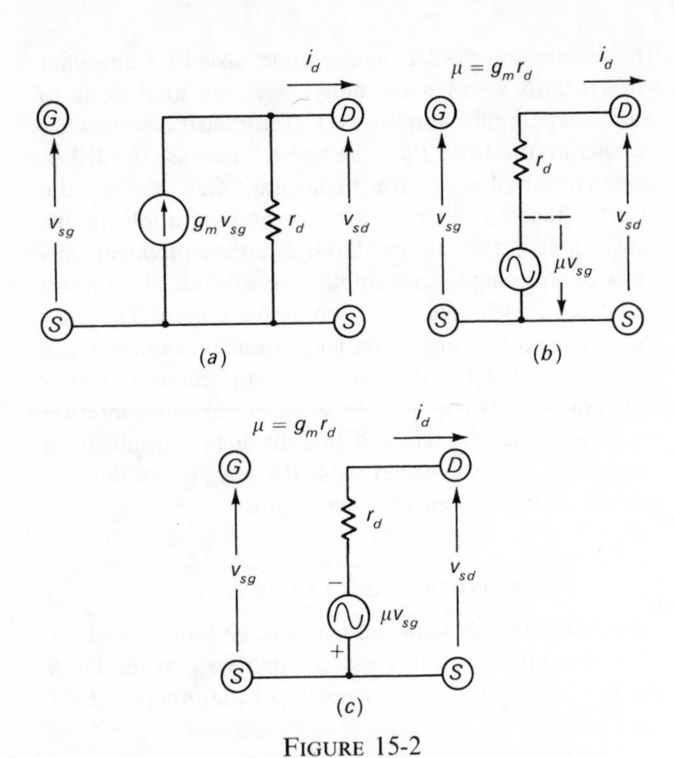

FIGURE 15-2

device, the actual dc drain current will increase due to the added increment of i_d caused by the $g_m v_{sg}$ current source. On the other hand, if we are talking about a *P*-channel device and v_{sg} swings positive causing an increment of i_d to flow out of the drain as shown in Fig. 15-2*a*, this increment opposes the nominal direction of drain current flow which is into the drain (remember we are talking about electron flow). Thus the net drain current decreases as would be expected for a positive swing in gate voltage for a *P*-channel device. In either case this incremental or ac model is valid. Be sure you can accept that.

Note that in the interest of making the illustrations clearer, the equivalent circuit is drawn as if there were two separate source terminals, whereas in actuality the device only has one source terminal. This should not prove confusing since both terminals marked *S* (source) are connected with a solid wire and hence may be considered as one single terminal.

Usually we can obtain a better feel for the properties of the basic amplifiers if we convert the current source model of Fig. 15-2*a* to the voltage source model of Fig. 15-2*b* by applying Thévenin's theorem where

1) $$\mu = g_m r_d \tag{15-1}$$

Note the reference direction of the μv_{sg} voltage source. It is exactly opposite to the reference direction of v_{sg} in relation to the source terminal. Thus, if terminal *G* is made positive relative to terminal *S* it means v_{sg} will be a positive quantity since the head of the arrow is swinging positive relative to its tail. Therefore, the head of the μv_{sg} arrow will also swing positive relative to its tail since this is a dependent voltage source. This means that when *G* swings positive relative to *S* the drain *D* swings negative relative to *S*. Conversely, if v_{sg} is a negative quantity, v_{sd} will be a positive quantity due to the μv_{sg} generator. In other words v_{sg} and v_{sd} are 180° out of phase. The μv_{sg} voltage source in Fig. 15-2*b* has to be directed opposite to the reference direction of v_{sg}, because back in Fig. 15-2*a* when v_{sg} swings positive, the open-circuit voltage between source and drain swings negative; and hence, if the current and voltage source models are to be equivalent, v_{sd} must also swing negative in Fig. 15-2*b*.

Another and perhaps more common way of denoting the phase relationship between v_{sg} and the μv_{sg} voltage source is shown in Fig. 15-2*c*. Thus if v_{sg} is shown in the indicated reference direction, the reference polarity for the dependent μv_{sg} source will appear as shown in Fig. 15-2*c*. This does not mean that the source is always positive in actual polarity with respect to the drain. It simply indicates the reference polarity for the dependent voltage source.

Now this may not all be too clear, and it is one of the biggest stumbling blocks for the student in dealing with the circuits involving active devices with dependent sources. Consequently it is imperative that you understand the notation we are using. It has been developed in the first volume of this series, "Outline for DC Circuit Analysis with Illustrative Problems."[1] Now do problem PS 15-1 before proceeding.

15-2 The common-source amplifier

Figure 15-3*a* illustrates a common-source amplifier This basic amplifier configuration is characterized by a

[1] See also Phillip Cutler, "Electronic Circuit Analysis," vol. 1, McGraw-Hill Book Company, New York, 1960.

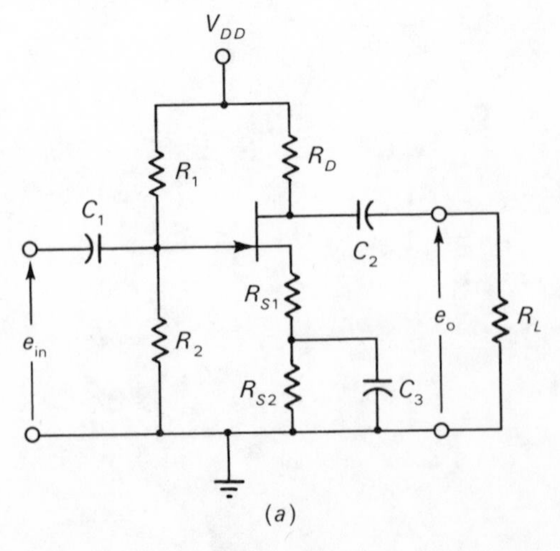

FIGURE 15-3

relatively high input impedance, relatively high output impedance, high voltage gain, and an output which is 180° out of phase with the input. Capacitor $C1$ serves as a blocking or coupling capacitor to remove any dc component of the input signal from the gate of the FET. Resistors $R1$ and $R2$ set the V_{GG} supply and resistors R_{S1} and R_{S2} also establish the bias point. Resistor R_{S2} is bypassed with capacitor $C3$ so that the ac signal is not completely degenerated. Resistor R_{S1} could be omitted to obtain maximum gain. However, R_{S1} tends to stabilize the ac gain. Resistor R_D is the drain load resistor and the voltage developed across it by the fluctuations in drain current are coupled through the coupling capacitor C_2 to the load R_L. Although this basic circuit may be modified in many ways, as long as the input signal is applied to the gate relative to the common ground and the output signal is taken from the drain relative to the common ground, we have basically a common-source amplifier.

The essential core of the common-source amplifier is shown in Fig. 15-3*b*. If we can analyze this particular circuit, we can analyze the circuit of Fig. 15-3*a* or its variations with relative ease since we need only add other circuit components around the core circuit. Resistor R_S in Fig. 15-3*b* represents any unbypassed source resistance such as R_{S1} in Fig. 15-3*a*.

Figure 15-3*c* illustrates an IGFET common-source amplifier. Resistor R_G in conjunction with R_D sets the bias in this circuit. Although it is not included here, quite often R_G is divided into two equal resistors with a bypass capacitor to ground from the midpoint. This bypass capacitor prevents ac signal degeneration and must be used if maximum voltage gain is desired at a sacrifice in gain stability. Figure 15-3*d* illustrates the basic guts of the common-source IGFET amplifier. A resistor R_S could have been included in the source lead to make Fig. 15-3*d* identical with Fig. 15-3*b*. We will include R_S in our analysis as it presents a more general case. If R_S is absent in the actual circuit involved, it need be merely set equal to zero in the results we will obtain.

The quantities we are specifically interested in with regard to Fig. 15-3*b* are the input impedance z_{in}, the output impedance z_o, and the voltage gain from e_1 to e_2 which is designated as A_v. Our basic approach will be to develop a Thévenin's equivalent circuit seen looking into the output terminals of the core circuit. Once this Thévenin's equivalent circuit is developed in terms of the driving voltage e_1, we can readily determine the

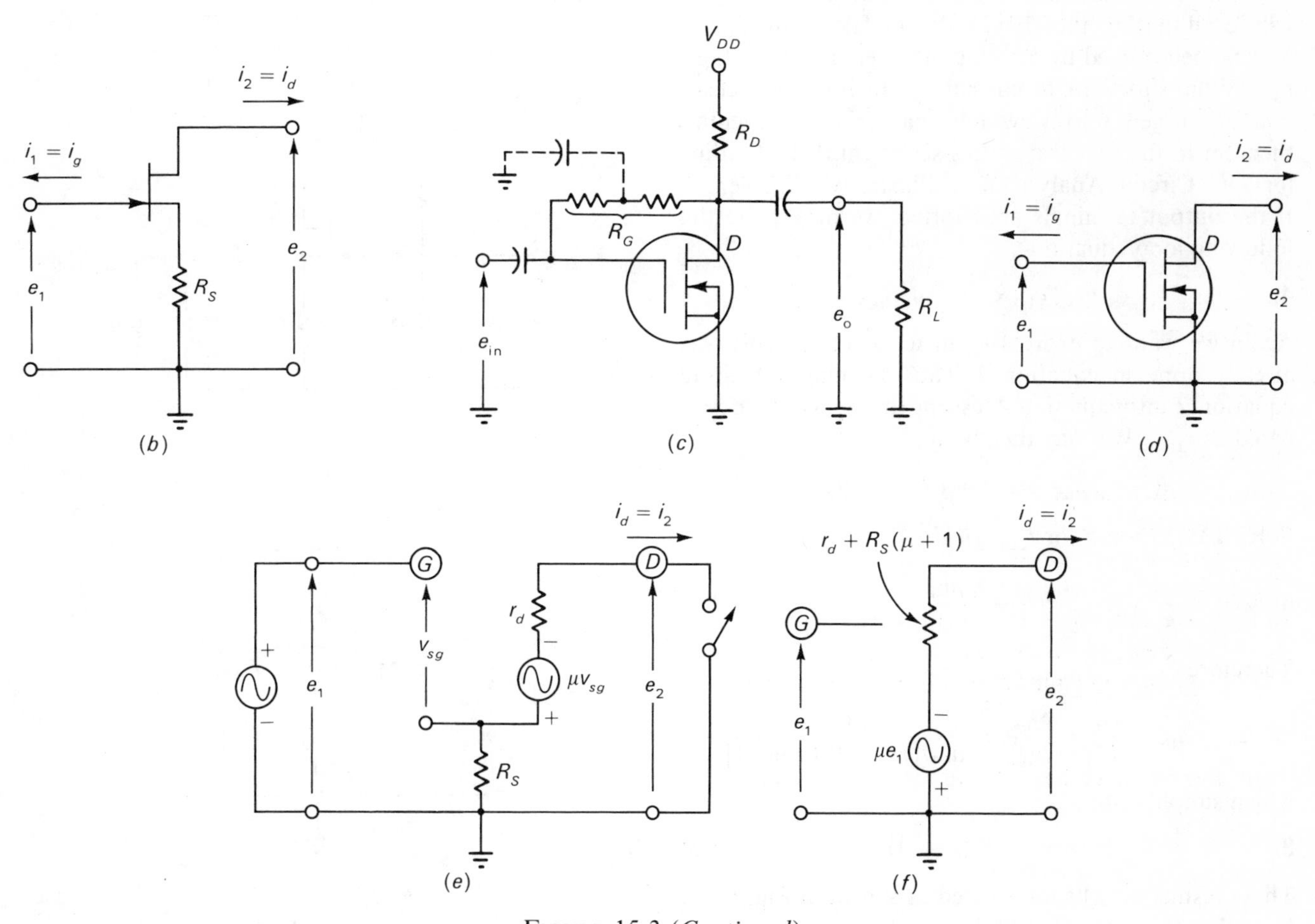

FIGURE 15-3 (*Continued*)

current, voltage, and anything else of interest that a load across the output terminals would be exposed to.

If we substitute the equivalent circuit of Fig. 15-2*c* into Fig. 15-3*b* and assume some voltage source e_1 connected across the input terminals, we have the equivalent circuit of Fig. 15-3*e*. When the switch on the output side is open, e_2 is the open-circuit voltage developed across the output terminals. Since with the switch open $i_2 = 0$, it follows that

1) $$e_{2(oc)} = -\mu v_{sg}$$

We do not, however, want e_2 in terms of v_{sg} which is an internal voltage; but instead we want e_2 expressed as a function of e_1 the effective driving or input voltage. Thus, to express v_{sg} in terms of e_1 we would write, starting at the tail of the v_{sg} arrow and proceeding towards the head,

2) $$v_{sg} = -R_S i_2 + e_1$$

But $i_2 = 0$, and therefore

3) $$e_{2(oc)} = -\mu(-R_S i_2 + e_1)$$
$$= -\mu(0 + e_1) = -\mu e_1 \qquad (15\text{-}2)$$

Now we must determine the Thévenin's equivalent impedance seen looking into the output terminals. You will recall that z_{th}, the Thévenin's equivalent impedance, may be determined by dividing the open-circuit voltage, v_{oc}, by the short-circuit current, i_{sc}. If this is not clear, you are urged to review the chapter on Thévenin's theorem in the first text of this series entitled, "Outline for DC Circuit Analysis with Illustrative Problems." If the output terminals are shorted, we may write the following loop equation

4) $$\mu v_{sg} = (R_S + r_d)i_{2(sc)}$$

Again we want to express v_{sg} in terms of e_1. This was already done in equation 2. Thus we may substitute equation 2 into equation 4, except that i_2 will be designated as $i_{2(sc)}$. We may then write

5) $$\mu(-R_S i_{2(sc)} + e_1) = (R_S + r_d)i_{2(sc)}$$

Solving equation 5 for $i_{2(sc)}$ *yields*

6) $$i_{2(sc)} = \frac{\mu e_1}{r_d + R_S(\mu + 1)}$$

Therefore

7) $$z_o = z_{\text{th}} = \frac{v_{oc}}{i_{sc}} = \frac{e_{2(oc)}}{i_{2(oc)}} = \frac{\mu e_1}{\mu e_1/[rd + R_S(\mu + 1)]}$$

which simplifies to

8) $$z_o = r_d + R_S(\mu + 1) \qquad (15\text{-}3)$$

These results may be interpreted as shown in Fig. 15-3*f*. Note that the output is 180° out of phase with the input.

Note also that the maximum possible voltage gain is

9) $$A_{v(\max)} = \frac{e_{2(oc)}}{e_1} = \frac{-\mu e_1}{e_1} = -\mu \qquad (15\text{-}4)$$

The minus sign in equation 9 is indicative of the fact that the output is 180° out of phase with the input. Note also that if $R_S = 0$, $z_O = r_d$.

Now do problem PS 15-2.

15-3 The common-drain amplifier (source follower)

Figure 15-4*a* illustrates a source-follower FET amplifier. Various biasing arrangements may give the circuit a somewhat different form, but the input is always ultimately applied to the gate and the output is taken from the source. The source follower is characterized by a high input impedance, low output impedance, and voltage gain less than unity. The output voltage is also in phase with the input. The source follower finds its greatest utility as an impedance-transforming buffer amplifier. Normally the source follower is operated without any impedance in the drain lead. However, in the interest of generality, we will analyze the circuit with some resistance R_D assumed in the drain lead.

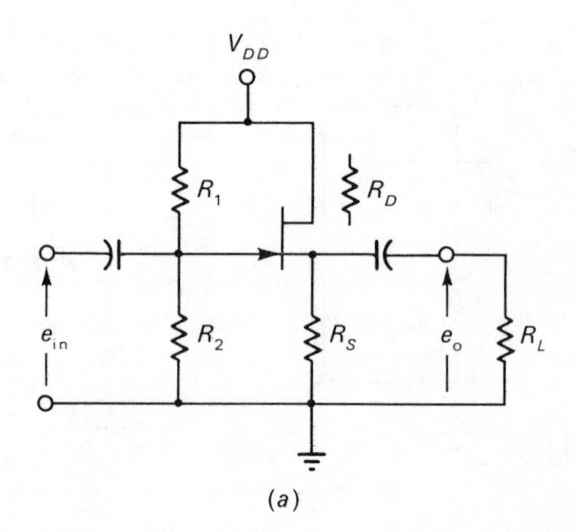

(*a*)

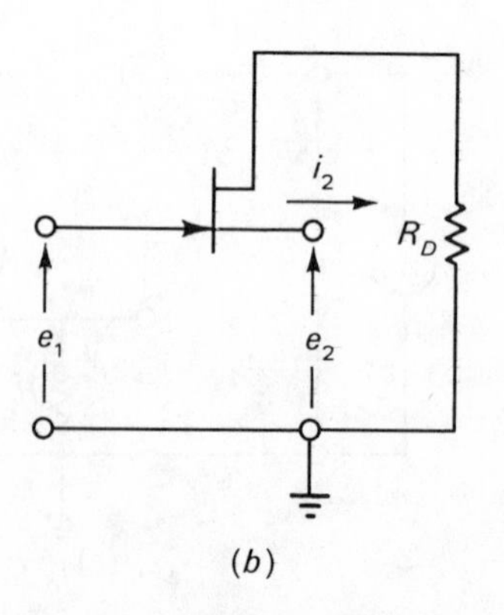

(*b*)

FIGURE 15-4

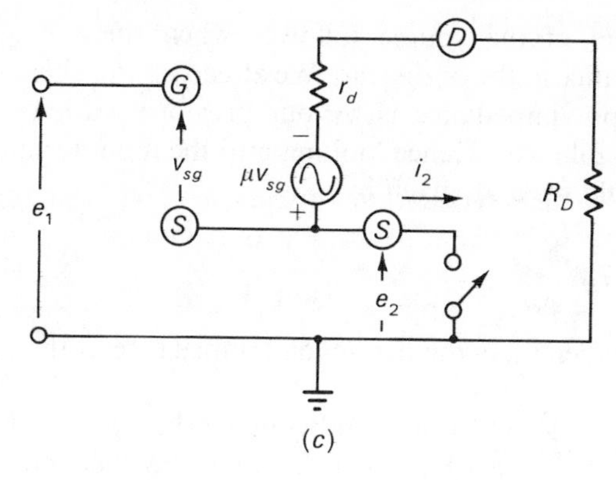

(c)

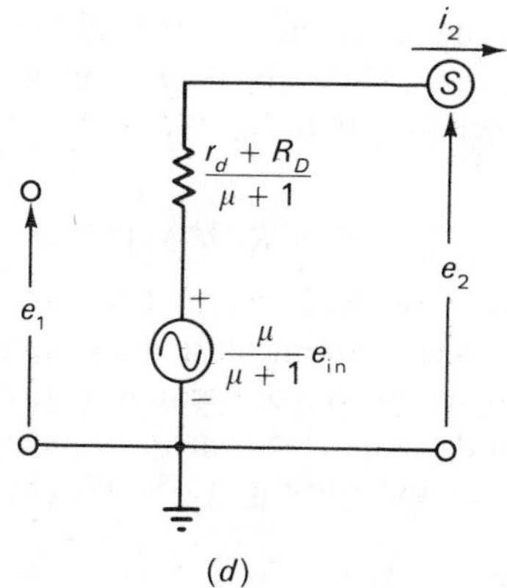

(d)

FIGURE 15-4 (*Continued*)

Figure 15-4*b* illustrates the essence of the source follower. We wish to determine the voltage gain and the input and output impedance. Figure 15-4*c* illustrates the equivalent circuit of the source follower. Again we will thevenize to determine the open-circuit voltage and Thévenin's equivalent impedance seen looking into the output terminals. The input impedance presented to e_1 in the low-frequency spectrum is essentially infinite.

On the output side with the switch open we may write

1) $$e_{2(oc)} = \mu v_{sg}$$

But

2) $$v_{sg} = -e_2 + e_1$$

Substituting equation 2 into 1, setting $e_2 = e_{2(oc)}$, and simplifying yields

3) $$e_{2(oc)} = \frac{\mu}{\mu + 1} e_1 \qquad (15\text{-}5)$$

Closing the switch to determine the short-circuit current we may write the following loop equation:

4) $$0 = (r_d + R_D) i_{2(sc)} + \mu v_{sg}$$

But with the switch closed

5) $$v_{sg} = e_1$$

Substituting equation 5 into 4 and solving for i_{2sc} yields

6) $$i_{2(sc)} = \frac{-\mu e_1}{r_d + R_D}$$

The negative sign in equation 6 is solely due to the choice of reference direction for i_2 in Fig. 15-4*c* and will be ignored in determining the Thévenin's equivalent impedance. Thus

7) $$z_o = z_{th} = \frac{v_{oc}}{i_{sc}} = \frac{e_{2(oc)}}{i_{2(sc)}}$$

Substituting equations 3 and 6 into 7 we obtain

8) $$z_o = \frac{[\mu/(\mu + 1)] e_1}{[\mu/(r_d + R_D)] e_1} = \frac{r_d + R_D}{\mu + 1} \qquad (15\text{-}6)$$

The physical significance of equation 15-6 is that when looking into the source terminal any impedance in the drain lead (internal and external) will appear as if it is divided by $\mu + 1$. This is in contrast to the common-source amplifier where any impedance in the source lead looks as if it is multiplied by $\mu + 1$ when looking into the drain terminal. The Thévenin's equivalent circuit of the source follower will then appear as shown in Fig. 15-4*d*.

Now you should carefully work your way through problem PS 15-3.

15-4 The common-gate amplifier

The common-gate amplifier, or grounded-gate amplifier as it is sometimes called, is shown in Fig. 15-5*a*. Although there may be variations in the appearance of the circuit

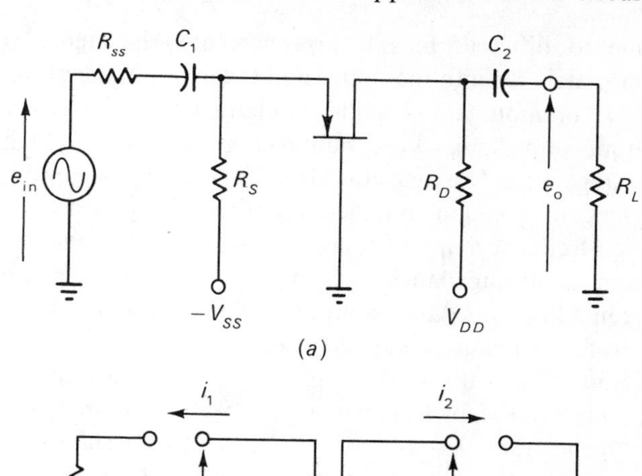

(a)

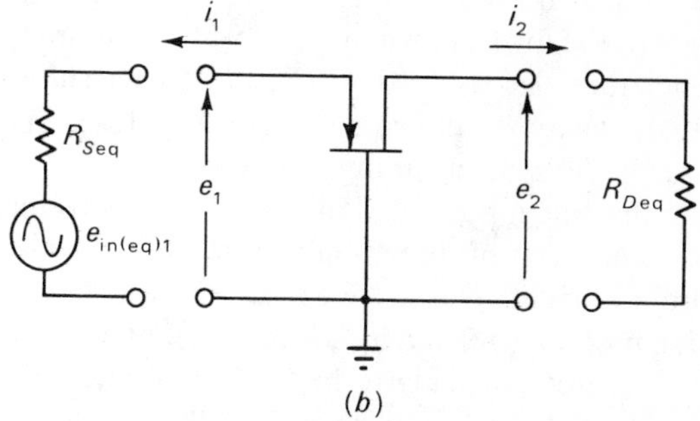

(b)

FIGURE 15-5

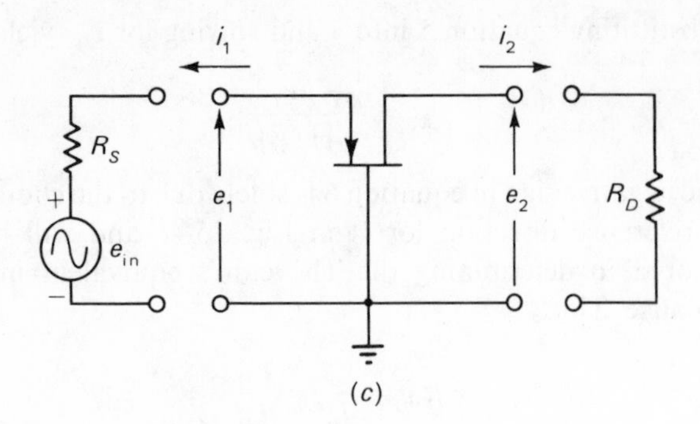

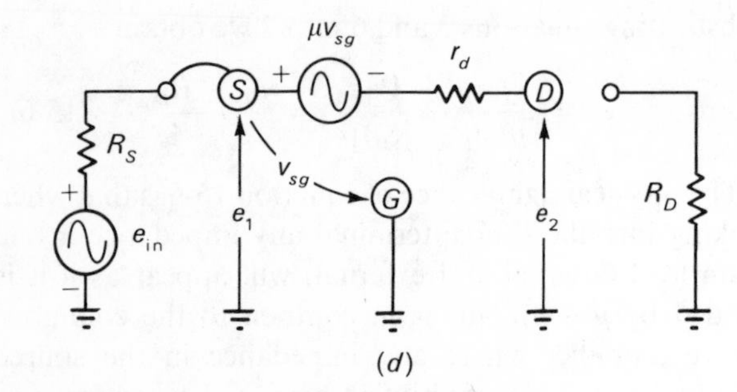

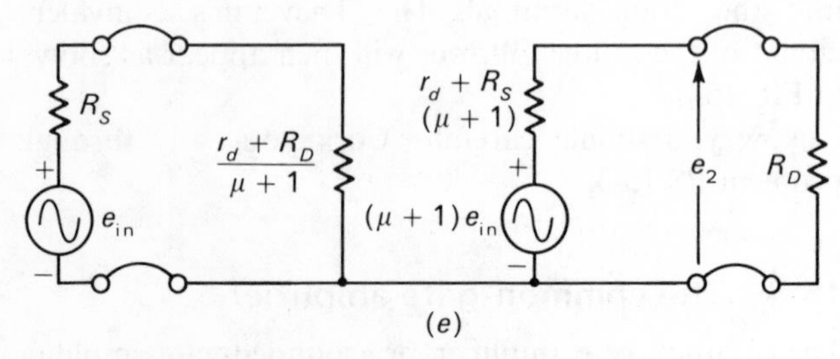

FIGURE 15-5 (*Continued*)

due to different biasing arrangements, the signal is ultimately fed into the source and taken out of the drain. The common-gate amplifier is characterized by a low input impedance, high output impedance, and high voltage gain. The output is also in phase with the input. The common-gate amplifier is particularly applicable in high-frequency applications, although it may also be used as an impedance-transforming device when going from a low-impedance source to a high-impedance load. Insofar as the ac signal is concerned, the circuit of Fig. 15-5*a* may, with the aid of Thévenin's theorem, be reduced to the form shown in Fig. 15-5*b*. To simplify the nomenclature, let us redraw Fig. 15-5*b* as shown in Fig. 15-5*c* where R_S and e_{in} and R_D in Fig. 15-5*c* are not necessarily the same as in Fig. 15-5a.

Note that the input signal in Fig. 15-5*c* has to furnish the ac component of source current which is the same as the drain current. Hence, since current will flow from the signal source we might suspect that the input impedance is not particularly high. As a matter of fact, looking into the source terminal of the common-gate amplifier is actually the same as looking into the output terminal of the source follower when there is some impedance in the drain lead. We already know then what the input impedance is by our previous work on the source follower. Hence looking into the input terminals, e_1 would see a z_{in} given by

1) $$z_{in} = \frac{r_d + R_D}{\mu + 1} \qquad (15\text{-}7)$$

Remember R_D is the net ac load impedance in the drain lead.

We also know from our previous work on the common-source amplifier that when looking into the drain terminal, any impedance in the source lead looks as if it is multiplied by $\mu + 1$. Hence by inspection, we may write the output impedance that R_D in Fig. 15-5*c* looks into as being

2) $$z_o = r_d + R_S(\mu + 1) \qquad (15\text{-}8)$$

The remaining problem is to determine the open-circuit voltage seen looking into the output terminals when the input signal source is connected. This may be done with the aid of Fig. 15-5*d* where we have substituted an incremental model into Fig. 15-5*c*. We may now write

3) $$e_{2(oc)} = e_{in} - \mu v_{sg}$$

but

4) $$v_{sg} = -e_{in}$$

Substituting equation 4 into 3 yields

5) $$e_{2(oc)} = (\mu + 1)e_{in} \qquad (15\text{-}9)$$

Equations 1, 2, and 5 may be given the physical interpretation shown in Fig. 15-5*e*.

Now do problem PS 15-4.

15-5 High-frequency considerations

As the input signal frequency increases, reactive affects come into play. A major problem which causes degradation of gain in the common-source amplifier is Miller effect. The Miller effect, you will recall from Chapter 11, refers to the sensitivity of an amplifier's input admittance to the gain. Furthermore, it developed that any feedback capacitance between input and output appeared across the input terminals as if it were multiplied by the stage gain. Let us review and explore the Miller effect with regard to the common-source FET amplifier of Fig. 15-6*a*. Our immediate goal is to develop a feeling for the input admittance.

Substituting the incremental model of Fig. 15-1 into Fig. 15-6*a* yields Fig. 15-6*b*. To keep the analysis simple let us redraw Fig. 15-6*b* as shown in Fig. 15-6*c*. This is permissible since everything on the right-hand side of C_{DG} is sitting at $e_o = A_v e_{in}$ volts relative to the common input (ground) terminal. For Fig. 15-6*c* we may write

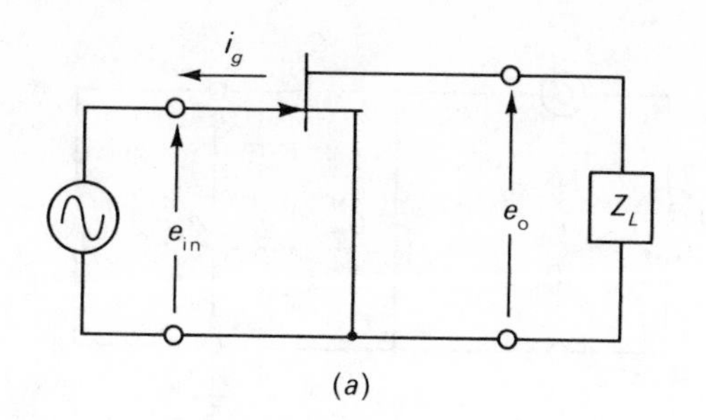

(a)

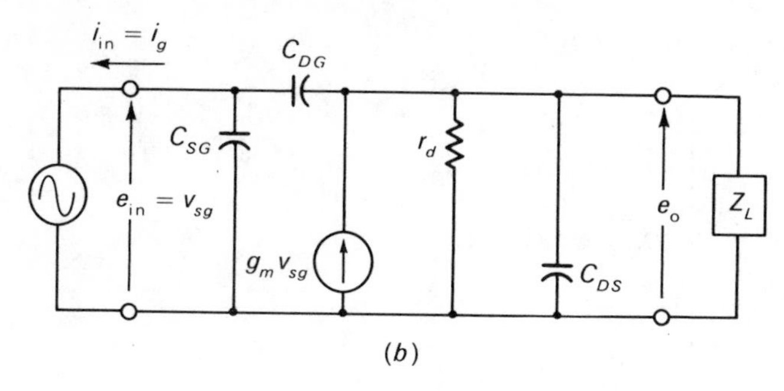

(b)

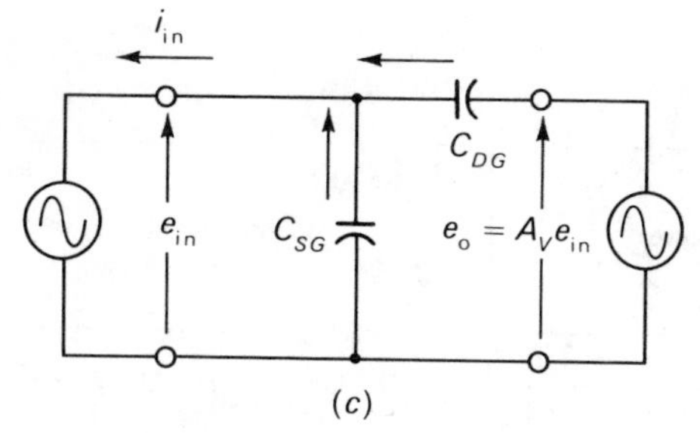

(c)

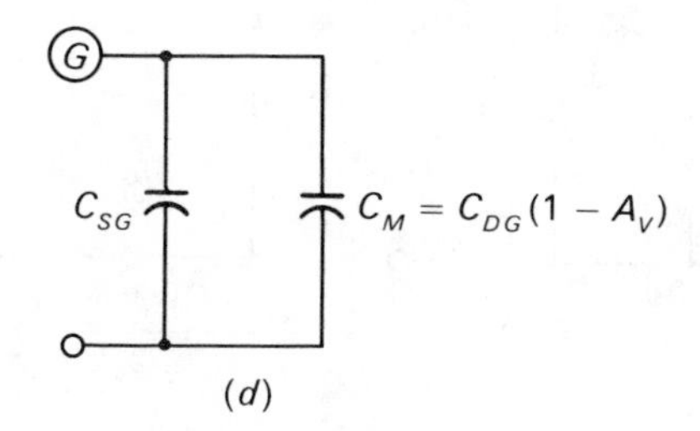

(d)

FIGURE 15-6

1) $$i_{in} = j\omega C_{SG} e_{in} + j\omega C_{DG}(e_{in} - A_v e_{in})$$

which simplifies to

2) $$i_{in} = [j\omega C_{SG} + j\omega C_{DG}(1 - A_v)]e_{in}$$

Therefore

3) $$y_{in} = \frac{i_{in}}{e_{in}} = j\omega[C_{SG} + C_{DG}(1 - A_v)] \quad (15\text{-}10)$$

or in terms of the parameters usually specified on a data sheet

4) $$y_{in} = j\omega(C_{iss} - A_v C_{rss}) \quad (15\text{-}11)$$

Equation 15-10 may be given the physical interpretation shown in Fig. 15-6*d*. Since A_v is usually a number much larger than unity and inverted in phase (a negative quantity) it follows that C_{DG} will actually appear as if it were multiplied by one plus the stage gain.

We have not specified A_v as yet. Actually A_v in a circuit such as Fig. 15-6*b* will be a function of C_{DS} and C_{DG} as well as g_m, r_d, and Z_L. An exact expression for A_v when the signal-source impedance $R_{ss} = 0$ as in the circuit of Fig. 15-6*a* or 15-6*b* is

5) $$A_v = \frac{e_o}{e_{in}} = \frac{e_o}{v_{sg}}$$

$$= \frac{-Z_L(g_m - j\omega C_{DG})}{1 + Z_L[y_d + j\omega(C_{DG} + C_{DS})]} \quad (15\text{-}12)$$

where $y_d = 1/r_d$. Equation 5 may also be expressed in terms of the device parameters usually given on a data sheet. Thus

6) $$A_v = \frac{e_o}{v_{sg}} = -\frac{Z_L(g_{fs} - j\omega C_{rss})}{1 + Z_L(g_{os} + j\omega C_{oss})} \quad (15\text{-}13)$$

Now it may be shown that for values of

7) $$\omega = 0.5\,\omega_D \quad (15\text{-}14)$$

where

8) $$\omega_D = \frac{1 + Z_L g_{os}}{C_{oss} Z_L} \quad (15\text{-}15)$$

the gain A_v is essentially given by

9) $$A_v = -\frac{Z_L g_{fs}}{1 + Z_L g_{os}} = \frac{-\mu Z_L}{r_d + Z_L} \quad (15\text{-}16)$$

For a typical JFET with Z_L being a resistance in the neighborhood of 10 kilohms or less, ω_D will be in the neighborhood of a few megahertz. Thus equation 15-16 is certainly a good approximation for A_v within the audio frequency spectrum.

It turns out that if Z_L is reactive, y_{in} contains a resistive component which may be positive if Z_L is capacitive in nature or negative if Z_L is inductive. This means that an inductive load can cause a component of negative input resistance to appear at the input terminals which may give rise to instability (oscillations).

Now work through problem PS 15-5.

To see why the source-follower circuit presents a high input impedance, consider Fig. 15-7*a*, which may be redrawn as Fig. 15-7*b* in the interest of simplicity. Note that this time C_{SG} rather than C_{DG} is the feedback capacitance. Evaluation of the input admittance for Fig. 15-7*b* yields

$$y_{in} = j\omega[C_{DG} + C_{SG}(1 - A_v)] \quad (15\text{-}17)$$

Equation 15-17 may be given the physical interpretation shown in Fig. 15-7*c*. Again we have a Miller-effect capacitance C_M which, in this case, is equal to $C_{SG}(1 - A_v)$. Usually C_{SG} is a few times larger than C_{DG} which would tend to make the Miller-effect capacitance large.

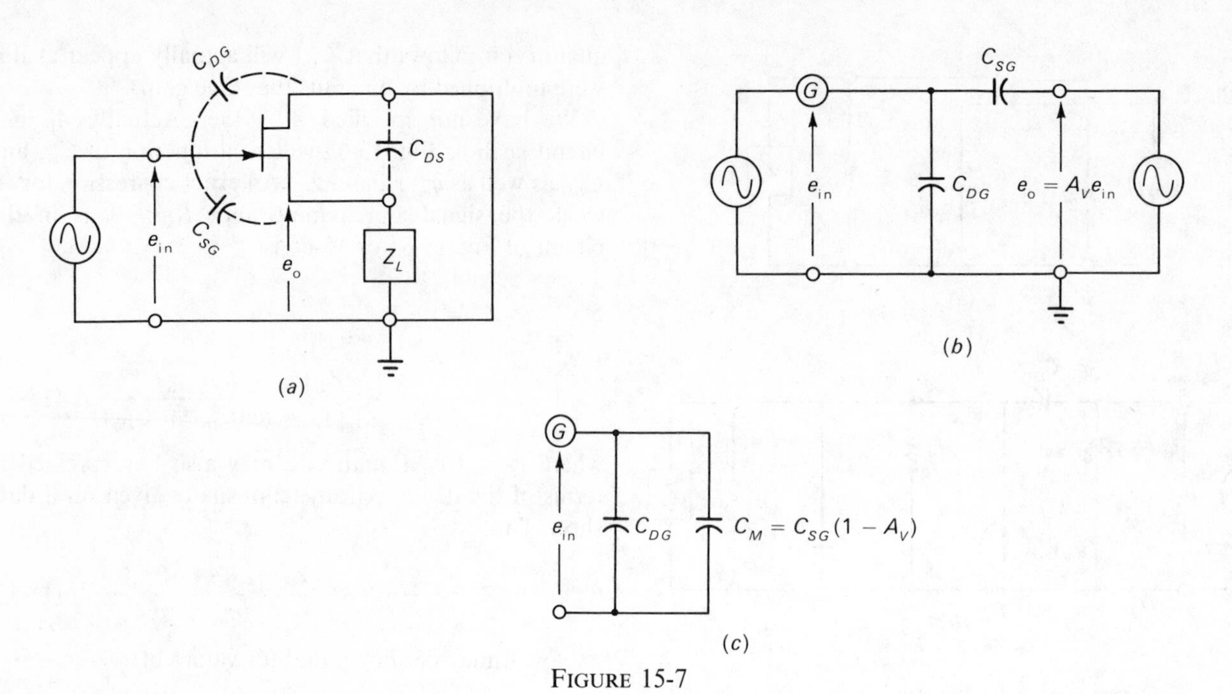

FIGURE 15-7

Table 15-1

$\mu = g_m r_d = \dfrac{g_{fs}}{g_{os}}$	Common source	Source follower	Common gate
Circuit	e_{in}, z_{in}, R_S, z_o, R_L, e_o	e_{in}, z_o, R_D, R_L, e_o	R_S, e_{in}, z_{in}, z_o, R_L, e_o
Equivalent circuit	G, D, $r_d + R_s(\mu + 1)$, e_{in}, μe_{in}, e_o, R_L	G, S, $\dfrac{r_d + R_D}{\mu + 1}$, e_{in}, $\dfrac{\mu}{\mu + 1} e_{in}$, e_o, R_L	S, D, $r_d + R_S(\mu + 1)$, R_S, $\dfrac{r_d + R_L}{\mu + 1}$, e_{in}, $(\mu + 1)e_{in}$, e_o, R_L
Voltage gain $A_v = \dfrac{e_o}{e_{in}}$	$\dfrac{-\mu R_L}{R_L + r_d + R_S(\mu + 1)}$	$\dfrac{\mu R_L}{R_D + r_d + R_L(\mu + 1)}$	$\dfrac{(\mu + 1)R_L}{R_L + r_d + R_S(\mu + 1)}$
z_{in}	$\cong \infty$	$\cong \infty$	$\dfrac{r_d + R_L}{\mu + 1}$
z_o	$r_d + R_S(\mu + 1)$	$\dfrac{r_d + R_D}{\mu + 1}$	$r_d + R_S(\mu + 1)$

However, you will recall that in the source follower A_v is in phase with the input so that A_v is a positive quantity and also the magnitude of A_v is fairly close to unity. This means that the quantity $1 - A_v$ will be equal to one minus some number which is fairly close to, but slightly less than one, so that the quantity in the parenthesis is quite small; and hence the multiplier for C_{SG} will be very small. Hence the effect of C_{SG} is considerably diminished by the Miller effect in the source follower. Physically it simply means that the voltage across C_{SG} tends to be small because the output follows the input. If the voltage across the capacitor is small, only a small amount of current can flow into it which reduces the input admittance.

Now do problem PS 15-6.

Table 15-1 summarizes some of the basic amplifier properties.

PROBLEMS WITH SOLUTIONS

PS 15-1 Determine v_{sd} for each of the three switch positions shown in Fig. PS 15-1*a* and *b*.

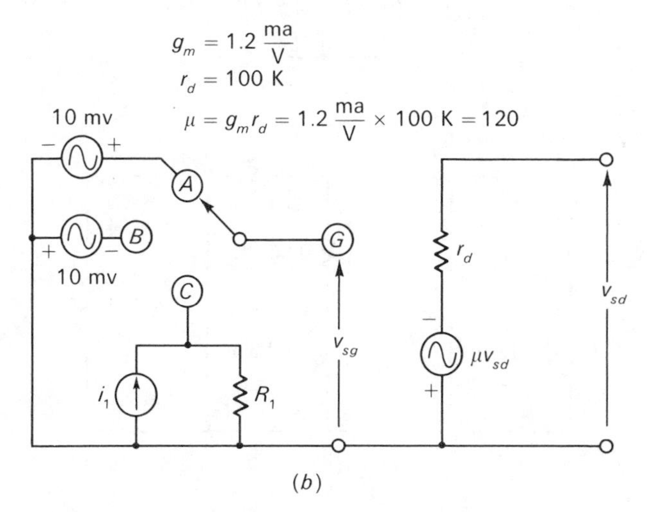

FIGURE PS 15-1

SOLUTION In general for Fig. PS 15-1*a*

$$1)\quad v_{sd} = -r_d g_m v_{sg} = -100 \text{ kilohms } (1.2 \text{ ma/volt}) v_{sg} = -120 v_{sg}$$

In position A, $v_{sg} = +10$ mv. Therefore, $v_{sd} = -120(+10$ mv$) = -1.2$ volts. In position B, $v_{sg} = -10$ mv. Therefore, $v_{sd} = -120(-10$ mv$) = +1.2$ volts. In position C, $v_{sg} = -R_1 i_1$. Therefore, $v_{sd} = -120(-R_1 i_1) = +120\ R_1 i_1$. In general, for Fig. PS 15-1*b*

$$2)\quad v_{sd} = -\mu v_{sg}$$

where

$$3)\quad \mu = g_m r_d = (1.2 \text{ ma/volt}) \ 100 \text{ kilohms} = 120$$

Substituting 3 into 2 we have

$$4)\quad v_{sd} = -120\ v_{sg}$$

In position A, $v_{sg} = +10$ mv. Therefore, $v_{sd} = -120(+10$ mv$) = -1.2$ volts. In position B, $v_{sg} = -10$ mv. Therefore, $v_{sd} = -120(-10$ mv$) = +1.2$ volts. In position C, $v_{sg} = -R_1 i_1$. Therefore, $v_{sd} = -120\ (-R_1 i_1) = +120\ R_1 i_1$.

Note the phase relationships and equivalent results obtained from the current (Norton's) and voltage (Thévenin's) models.

PS 15-2 Determine e_o in the circuit of Fig. PS 15-2*a*.

SOLUTION Constructing a simplified ac circuit we obtain Fig. PS 15-2*b*, which may be simplified further to yield Fig. PS 15-2*c* where the apparent input signal is now

$$e_{in}' \approx \frac{22 \text{ megohms}}{1 \text{ megohm} + 22 \text{ megohms}} (5 \text{ mv}) = 4.78 \text{ mv}$$

and the effective signal source impedance is

$$R_{ss}' \approx 1 \text{ megohm} \parallel 22 \text{ megohms} = 0.957 \text{ megohm}$$

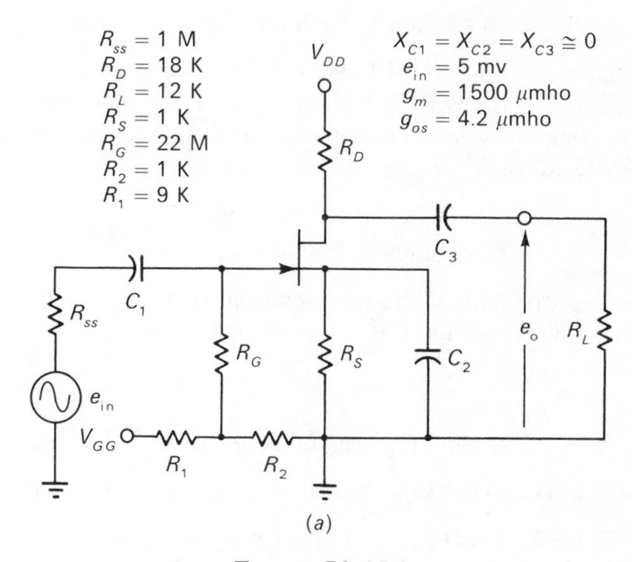

FIGURE PS 15-2

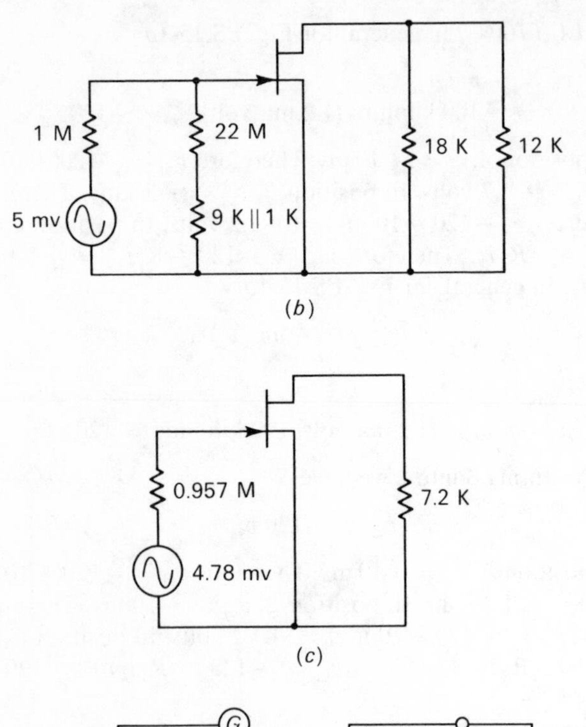

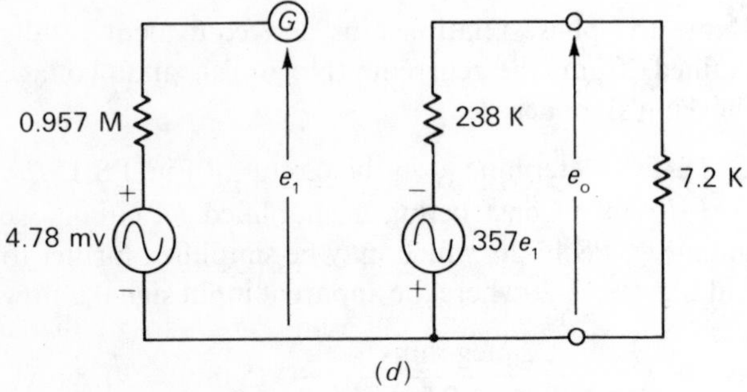

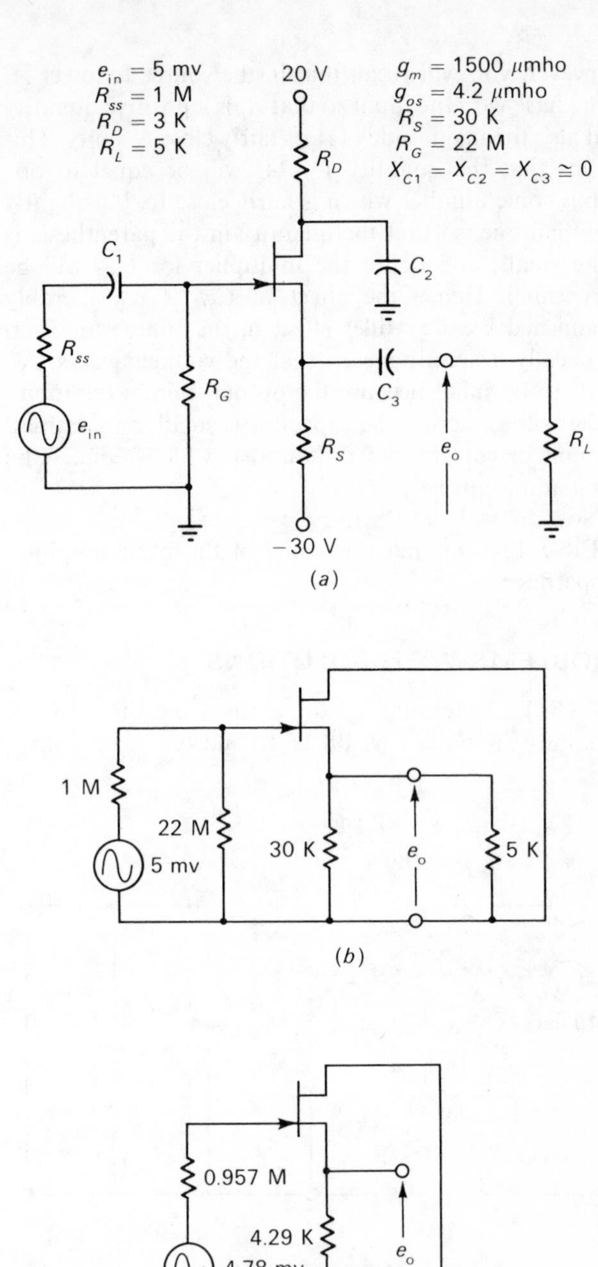

0.957 M
e_1
238 K
e_o
7.2 K
4.78 mv
$357e_1$
(d)

FIGURE PS 15-2 (*Continued*)

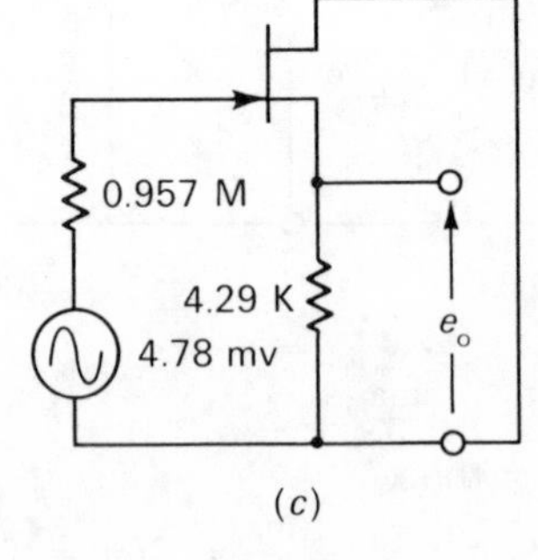

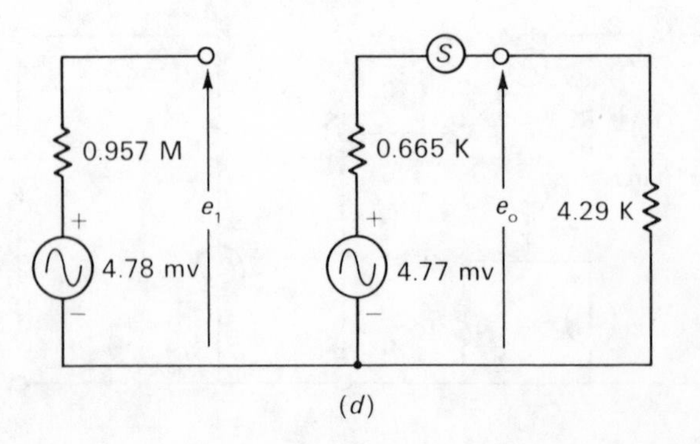

FIGURE PS 15-3

and the load impedance is

$$R_L' = 18 \text{ kilohms} \parallel 12 \text{ kilohms} = 7.2 \text{ kilohms}$$

Since $r_d = 1/g_{os} = 1/4.2\ \mu\text{mho} = 238$ kilohms and $\mu = g_m r_d = (1.5 \text{ ma/volt})(238 \text{ kilohms}) = 357$ we may, from the common-source Thévenin's equivalent circuit of Fig. PS 15-2*d*, write

$$e_o = -\frac{357\,(4.78 \text{ mv})\, 7.2 \text{ kilohms}}{7.2 \text{ kilohms} + 238 \text{ kilohms}} = -50.1 \text{ mv}$$

An approximate alternative solution to the gain when the source terminal is at ac ground potential and $R_L' \ll r_d$ is

$$\begin{aligned} e_o &\approx -R_L' g_m e_{in} \\ &= -7.2 \text{ kilohms}\,(1.5 \text{ ma/volt})(4.78 \text{ mv}) = -51.7 \text{ mv} \end{aligned}$$

which is close but slightly high because r_d was neglected.

PS 15-3 Determine e_o in the circuit of Fig. PS 15-3*a*.
SOLUTION First we draw Fig. PS 15-3*b* which is the

circuit as the input signal sees it. This figure is then simplified to Fig. PS 15-3c. Next we construct the Thévenin's equivalent circuit of a source follower as shown in Fig. PS 15-3d after determining the following

$$r_d = \frac{1}{g_{os}} = 238 \text{ kilohms}$$

$$\mu = g_m r_d = 357$$

$$v_{oc} = \frac{\mu}{\mu + 1} e_{in} = \frac{357}{358}(4.78\text{mv}) = 4.77 \text{ mv}$$

$$z_{th} = \frac{r_d}{\mu + 1} = \frac{238 \text{ kilohms}}{358} = 0.665 \text{ kilohm}$$

Consequently

$$e_o = \frac{4.29 \text{ kilohms } (4.77 \text{ mv})}{4.29 \text{ kilohms} + 0.665 \text{ kilohm}} = 4.13 \text{ mv}$$

Note that the gain of the source follower is slightly less than unity as might be expected.

PS 15-4 Determine e_o in the circuit of Fig. PS 15-4a.

SOLUTION Drawing the circuit as the ac signal sees it, we obtain Fig. PS 15-4b which clearly reveals we have

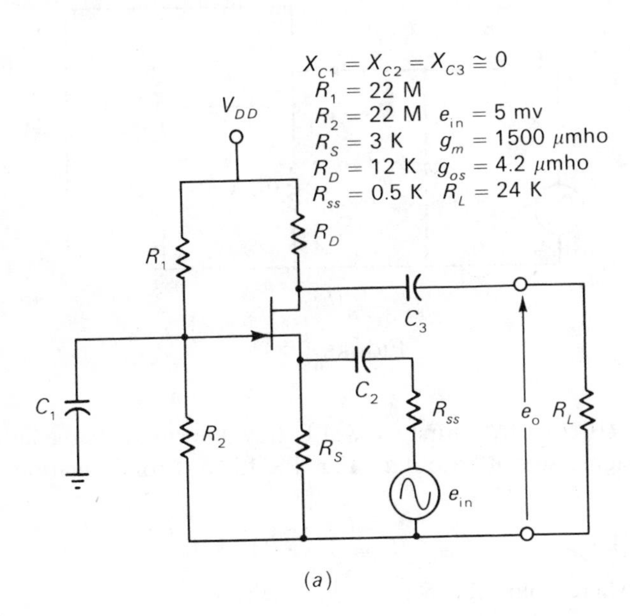

(*a*)

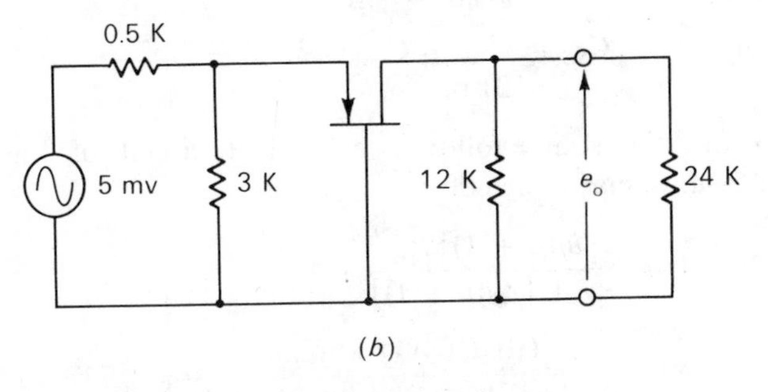

(*b*)

FIGURE PS 15-4

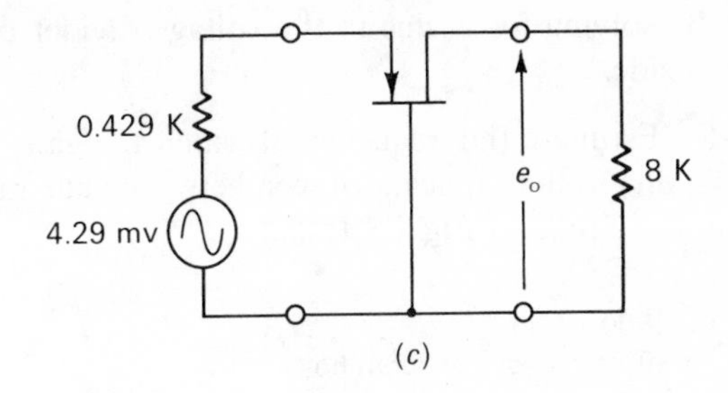

(*c*)

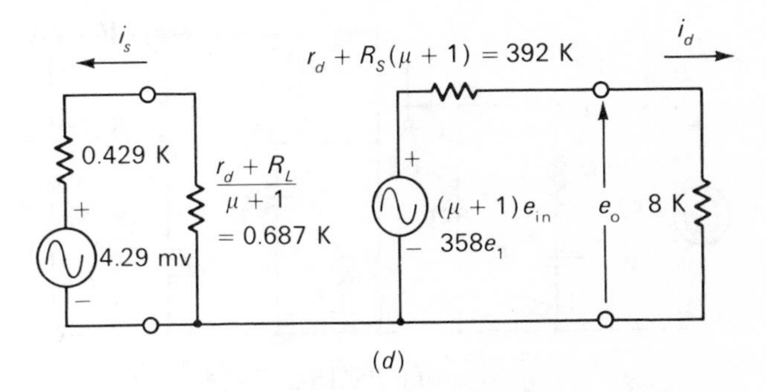

(*d*)

FIGURE PS 15-4 (*Continued*)

a common-gate amplifier. Thevenizing and simplifying Fig. PS 15-4b we obtain Fig. PS 15-4c. Now

$$r_d = \frac{1}{g_{os}} = 238 \text{ kilohms}$$

$$\mu = g_m r_d = 357$$

The input impedance of the common-gate amplifier in Fig. PS 15-4c is

$$z_{in} = \frac{r_d + R_L}{\mu + 1} = \frac{238 \text{ kilohms} + 8 \text{ kilohms}}{358}$$

$$= 0.687 \text{ kilohm}$$

The open-circuit voltage on the output side is

$$v_{oc} = (\mu + 1)e_{in} = 358e_{in}$$

The output impedance is

$$z_o = r_d + R_S(\mu + 1)$$
$$= 238 \text{ kilohms} + 0.429 \text{ kilohm}(358) = 392 \text{ kilohms}$$

The common-gate equivalent circuit of Fig. PS 15-4d may then be constructed. The easiest way to evaluate e_o is to determine i_s from

$$i_s = \frac{4.29 \text{ mv}}{0.429 \text{ kilohm} + 0.687 \text{ kilohm}} = 3.84 \ \mu a$$

Since $i_d = -i_s$ and $e_o = -8Ki_d$ for the reference directions indicated we have

$$e_o = -8 \text{ kilohms}(-3.84 \ \mu a) = 30.8 \text{ mv}$$

Note the output is in phase with the input which is correct for the common-gate amplifier. Verify this solution by solving for e_o due to the voltage divider on the output side.

PS 15-5 Estimate the frequency at which the gain is 3 db down due to the capacitance seen between gate and ground in the circuit of Fig. PS 15-5.

$g_{fs} = 3600\ \mu$mho
$C_{iss} = 39\ pf$ $\quad g_{oss} = 35\ \mu$mho
$C_{rss} = 8\ pf$ $\quad C_{oss} = 21\ pf$

FIGURE PS 15-5

SOLUTION From equation 15-11 which includes the Miller effect

1) $$C_{\text{in}} = C_{iss} - A_v C_{rss}$$

2) $$C_{\text{in}} = 39 \text{ pf} - A_v(8 \text{ pf})$$

3) $$A_v \cong -\frac{\mu R_L}{r_d + R_L}$$

where

4) $$r_d \approx 1/g_{oss} = 1/35\ \mu\text{mho} = 28.6 \text{ kilohms}$$

and

5) $$\mu = g_m r_d = (3.6 \text{ ma/volt})(28.6 \text{ kilohms}) = 103$$

Substituting 4 and 5 into 3 yields

6) $$A_v \approx \frac{-103(8 \text{ kilohms})}{28.6 \text{ kilohms} + 8 \text{ kilohms}} = -22.5$$

Substituting into equation 2 yields

7) $$C_{\text{in}} = 39 \text{ pf} - (-22.5)(8 \text{ pf}) = 219 \text{ pf}$$

The Thévenin's equivalent impedance driving C_{in} is due to $R_{ss} \parallel R_G$ = 5 megohms $\parallel$ 20 megohms = 4 megohms. Thus the time constant is T = 4 megohms (219 pf) = 876 μsec. The corresponding $f = 1/2\pi T$ = 182 H_z. Note that with a high source impedance the frequency response may roll off at low audio frequencies. This is a serious problem in the common-source amplifier.

PS 15-6 In the previous problem it was shown that driving a common-source amplifier from a high signal source impedance can cause the high-frequency response to actually deteriorate in the audio frequency spectrum. Since we usually need the high input impedance of a FET over a wider frequency range it is more advisable to use a source follower. To illustrate the advantage of the source follower in this respect, estimate the frequency at which the gain (although it is lower than in the common-source amplifier) is 3 db down due to the capacitance seen between gate and ground in the circuit of Fig. PS 15-6*a*. Assume the FET parameters are the same as in the previous problem.

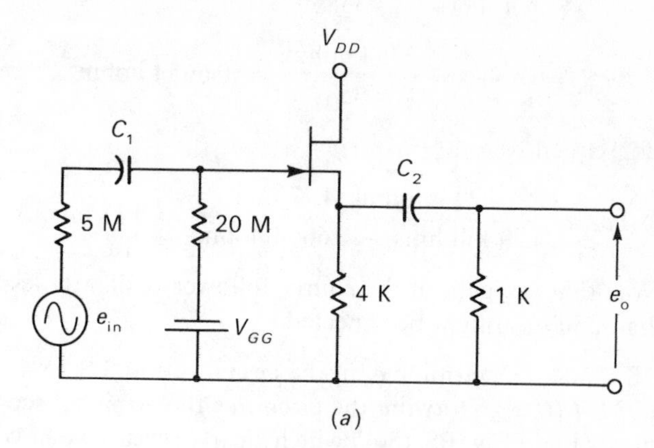

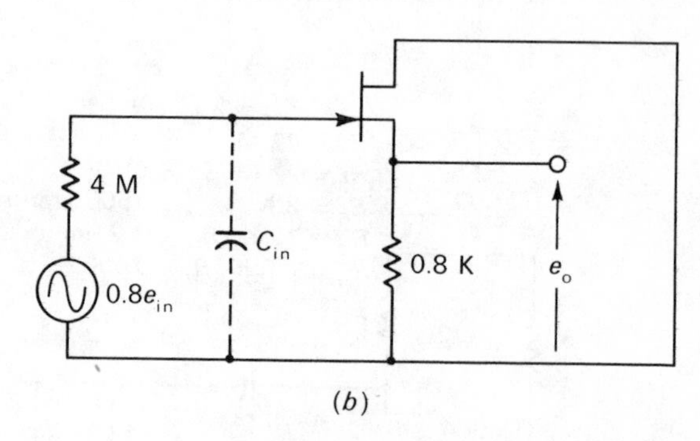

FIGURE PS 15-6

SOLUTION First we will redraw the circuit as the ac signal sees it to obtain Fig. PS 15-6*b*. From equation 15-17

1) $$C_{\text{in}} = C_{DG} + C_{SG}(1 - A_v)$$

where from Fig. 15-1

2) $$C_{DG} \cong C_{rss} = 8 \text{ pf}$$

3) $$C_{SG} = C_{iss} - C_{rss} = 39 \text{ pf} - 8 \text{ pf} = 31 \text{ pf}$$

From the source-follower equivalent circuit of Fig. 15-4*d* we may conclude

4) $$A_v = \frac{[\mu/(\mu + 1)]\, R_L}{R_L + [r_d/(\mu + 1)]} = \frac{(103/104)0.8 \text{ kilohm}}{0.8 \text{ kilohm} + 28.6 \text{ kilohms}/104} = 0.737$$

Substituting equations 2, 3, and 4 into 1 yields

5) $$C_{in} = 8 \text{ pf} + 31 \text{ pf}(1 - 0.737) = 16.2 \text{ pf}$$

Therefore

6) $$T = RC = 4 \text{ m}(16.2 \text{ pf}) = 64.6\ \mu\text{sec}$$

7) $$f = \frac{1}{2\pi T} = \frac{0.159}{64.6\ \mu\text{sec}} = 2460 \text{ hertz}$$

Although 2460 hertz for the source follower is a considerable improvement over 182 hertz for the common-source configuration, the roll-off due to input capacitance is still in the audio spectrum. Other JFETs and IGFETs may have smaller input capacitances but generally at the expense of transconductance. C_{in} is still a serious problem when the input signal emanates from a high source impedance. Later we will see how feedback alleviates this problem.

PROBLEMS WITH ANSWERS

PA 15-1 Determine e_o in Fig. PA 15-1.

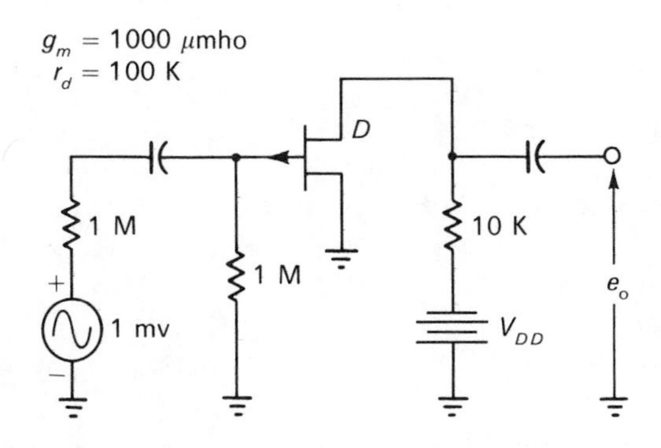

ANSWER −4.55 mv

PA 15-2 Determine e_o in Fig. PA 15-2.

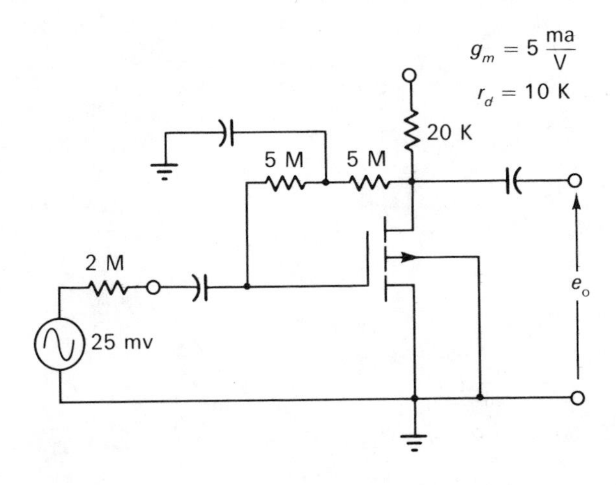

ANSWER −569 mv

PA 15-3 Determine e_o in Fig. PA 15-3 with the switch across R_{S1} closed and open.

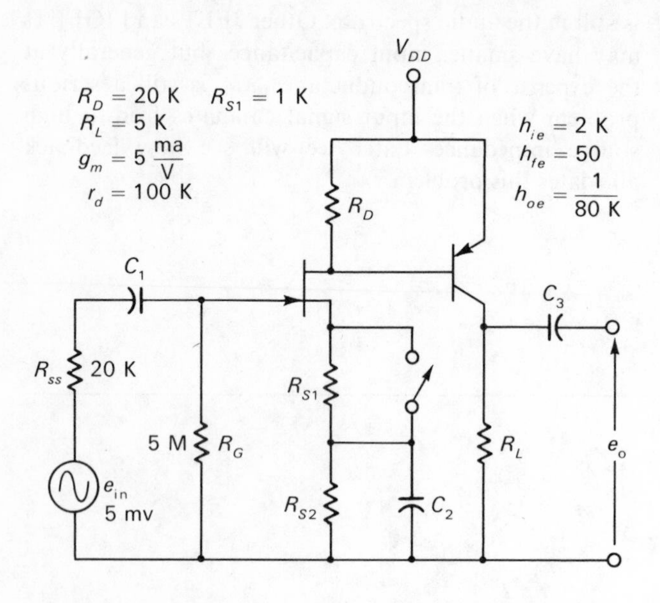

ANSWER 5.24 volts; 0.888 volt

PA 15-4 Estimate e_o in Fig. PA 15-4.

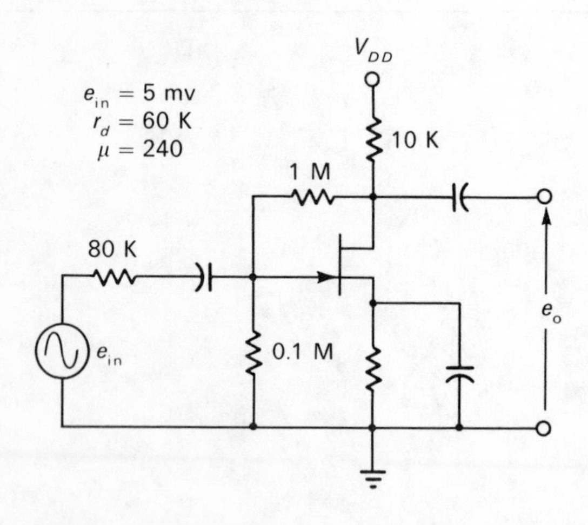

ANSWER -37.1 mv

PA 15-5 Determine e_o in Fig. PA 15-5 at 20 kilohertz.

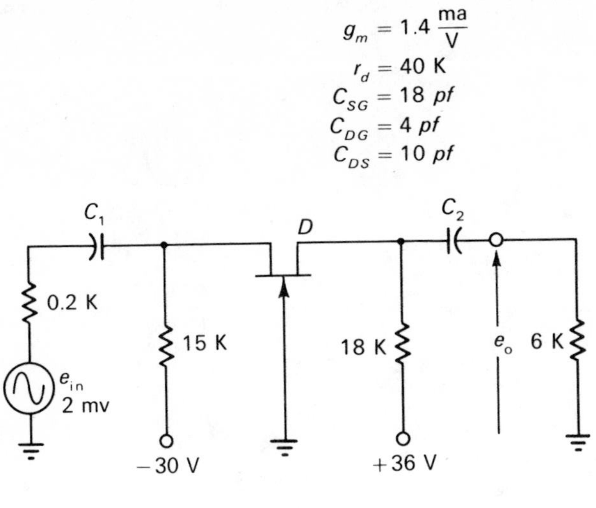

$g_m = 1.4 \frac{ma}{V}$
$r_d = 40$ K
$C_{SG} = 18$ pf
$C_{DG} = 4$ pf
$C_{DS} = 10$ pf

C_1 D C_2
0.2 K
15 K
18 K
e_o 6 K
e_{in} 2 mv
−30 V
+36 V

ANSWER 9.06 mv

PROBLEMS WITHOUT ANSWERS

P 15-1 For any given V_{DD} supply and choice of FET types, would you choose a low or high V_P unit if maximum voltage gain were desired?

P 15-2 What is the voltage gain in the circuit of Fig. P 15-2?

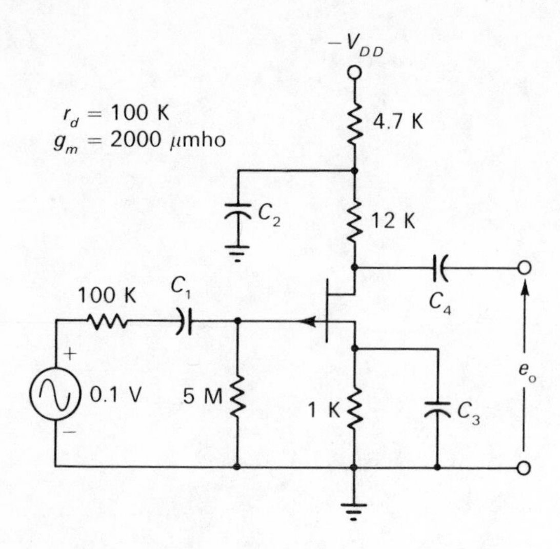

P 15-3 Determine $C1$ for a low-frequency breakpoint (3 db down) at 10 hertz.

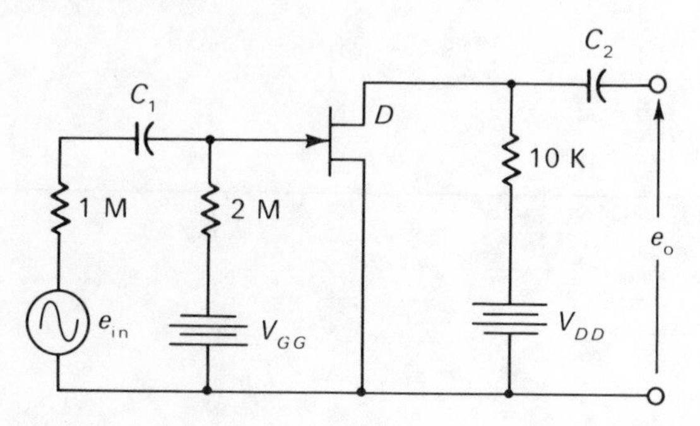

P 15-4 If in the circuit of Fig. P 15-4 there is 1 volt of ripple on the V_{DD} supply, determine the ripple component of drain current.

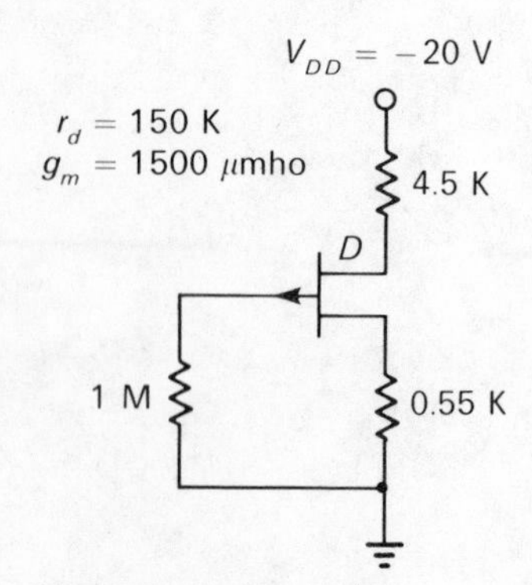

P 15-5 Prove that the output conductance in the circuit of Fig. P 15-5 is given by

$$g_o = \frac{g_{os1}g_{os2}}{g_{os1} + g_{os2} + g_{m2}}$$

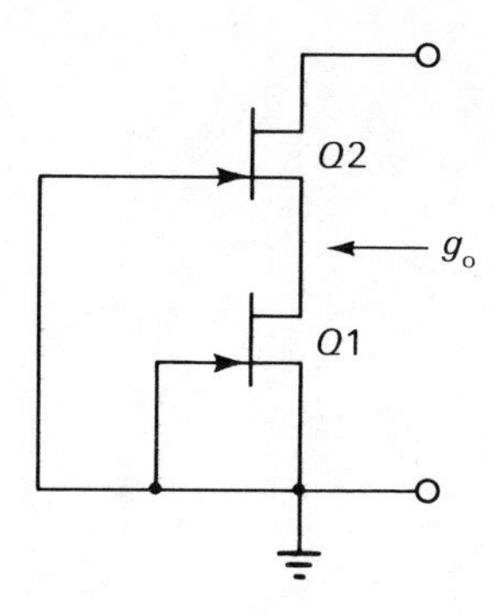

P 15-6 The FET in the circuit of Fig. P 15-6 is to operate at a Q point where the temperature coefficient of drain current is zero. If $V_p = -4.5$ volts, $I_{DSS} = 2.8$ ma, and V_{SDQ} is to equal $2|V_p|$, determine R_S, R_D, and A_v.

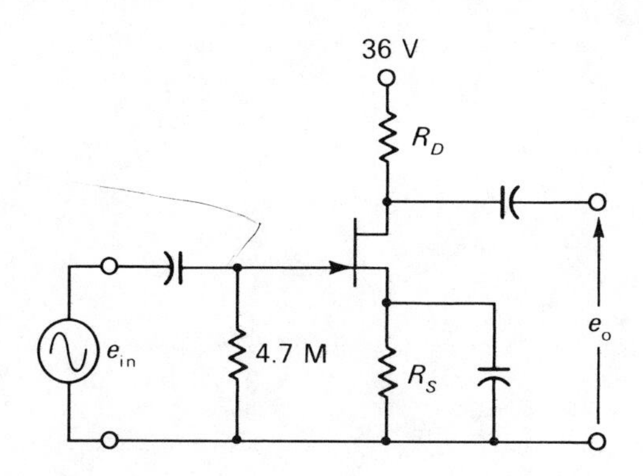

A current notation

The notation used in this text to represent an electric current is based upon the following considerations:

1) Electric current flow is designated with an upper- or lowercase symbol I and an appropriate subscript if required.
2) The assumed positive reference direction of current flow is indicated with an arrow as shown in Fig. A-1.

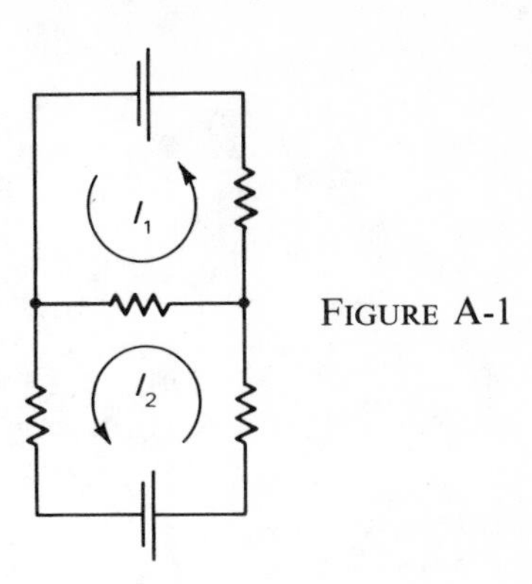

FIGURE A-1

3) Current flow in this text will, in general, imply electron or negative charge flow. Thus the current I in flowing through a resistor R polarizes the terminal which it first enters negative with respect to the terminal from which it emerges. This is illustrated in Fig. A-2.

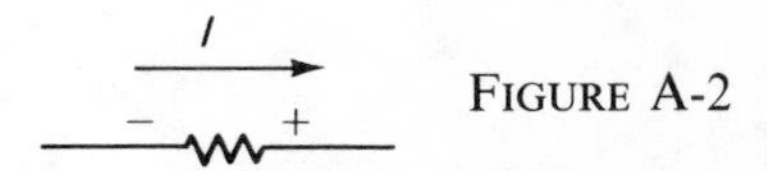

FIGURE A-2

If your previous training is such that you are more comfortable with so-called conventional current flow which implies the transfer of positive rather than negative charge, you need only reverse the arrow associated with I in this text and proceed with the development presented or any problem you might be attempting to solve. As you can see from Fig. A-3 the resultant polarity of the voltage developed across an impedance is still the same since the end of the impedance that the conventional current first enters is polarized positive with respect to the end from which it emerges.

For additional reference material regarding the notation used in this test, the reader is referred to chapter 1 of volume 1, "Electronic Circuit Analysis," by Phillip Cutler, published by McGraw-Hill.

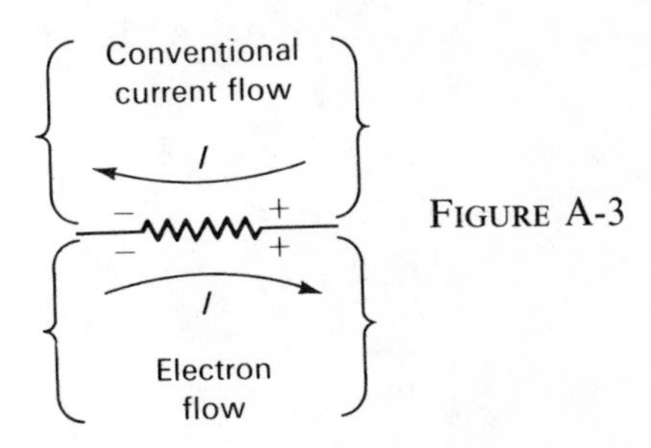

FIGURE A-3

B voltage notation

The notation used in this text to describe a voltage is intended to avoid the confusion that sometimes arises in the students mind with regard to a voltage drop or a voltage rise. Instead, this text merely refers to a voltage which is rigorously described according to the following convention:

1) The voltage between a pair of terminals is to be described by the symbol E, or V, in association with an arrow or double subscripts to indicate the terminals between which this voltage exists.
2) It will be agreed that the reference or starting point from which the voltage is reckoned, is the tail of the arrow in arrow notation, or the first subscript in double subscript notation.
3) To evaluate V or E you start at the tail of the arrow (or the first subscript) and follow some path through the circuit while taking note of the various potentials encountered, until you arrive at the head of the arrow (or the second subscript). The resultant value of the expression for V or E will yield the voltage at the head of the arrow (or second subscript) relative to the reference (starting) point.
4) In evaluating (or expressing) V or E as you proceed from the reference point to the point in question, you will encounter various circuit elements. If in traversing this path you enter a circuit element at one terminal and emerge at the other terminal at a potential more positive than the point of entry, you have gone in the positive direction or up in potential, and this is indicated with a positive sign. On the other hand if you emerge at a potential more negative than the point of entry you have gone down in potential and this is indicated with a negative sign.

Some illustrative examples are shown in Figs. B-1 and B-2. Remember, the current arrows indicate electron

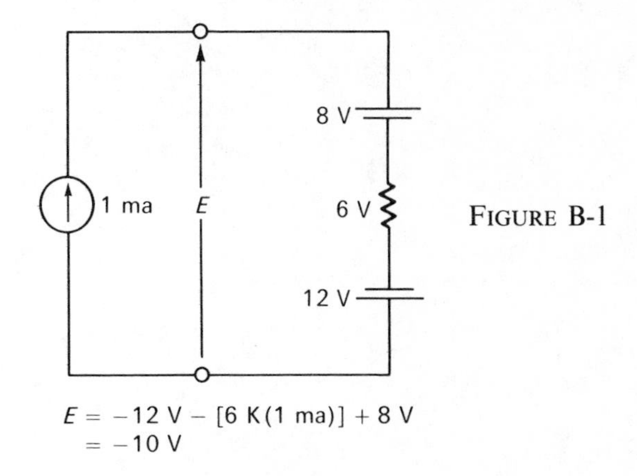

$$E = -12\text{ V} - [6\text{ K}(1\text{ ma})] + 8\text{ V}$$
$$= -10\text{ V}$$

FIGURE B-1

flow and hence the end of the resistor that the current first enters is polarized negative with respect to the end from which it emerges.

For additional reference material regarding this form of voltage notation see chapter 1 of volume 1, "Electronic Circuit Analysis," by Phillip Cutler, published by McGraw-Hill.

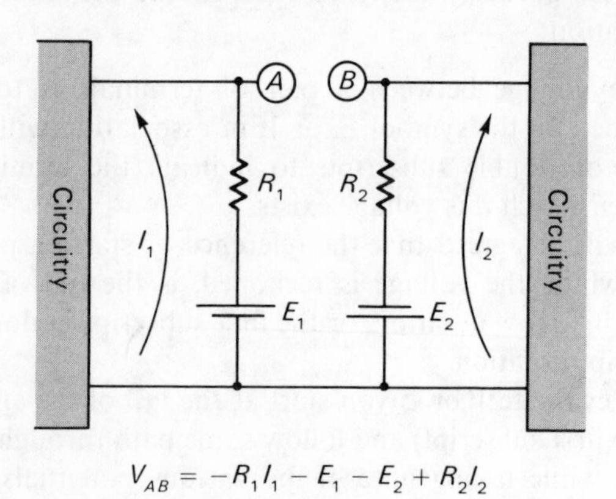

$$V_{AB} = -R_1 I_1 - E_1 - E_2 + R_2 I_2$$

FIGURE B-2

index